10.99

PENGUIN BOOKS

SO SHALL WE REAP

'Compelling . . . a devastating analysis of the arrogance and wrong-headedness of the food-production business' *The Times*

'A frightening reckoning of the damage being done by our appetite for cheap food' *Daily Mail*

'Were it possible to mandate required reading for all the politicians, civil servants, bio-technologies, food processors and retailers with a hand in agriculture today, this would be top of my list' Jonathon Porritt

'A well-thought out, detailed book . . . Although full of scientific rationale, it also appeals to common sense . . . he gives a great deal of intellectual weight to many of the arguments he presents' Scarlett Thomas, *Independent on Sunday*

'This is a journey all the way from the hunter-gatherers who first tried farming to the agri-businessmen who clear the land for industrial food production. The byways through science, nutrition, genetics, economics and even morality are adorned by learning worn lightly . . . Colin Tudge persuasively lays out the needs and prospects for enlightened agriculture if future generations are to survive and prosper. It is a tract for our times' Sir Crispin Tickell, Director of the Green College Centre for Environmental Policy and Understanding, University of Oxford

'Unlike many opponents of GMOs Tudge can argue the detailed scientific case, but he also has a wonderful eye for anecdote' Felicity Lawrence, *Guardian*

'A blueprint for mankind that not only shows us how to feed everyone, but to feed them well without compromising the ethics required to nurture a responsible stewardship of this planet' *Countryside Voice*

ABOUT THE AUTHOR

Colin Tudge read zoology at Cambridge and then became a writer, first for magazines, including *New Scientist*, and then for the BBC. Since then he has focused increasingly on books, writing on agriculture and conservation and on genetics and evolution. His publications include *The Variety of Life*. Colin Tudge is a Fellow of the Linnean Society of London and visiting Research Fellow at the Centre of Philosophy at the London School of Economics.

COLIN TUDGE

So Shall We Reap

What's Gone Wrong With the World's Food – and How to Fix It

PENGUIN BOOKS

PENGUIN BOOKS

Published by the Penguin Group
Penguin Books Ltd, 80 Strand, London WC2R 0RL, England
Penguin Group (USA), Inc., 375 Hudson Street, New York, New York 10014, USA
Penguin Books Australia Ltd, 250 Camberwell Road,
Camberwell, Victoria 3124, Australia
Penguin Books Canada Ltd, 10 Alcorn Avenue, Toronto, Ontario, Canada M4V 3B2
Penguin Books India (P) Ltd, 11 Community Centre,
Panchsheel Park, New Delhi – 110 017, India
Penguin Group (NZ), cnr Airborne and Rosedale Roads,
Albany, Auckland 1310, New Zealand
Penguin Books (South Africa) (Pty) Ltd, 24 Sturdee Avenue,
Rosebank 2196, South Africa

Penguin Books Ltd, Registered Offices: 80 Strand, London WC2R 0RL, England

www.penguin.com

First published by Allen Lane 2003
Published in Penguin Books with a new Afterword 2004
4

Copyright © Colin Tudge, 2003, 2004
All rights reserved

The moral right of the author has been asserted

Printed in England by Clays Ltd, St Ives plc

Except in the United States of America, this book is sold subject
to the condition that it shall not, by way of trade or otherwise, be lent,
re-sold, hired out, or otherwise circulated without the publisher's
prior consent in any form of binding or cover other than that in
which it is published and without a similar condition including this
condition being imposed on the subsequent purchaser

ISBN-13: 978–0–141–00950–6
ISBN-10: 0–141–00950–0

Contents

B.C.F.T.C.S.

IV
Enlightened Agriculture

B.C.E.T.C.S

Acknowledgements

I am always aware of my debt to everyone who has ever taught me anything; and even on the particular matter of food and farming, the people all around the world who have expanded my thoughts are too numerous to mention. But among them I should single out the late (and much missed) Anil Agarwal, who fought for many years in India for the environment and for justice; Michael Allaby, who thought long and hard about self-sufficiency in particular, before I got around to it, and with whom I once co-authored a book; to Sir Kenneth Mellanby, to whom Mike Allaby introduced me; and to Dr Paul Richards, of University College London, who recognized the strengths of indigenous farming long before most people, and opened my eyes to them. I have also benefited enormously from discussions with Peter Bunyard, who has for many years worked for *The Ecologist*; Dr Matt Ridley, a fine writer on biological matters, who also runs an estate in exemplary fashion; Dr Bernard Dixon, who has guided me in matters relating to animal disease; and Dr Jeremy Cherfas, zoologist turned seedsman, now with IPGRI (the International Plant Genetics Research Institute, Rome). Recently, too, I have come to know Satish Kumar, at Schumacher College, who has helped me to refine my ideas on ethics, and in particular on the philosophy of Gandhi. Among the many scientific institutions I have visited this past three decades, I remember in particular the Rowett Research Institute, Aberdeen, in the 1970s, when it was directed by Sir Kenneth Blaxter; Rothamsted, the oldest dedicated agricultural research institute in the world, where among many others I was privileged to talk at length to N. W. Pirie – another pioneer thinker from the 1970s; ICRISAT (the International Crops Research Institute for the Semi-Arid Tropics, Hyderabad, India); and IPGRI. Of all the agricultural scientists I have met, the ones I spent most time with and so learnt most from are Dr Bob Orskov, formerly at Rowett and now at the Macauley Research Institute, Aberdeen; and Dr Mike Gale, FRS, at what was then the Plant Breeding Institute, Cambridge. In recent years I have learnt an enormous amount from conversations at the Centre for the Philosophy of the Natural and Social Science, and will always be grateful to Dr Helena Cronin for inviting me to become a Visiting Fellow. I am especially grateful, in the present context, to Max Steuer at the London School of Economics (LSE) for his input on economics in general; Dr Richard Webb, for his views on globalization (on which we seem to disagree more or less completely); and Professor Tim Dyson, for his particular insights into demography. Through the LSE I also met Professor Aubrey Sheiham, at University College London, who has done his best to keep me up to date in nutritional science. Since I moved to Oxford, too, I have received great help and encouragement from Sir Crispin Tickell, who has added a slice of my own concerns to his already prodigious workload. To all: much thanks.

Finally, I owe a great deal both to John Brockman and to Felicity Bryan for helping me to get the initial outline of this book into shape; and to Stefan McGrath at Penguin who took the book on. Finally, I am in endless debt to my wife, Ruth West, for putting up with me in general, making it easy to focus on the work in hand (at the expense of her own work), and for her very significant intellectual input (for she too has pondered at great length the issues raised in this book).

Acknowledgements for the Paperback Edition

Two people in particular have influenced my thoughts since I submitted the original text for *Reap* in the spring of 2003 (as now reflected in the Afterword). Professor Norman Myers, who has various posts but is based mainly at Oxford, clarified my thoughts on the (very uncertain and often negative) relationship between money and human well-being. In the Summer of 2003, too, my wife Ruth and I had the privilege of staying with Robin and Binka Le Breton at their wonderful dairy farm *cum* teaching centre at Iracambi, on the Atlantic coast of Brazil (where Robin is also helping to restore the sadly depleted rainforest). Over several days, and in continuing correspondence, we continue to the discuss the need for and the realities of agrarian economies – which perhaps is the route to pursue in the much-needed reform of world food production as a whole.

February, 2004

Introduction

Be not deceived . . . whatsoever a man soweth, that shall he also reap.

Galatians 6: 7

This book is about food, and in particular the production of it, which mainly means farming. It argues that if we – humanity – get farming right, then our future could be glorious. Our descendants should still be here in ten thousand years, or indeed in a million years, and everyone who is liable to be born from now on and for ever could be well fed: not simply according to the tenets of nutritional theory, but to the highest standards of gastronomy. Our descendants could be surrounded, too, if they chose (as they surely would), by other creatures: butterflies, elephants, all manner of plants – there should be room for all; or at least for a very fair proportion of those that are with us now. Health, environment, good relations between nations; everything follows, or can follow, from good farming. But if we get it wrong, then the future will be dire; as Thomas Hobbes said in the seventeenth century in a slightly different context, 'poor, nasty, brutish', and a great deal shorter than it should be.

There are good reasons for thinking we are getting it wrong. Most obviously, famine is common, primarily these days in sub-Saharan Africa but also, from time to time, elsewhere. At the time of writing (November 2002) people are starving even in Argentina, once a brave new world to which poor Europeans emigrated, and only recently held up as the model of economic development for all the rest to follow. But over-nutrition is at least as common as starvation, and far more conspicuous: the World Health Organization now describes obesity as a 'global epidemic'. At the same time, of course, we are wiping out our fellow creatures at an unprecedented rate while at least in some parts of the world the relationship between nations is again as Thomas Hobbes described, 'a condition of war of everyone against everyone'.

This book asks – or rather shows – what we need to do to get it right. The thinking required is radical but not revolutionary. We need not contemplate anything so dramatic or unlikely as the overthrow of capitalism: just a different model of capitalism from the abstracted, overheated, aggressive form that now demands the maximization of cash efficiency on a global scale. We need not become ascetics. Vegetarianism is not called for. The opposite is the case: traditional, regional cuisines beautifully meet the requirements both of nutritional theory and of sound farming. People who really care about food – gourmets – are much easier to feed than faddists, or those who are wedded to lean meat and junk.

We need not, and indeed should not, be anti-science. Science provides the sharpest insights into the mechanisms of life that have yet been devised, and perhaps ever can be devised, and we need much more of it rather than less. But we need to liberate science. We need it to suggest new ways of looking at the world, of course. But primarily we need to identify the problems that really need solving, and then direct science at them. At the moment it seems to be assumed worldwide that the proper role of agricultural science is simply to help us to do whatever needs doing more cheaply, and on a larger scale: primarily to replace human labour with machines and industrial chemistry and now with biotechnology. As we lose people from the land we also lose ingenuity and attention to detail – plus, of course, the rural communities that until recently accommodated most of humanity. Out of traditional farming these past ten thousand years or so have emerged principles and techniques that can collectively be called 'good husbandry'. They include the conservation of diversity, of crops, livestock and landscape, and the interruption of chains of infection. Agricultural science is one of the world's greatest assets when it is used to abet and enhance good husbandry. But at present, increasingly, it is used to override the common-sense principles of farming, simply because that is the cheaper course. Then, despite the innate brilliance of many scientists and the good intentions of the great majority, science can be numbered among the world's most serious enemies.

Why, though, has the world drifted so seriously off beam? Why do we do things badly and destructively when we have the power – easily – to look after ourselves very well indeed? It is easy to moralize, to suggest that people in power are greedy, or in some general way are 'wicked'. There are evil people about beyond doubt, and of course they make a difference, but simple moral turpitude is not the main cause of our present ills. Most politicians would rather do good than harm. Scientists are commonly motivated by a desire to help humanity. Most industrialists that I have met are very civilized people, convinced that well-run, efficient industry alone can keep the world on an even keel – including what is now called 'agro-industry'.

Yet, astonishing though it may seem in this age of experts and think-tanks, I believe that the people in the highest places have misconstrued the nature of the problem. They do not see (because it has never occurred to them, and because their education has led them in quite different directions) that at bottom, the problems of humanity as a whole are those of biology. If we really want to survive in the long term (and ten thousand years is 'the long term'; not the thirty-year projections of conventional economics) then we have to begin by thinking of ourselves as a biological species, *Homo sapiens*, and the earth as our habitat; not simply a stage, or a tabula rasa, on which we can impose any manner of fantasy and whim. We need to see that farming must march to its own drum – that of 'good husbandry', founded in sound biology, and steered by respect for human values; and that this in many practical ways runs totally counter to the modern mantra which says, in the chill phrase I have heard so often these past three decades, that 'agriculture is just a business like any other'. Actually, all businesses are different, but farming is more different than all the others: indeed we could sensibly place farming in one panel of a diptych, and everything else we do in the other. Agriculture is properly seen as the counterpoise of all other human activities.

We need again to see farming as a major employer – indeed to perceive that to employ people is one of its principal functions, second only to the need to produce good food and maintain the landscape. Yet modern policies are designed expressly to cut farm labour to the bone, and then cut it again. This is seen as 'modernization', which in turn is seen to be both necessary and good. Thus Britain's government recently agreed to finance a scheme in Andhra Pradesh, India, which would drive 20 million farmers from the land. It's not just me, a comfortable European, who says that this is misguided. We need also to acknowledge on the one hand that science is wonderful, and can solve problems; but that it does not and never can provide omniscience, and in the creation of policy must always be secondary to human values. It is a huge mistake to suppose, as so many influential people these past few centuries have supposed, that science can create a qualitatively different world, in which all problems are solved, and that all we have to do is follow its lead. Absent from this line of thinking is any appreciation of human values, or indeed of the philosophy of science.

It may seem odd to suggest that so many people in high places have misconstrued the problems of humanity so absolutely, but no other explanation holds water. Yet if we look more closely we see it is not so implausible. Most of the world's leaders have been educated in law or economics or history, none of which take any serious account of life's physical realities: of the fabric of the earth itself, or the physiology of human beings and other animals, or of plants. For most people, even or perhaps especially for the most expensively educated, life itself and the things that support it – flesh, atmosphere, soil – are merely a 'given': the background, the setting, in which human beings can do exclusively human things. Only recently has it occurred to politicians in high places that 'the environment' itself must be taken care of. The first world summit that specifically addressed its problems was the United Nations Conference on the Human Environment in Stockholm in 1972, which is commonly perceived as a 'watershed'. Scientists ought to be more alert to the physical realities of life than politicians are, and some of them are. But scientists are brought up in an increasingly bullish atmosphere, effectively to believe that they can do anything that might be conceived and even, nowadays, that it is morally right to do whatever the market will pay them to do. It isn't that difficult to see how the world could have deceived itself; and when we contrast what is theoretically possible with what is actually happening we see that this self-deception, unlikely though it may seem at first, must indeed be the case. Often in the past, after all, world leaders have got things horribly wrong. Anyone who spent any time at all in the twentieth century would know that.

For my part, I effectively began writing this book in the early 1970s when I spent several years at *Farmer's Weekly*, which was (and I believe still is) Britain's leading agricultural magazine. It was enormous fun (my best years as a magazine journalist) but I also took it very seriously: treated my time there as an undergraduate course, to flesh out the biology I had learnt at university. I also took a great interest in nutritional theory, which (as this book will discuss) took some radical turns in the 1970s, and in some highly significant ways was turned on its head. I liked cooking, too, and (after *Farmer's Weekly*) worked for a time in London's West End, among the best restaurants; and (another perk of journalism) had an expense account that enabled me to sample just about everything. I even wrote a few restaurant reviews, including one of Britain's first ever tandoori restaurant. Now, of course, there is a tandoori in every high street (and a balti and a Thai takeaway as well); proving, if proof were needed, that people's tastes even in conservative Britain are nothing like so conservative as is often supposed.

Then in 1974 I attended a couple of key conferences. The first was the World Food Conference in Rome that was convened by Henry Kissinger, following famines not least in India and Pakistan. In Rome, shockingly, I saw how the representatives of nations at big meetings do not for the most part address the central issues, but contrive primarily to ensure that whatever is agreed does not cost them too much, and to show that whatever might have gone wrong in the past was not their fault. The few nations at Rome that really did seem to focus on the issue of world famine – Canada, Norway, and indeed China – were those who were outside the mainstream of world power politics or (like Algeria and China) had their own fish to fry. Only the NGOs – the 'non-governmental organizations', which for example include charities such as Oxfam, and religious groups such as the Roman Catholic Church – focused consistently on the matter that was ostensibly in hand, which is that people were starving. 'Twas ever thus, at world summits; including, most recently, at the World Summit on Sustainable Development, held in Johannesburg, South Africa, in August 2002. Nonetheless, most who have anything to do with these summits agree that they are a good thing. Among other things, good institutions come out of them: Stockholm, for example, gave rise to UNEP, the United Nations Environment Programme. Recent mutterings within the British government, following Johannesburg, that summits are a waste of time, are surely pernicious. If we didn't have them, then the only grand global meetings to be held at all would be those to do with world trade. Even as things are, everything besides world trade is considered secondary. The loss of summits, for all their obvious deficiencies, would make this neglect official.

The other pivotal meeting of 1974, at least for me, was on 'Food Technology in the 1980s'. It was held at the Royal Society in London, widely acknowledged (even in France, Russia and America) as the world's most prestigious scientific club. At this meeting, scientists both in industry and in academe urged the need in particular for 'textured vegetable proteins', or TVPs: artificial meat spun from protein extracted from soya, or from fungi, or indeed from bacteria raised on oil – which, in those still pre-OPEC days, was perceived as the earth's inexhaustible benison. The mantra of the day had it that people in general need large amounts of protein, and that they (we) have a particular predilection for meat. The supply chain (the passage of food from field to table) was in those days dominated by the food processors, who took it as read that farms merely produced raw materials which then had to be manipulated in various ways before people at large (peremptorily known as 'the public') could be allowed to eat them.

After these two meetings I experienced what I think should probably rank as a revelation. I remember sitting in the garden one bright June dawn – it must have been about 5 a.m. – on our children's swing, contemplating famine in general, and the particular claims of the food processors that the problems could be solved only by their own ministrations, for example through the provision of TVPs. I felt in my bones that their claims made no sense. If people really do need TVPs, then we should all be dead already – since, at that time, TVPs were still novel, and available only to a few.

The revelation that came to me was that the products of good, basic farming precisely match the requirements of the nutritional theory that was then unfolding: in all but the most extreme latitudes traditional farming provides precisely that combination of plants and animals that human physiology requires. It also came to me as if from on high that all the world's great cooking – Chinese, Indian, peasant French and Italian, Turkish, North African, South-East Asian, East European – is based on the products of traditional farming, for how could it be otherwise? In other words, the principles of good, basic farming, of the most up-to-date (and convincing) nutritional theory, and of great gastronomy, work perfectly in harmony. Indeed, it could not be any other way, since human physiology is adapted to the produce of wild nature; and traditional farming reflects wild nature; and great cooking has evolved over hundreds or indeed thousands of years to make best use of what wild nature, and traditional farming, provide. It occurred to me that the efforts of the food-processing industry were the most absolute nonsense: that here was a multimillion (or billion) dollar world industry with all the substance of the South Sea Bubble (a wild speculation based on nothing at all that in 1720 caused many of the most prominent Englishmen, including King George I, to lose fortunes). I wrote a book about food and the food industry, called *The Famine Business*, first published in Britain by Faber & Faber in 1977. Some people liked the book and some are still kind enough to remember it, but on the whole it went down like a lead balloon. Twenty-five years later the world has moved on. So have I: among other things, I have had the opportunity to talk to farmers and scientists in at least a score of countries on five continents, and have written about half a dozen scientific reports for various agricultural and related institutions, including those of what was then called the Agricultural and Food Research Council (AFRC) in the late 1980s. So I have learnt a lot more science, and seen a lot more farms in far-flung and sometimes unlikely places, and done a lot more thinking about politics and matters of morality. This, then, is an update of *The Famine Business*, but much broader in scope. I now regard *The Famine Business* as an apprentice work.

I

The Nature of the Problem and the Meaning of Agriculture

Every moving thing that liveth shall be meat for you; even as the green herb have I given you all things. Genesis 9: 3

In the sweat of thy face shalt thou eat bread, till thou return unto the ground. Genesis 3: 19

I

The Nature of the Problem

The prospects for humanity and for the world as a whole are somewhere between glorious and dire. It is hard to be much more precise.

By 'glorious' I mean that our descendants – all who are born on to this earth – could live very well indeed, each as comfortably and a great deal more securely than any ancient potentate, and could continue to do so for as long as the earth can support life, which should be for a very long time indeed. We should at least be thinking in terms of the next million years. Furthermore, our descendants could continue to enjoy the company of other species – establishing a much better relationship with them than we have now. Other animals need not live constantly on their beam-ends, and in perpetual fear of us. Many of those fellow species now seem bound to disappear but a significant proportion, enough to be well worth saving, could and should continue to live alongside us. Such a future may seem idyllic, and so it is. Yet I do not believe it is fanciful: no mere Utopia. There is nothing in the physical fabric of the earth or in our own biology to suggest that this is not possible.

'Dire' means that we, human beings, could be in deep trouble within the next few centuries, living but also dying in large numbers in political terror and from starvation, while our fellow creatures, a huge and random swathe of them, would simply disappear, leaving only the ones that we find convenient, or that we can't shake off: our chickens, our cattle, and the flies and mice that come along for the ride. I'm taking it to be self-evident that glory is preferable.

Our future is not entirely in our own hands because Earth has its own rules, and is part of the solar system, and is neither stable nor innately safe. Other planets in the solar system are quite beyond habitation, deep frozen or melting hot, and ours too in principle could tip either way. Recondite and unspectacular changes in the atmosphere could do the trick. The core of Earth is hot, which in many ways is good for living creatures, but the surface is pocked with volcanoes, and every now and again the melted rock beneath bursts through. Among the biggest volcanic eruptions in recent memory was Mount St Helens, in Washington State, which threw out a cubic kilometre of ejecta – fortunately in an area where very few people live. Vesuvius in AD 79 was no bigger, but it wiped out Pompeii. The eruption of Santorini on the island of Thira to the north of Crete in 1470 BC was ten times larger than either – and the aftermath destroyed the civilization of the Minoans (and left the Mediterranean to the Greeks). The eruption of Krakatoa in Indonesia in 1883 matched Santorini for violence and affected the weather worldwide. Between April and July 1815 Tambora (on the island of Sumbawa east of Java) threw ten times more ejecta into the upper atmosphere than either Santorini or Krakatoa and disrupted the climate for season after season. The people of New England referred to the winter of 1816 as 'Eighteen hundred and froze to death'. Mary Shelley, on holiday in the Alps with Percy Bysshe Shelley and Byron, wrote *Frankenstein* while they all sheltered from the rain.

Yet none of these show what volcanoes can really do. Almost all of India sits on the consolidated lava of the ancient Deccan volcanoes that apparently erupted soon before the dinosaurs disappeared. Yellowstone, the biggest national park in the US outside Alaska, occupies the caldera of an exceedingly ancient volcano of extraordinary magnitude. Modern surveys show that its centre is now rising. Sometime in the next 200 million years Yellowstone could erupt again and when it does the whole world will be transformed. Yellowstone could erupt tomorrow. But there's a very good chance that it will give us another million years, and that surely is enough to be going on with. It seems sensible to assume that this will be the case.

The universe at large is dangerous too: in particular, we share the sky with a swarm of wayward asteroids, and every now and again they encroach upon us. An asteroid the size of a small island, hitting Earth at 10,000 miles an hour (a modest relative speed by the standards of heavenly bodies) would strike the ocean bed like a rock in a puddle, as if the water was not there at all, send a tidal wave around the world as high as a small mountain and as fast as a jumbo jet, and propel us into an ice age that could last for centuries. There are plans to head off such disasters (including heroic and hugely expensive rockets to shove approaching asteroids into new trajectories), but in truth it's down to luck. On the other hand, the archaeological and the fossil evidence tells us that humanity's luck has held reasonably enough for the past 5 million years and that no truly devastating asteroid has struck since the one that seems to have accounted for the dinosaurs, 65 million years ago. So again, there seems no immediate reason for despair. The earth is indeed a precarious place, in an uncertain universe, but with average luck it should do us well enough. If the world does become inhospitable in the next few thousand or million years, then it will probably be our own fault. In short, despite the underlying uncertainty, our own future and that of our fellow creatures is very much in our own hands. Or that, at least, is the sensible way to look at it.

Given average luck on the geological and the cosmic scale, the difference between glory and disaster will be made, and is being made, by politics. As this book unfolds I want to discuss the kinds of political systems and strategies that would predispose us to long-term survival (and indeed to comfort and security and the pleasure of being alive), and also the kind that will take us more and more frenetically towards collapse. The broad point is, though, that we need to look at ourselves – humanity – and at the world in general in a quite new light. We need in particular to perceive that our material problems are, at the most fundamental level, those of biology. We need to think, and we need our politicians to think, biologically. Do that, and take the ideas seriously, and we are in with a chance. Ignore the great beating drum of biology and we and our fellow creatures haven't a prayer. To begin with, we might reasonably turn the biological spotlight on ourselves.

The Importance of Thinking Biologically

Aristotle said that human beings are 'political animals'. I remember from my school choir days a medieval chant that seemed somewhat prematurely to celebrate death, and the consequent liberation of the soul: 'That from clod of earth set free, winged with zeal flies up to thee!' In these two sentiments is encompassed the classical, Western view of humanity: we see ourselves as cerebral and secular creatures on the one hand, jockeying for power, and as spiritual beings on the other, destined for heaven; or at least preferring the world of the imagination to what we perceive as the crude realities of life. In general we share Jaques's view in *As You Like It*, that 'All the world's a stage' for us to act out our fantasies upon; or to put the matter more fiscally, that the world is simply raw material, for us to mine and to trade.

We think of ourselves as British or French or American or whatever, or as doctors or bricklayers, as frustrated schoolchildren or harassed parents, each with our own particular problems and claiming our particular 'rights', but we don't as a matter of habit think of ourselves in our raw biological form – as animals: collectively as the species *Homo sapiens*. We don't treat the world at large as our habitat and no longer think, as many other societies have often thought, of other species as our fellow creatures. Anthropologists tell us that some other peoples through history and prehistory saw themselves as a part of a Creation, alongside other species: they might see themselves primarily as Cherokee or Algonquin but also as the brothers and sisters of bears and beavers. There are hints of such a tradition in the Western world, for example in St Francis of Assisi, but in the modern West this is not the common view. Rational and materialist on the one hand, and effetely 'spiritual' on the other, we like to think we are above mere flesh, and that with our rational minds and our technologies we can do whatever we want. There is no need to take other life forms seriously, or the fabric of the earth itself, because we can make of them what we will. The men and women who run the world's affairs and spend so many hours around the conference table seem to think exclusively in these political and bureaucratic terms. Until very recent years, when 'environment' became a slogan that could attract votes, and 'biodiversity' has become the subject both of diplomacy and of law, most politicians (I can attest from experience) found any mention of either to be slightly ludicrous. Even now, the world's leaders typically seem to suppose that 'environment' means 'golf course' and that 'biodiversity' is simply another resource, vaguely linked to tourism on the one hand and to biotechnology on the other, but in either case far less interesting than oil.

But whatever else we may be, whatever our aspirations and pretensions, in the end we are animals, and big and voracious animals at that; and the greatest mistake humanity has made these past few thousand years is to forget this most elementary fact. Earlier societies might be forgiven for messing up their environments (as the archaeological and historical records show they often did) because, at least in some cases, they did not know enough biology to keep out of trouble. We have much less excuse. Present science is far from perfect but we should do better than, say, the ancient Greeks or the Mayans. On the whole, though, we don't. Indeed, because we act on a much bigger scale and with much more vigour, we do a great deal worse. The trouble is, of course, that we don't use our science for general human comfort and long-term survival; and the basic reason for that, I suggest, is not that the people in charge are evil but that they, along with most of the rest of us, have misconstrued the nature of the problem. We don't think of ourselves as animals. We don't think of the world as our habitat, but as modelling clay to be shaped as we choose. Nature is remarkably flexible and astonishingly adaptable, and will lend itself to a huge variety of manipulations. But it is not infinitely forgiving. If for commercial or political or ideological reasons our manipulations drift too far from what is biologically permissible ('sustainable' is the fashionable term) then the biological systems collapse, spin off in quite new directions, and humanity and our fellow creatures must collapse with them. The archaeological and historical record is crammed with examples of policy outstripping biological possibility, or sometimes of plain apathy; and as the Spanish-American philosopher George Santayana observed, 'Those who do not learn from history are doomed to repeat its mistakes'.

So what does it mean to think of ourselves in biological terms?

What Kind of Creatures Are We?

Theologians and poets are right to extol the virtues of human beings, for we are indeed extraordinary. We are not the only creatures with claims to consciousness but we are clearly *more* conscious than, say, dogs or elephants. We are not the only animals with language, but we alone have verbal language, syntactically based, that enables us not only to frame our thoughts more subtly than other creatures but also – crucially – to share them with each other, rapidly and in detail, so that each of us becomes party to a universal human collective intelligence that in principle stretches across the whole world, and back through all of history.

But our preoccupation with our own brilliance tends to blind us to our underlying corporealness – that we are also flesh and blood, with flesh-and-blood needs. Crude survival is not about theology and poetry (or politics and philosophy), but about the need to keep our physiology intact, in tolerable habitats. It is perilous to lose sight of our biological roots.

The roots of our own particular 'hominid' lineage evidently began in Africa about 5 million years ago, with an ape-like creature, less than a metre tall, known as *Ardipithecus*, who soon evolved into the better-known *Australopithecus*. Both had chimp-sized brains, around 400 ml in size. About 2.5 million years ago the global climate became drier and cooler, forests worldwide retreated, and *Australopithecus* became more and more committed to the spreading savannah, and left the shrinking woods to its closest relatives, the chimpanzees and gorillas. The savannah evidently tested the ingenuity of *Australopithecus* – and so promoted the growth of its brain. By 2.2 million years ago *Australopithecus* had given rise to the first members of our own genus, *Homo*. This human debutant, *Homo habilis*, had a brain of around 700 ml and, it seems, was the first bona fide maker of stone tools.

Members of the genus *Homo* went on getting taller and brainier. The people broadly known as *Homo erectus* had appeared by about 1.6 million years ago. They were up to six feet tall, had brains of around 1,100 ml, and were the first to migrate out of Africa, across Europe and as far as South-East Asia. *Homo erectus* developed fire, possibly as long as 1.4 million years ago; and fire is extremely important, first as a means of influencing habitat (by burning vegetation) and to extend the range of the diet, by cooking. By 500,000 years ago people had assumed the form traditionally known as 'archaic *Homo sapiens*'. They had brains as big as ours (around 1,450 ml) but also had big 'caveman' faces and bodies. One group of 'archaics' went on to become the Neanderthals (*Homo neanderthalensis*), who became even brawnier (though still brainy) but died out around 30,000 years ago; and another group evolved into us – modern *Homo sapiens*. Clearly, the hominid lineage has evolved enormously in the past 5 million years, and particularly – spectacularly – in the past 2 million. Our bodies have doubled in height. Our brains have tripled in volume.

To a significant extent and in various ways, so many biologists argue, the rapid and unparalleled increase in human brain size depended upon diet, which had to be both various and reliable (see for example Michael and Sheilagh Crawford, *What We Eat Today*, Neville Spearman, London, 1972). Nervous tissue uses a lot of energy, and biochemically it is complex. In short, the brain is both pernickety and expensive. To foster and sustain such an organ, our ancestors needed in particular to consume a significant proportion of meat. They were never out-and-out carnivores, like lions. But they were very definite omnivores; unable to thrive exclusively on wild vegetation, and in general inclined to sample anything that grows and a few things (like salt) that do not.

The concentration of energy in vegetation is in general much less than in meat. For instance, as recorded in the nutritionist's bible, McCance and Widdowson's *The Composition of Foods* (revised by A. A. Paul and S. A. T. Southgate, Her Majesty's Stationery Office, London, 1978), 100 g of cabbage provides only 20 calories* – yet cabbage is a domestic plant, less fibrous and so more rich in energy than most wild leaves. A hundred grams of potato (a typical tuber) gives around 80 calories. Seeds are richer – 100 g of wheat flour offers around 350 calories. Pure sugar offers around 400 calories per 100 g, but sugar qua sugar is rare in nature except in the form of honey. But little compares with animal flesh, and especially with fat. Lean raw steak provides around 120 calories per 100 g but beef fat provides more than 600 calories; thirty times more, weight for weight, than cabbage. If the water is removed from fat, it provides around 900 calories per 100 g. Meat (and some plant materials) also supplies other nutrients that are of enormous significance for brains, including a range of recondite fatty acids used in the construction of nerve-cell membranes. Herbivores can sustain big brains if they eat enough of the right kinds of plants, as elephants demonstrate. But the general metabolic rate of a warm-blooded animal is inversely related to its body size, so that elephants have a very slow metabolic rate, which means they need much less energy per unit of weight than you or I do and many times less than, say, a mouse. So elephants can eat enough vegetation to sustain big brains within their vast bodies, but to do so must eat through most of the day and night.

Smaller animals that aspire to maintain big brains need richer fare – which in general means eating seeds and/or meat. Extreme specialist leaf-eaters, like the leaf-eating monkeys of Africa and South-East Asia and the koala of Australia, tend to have smaller brains than other animals of the same general kind and of comparable size. The difference in the energy content of leaves and meat is reflected in the wonderfully contrasting lifestyles of elephants and lions. Both are social and intelligent. But whereas elephants in the wild must eat for seventeen hours a day, lions

* The unit 'calorie' as used by nutritionists is 1,000 times bigger than the calorie which is the official unit of heat used by physicists. Strictly speaking, the nutritionists' calorie should be called a kilocalorie or Calorie. Throughout this book I will use calorie to mean Calorie.

sleep for around twenty hours a day and sit around for another two, and are commonly content to hunt about twice a week. In general, the more flesh you eat the lazier you can afford to be. None of this implies that modern people need or indeed should eat anything like as much meat as we in practice aspire to do, as discussed in Chapter 4. A little goes a long way. But in general, unless we have access to a modern health-food shop, a little animal flesh is advisable.

In the vocabulary of cattle farming, vegetation in general is classed as 'fodder', and meat as 'concentrate'; and it is worth taking trouble, and running risks, to acquire concentrate. Until the 1960s, biologists commonly believed that our fellow great apes – gorillas and chimps – were vegetarians. Indeed, they generally supposed that primates in general are primarily herbivorous, and many a zoo monkey has languished in consequence, wanly confronting its twice-daily bowl of oranges and chopped apple. In truth, most primates do like fruit and some are more or less vegetarian. But most are omnivores, and take meat when they can get it. Gorillas spend most of their time eating leaves, but they certainly do not refuse eggs in the wild, and John Aspinall used to give joints of roast lamb to his excellent troops of gorillas at Howlett's Zoo in Kent, south-east England, which they fell upon with relish. Jane Goodall showed in the 1960s to everyone's surprise (not to say horror) that chimpanzees can be fanatical hunters, particularly of monkeys, which they tear limb from limb. Chimps in full hunting flight are terrifying beasts. The cute little quasi-humans of sickly Hollywood movies and TV ads are only the infants.

Our own ancestors, from earliest times, must have been accomplished hunters. This surely influenced their psychology – though not in the way that it has become fashionable to suppose. Thus in the 1960s and '70s Robert Ardrey argued in books such as *African Genesis* and *The Hunting Hypothesis* that the earliest humans (then thought to be the Australopithecines) were indeed hunters, and that hunters perforce are innately aggressive, and that we moderns must have inherited our ancestors' aggressiveness. This idea had the ring of folk wisdom and it caught on. But Ardrey's biology always ran somewhat obliquely to the mainstream, and in any case scientific theory has moved on a great deal since then.

In the first place, there is no clear relationship between aggressiveness and diet. For my part, I would as soon be confined with a wolf as with, say, a stallion; and among 'big game', African buffalo have the most fearsome reputation of all although they are 100 per cent herbivores. Then – perhaps paradoxically (and in absolute contradiction of Ardrey) – the fact that our ancestors ate at least some meat probably enhanced their sociality, and ours too. Thus when chimpanzees have fruit, they keep it very much to themselves. But when they make a kill, they share the meat. We can easily see how natural selection would have favoured such discrimination: if a chimp eats more meat at any one sitting than it needs, the surplus is simply wasted (the body does not store surplus protein). In our own societies we put the surplus meat in the fridge for another day. Some wild animals also store their provender, including dogs, which bury bones. Ancient Icelanders traditionally stored large amounts of meat (notably that of sharks, some of which live in northern waters) by burying it on the beach, and there is evidence that Ice Age people in Britain stored meat in cairns. Out-and-out specialist predators like lions and pythons simply stuff the lot at a sitting, and then sleep it off. But chimps cannot do this, and for social animals that live in tropical forest there is little opportunity for safe storage. The most economical strategy for predatory chimps is to share the meat while it's there, so that the whole troop gets some benefit. After all, the troop must cooperate to make a kill in the first place. Thus for animals that are innately gregarious to begin with, meat-eating further encourages cooperativeness. As Homer Simpson observed (to Lisa's disgust) when he invited the neighbours to his burger-and-steak barbecue: 'You don't make friends with salad!'

Food chains are always pyramids: there are always many more creatures waiting to be eaten than there are to eat them. If predators consume more of their prey than the prey can replenish, then their food supply dwindles and they too begin to die off. Thus individual herbivores need great swathes of vegetation to graze and browse upon, while carnivores need entire herds to feed from and generally are content to pick off the stragglers. The food chain is also inefficient. As a very rough rule of thumb, for every gram of protein an animal adds to its body, it needs to eat 10 grams. The protein in plant material tends to be diluted by water and fibre, so herbivores need to consume an enormous amount of it, relative to their own body weight. The protein in meat is generally far more concentrated, but even so the meat-eater needs to eat far more than it can convert into its own body flesh. Thus the makers of myth may see the big carnivores as 'the kings of the jungle' but their total biomass will always be far less than that of the humbler herbivores on which they prey. In short, as Paul Colinbaux so succinctly put the matter in the title of his excellent book: 'big, fierce animals are rare' (*Why Big Fierce Animals are Rare*, Allen and Unwin, London, 1980).

Indeed, 'big, fierce animals are rare' might reasonably be seen as the most fundamental law of ecology. I know of only two exceptions. One is the crabeater seal of Antarctica, of which there are still many millions. But crabeaters live on krill, planktonic shrimp-like crustaceans which in turn live on diatoms and other single-celled organisms which grow by the million ton in the unshaded sunlight of the Southern Ocean. The crabeaters effectively graze on krill like cows on grass.

The other exception to this bedrock law of ecology is, of course, ourselves.

Population

Crabeater seals have broken the bedrock law because they live in very special circumstances. Human beings have broken the law because we have developed the skills of farming. What this entails I will discuss in the next chapter. Here I just want to show the consequences: how, in practice, human numbers have grown as farming has developed.

Convention has it that farming began 10,000 years ago in the Middle East, in what is called the 'Neolithic Revolution'. That, at least, is what the archaeological record seems to show. I will argue in the next chapter that in truth, human beings must have been managing the land and their fellow creatures to a significant extent long before 10,000 years ago – that they were not simple hunters, like lions, or straightforward, opportunist gatherers, like monkeys. But for two reasons, 10,000 years ago is a watershed nonetheless. First, it fell at the end of the last ice age, and except for the odd cold snap (as in Europe in the early fourteenth century and again in the late seventeenth) the climate seems to have given us a fairly clear run since then. Second, the archaeological changes of the Neolithic Revolution probably represent the beginnings of settled, arable farming and this as we will see was a very significant step indeed.

By 10,000 years ago people already occupied most of the world: they were present all through Eurasia and Africa, both Americas, and Australia. Of big land masses (apart from Antarctica) only Madagascar and New Zealand remained free of humans, the former until the first few centuries after Christ, and the latter until around AD 1000 (estimates vary in both cases). Yet the total human population 10,000 years ago is thought to have been only around 10 million (with guesses ranging from around 5 million to 15 million or more).

The populations of creatures like flies (known as r-strategists – rapid breeders, short-lived) progress by boom and bust: one week there seem to be very few, then there are billions, and then they might collapse again. But when long-lived, slow-breeding creatures (K-strategists) like us are doing well then their populations rise steadily (give or take a few blips), year by year and century by century, for as long as the good times last. The rise is by compound interest: a small percentage increase on what was there before. Because the initial population is small, the rise at first seems slow. A 1 per cent increase on 100 is only one, after all. But as the population grows the rise in real terms soon starts to seem very rapid, even if the percentage increase stays the same. A 1 per cent increase in a population of 1 billion is an extra 10 million. So a steady percentage rise translates into a faster and faster increase in absolute terms.

Thus, although in the first few millennia of farming the yearly percentage increase in world population was probably less than 1 per cent, total numbers by the time of Christ are thought to have reached anywhere between 100 million and 300 million. Three hundred million is probably nearer the mark and this was enough to support many advanced and diverse civilizations, in all the continents. By AD 1500 world numbers reached 500 million: half a billion. By 1825, 1 billion; by 1927, 2 billion. The 3 billion mark was passed in 1960. We got to 4 billion by 1975. Five billion was reached in 1986. A few years ago, 2,000 years after the birth of Christ and about 12,000 years after the Neolithic Revolution and the end of the Ice Age, we reached 6 billion. At the time of writing, in the midsummer of 2002, world numbers have reached around 6.15 billion: 6,150 million.

In 1798 the English cleric and economist Thomas Robert Malthus (he preferred to be known as Robert) observed in his first *Essay on the Principle of Population* that agricultural output could not keep pace with the inexorable rise of human numbers. Sooner or later, humanity was bound to be overtaken by famine. The rise in population was particularly conspicuous in Britain at that time, as improved agriculture encouraged an increasing birth rate and the new industrializing cities brought masses of people together. Darwin later perceived that Malthus's principle must apply to all living creatures. All have an innate tendency to out-breed their resources. Hence there was bound to be competition between them. The individuals that survived in each generation were the ones that were the best adapted to the particular circumstances that happened to prevail. Thus Malthusian competition drives Darwinian natural selection.

Many a politician and economist has been influenced by Malthus. In general, he seems self-evidently to be right. Any population of animals, including the human animal, clearly has the reproductive potential to outstrip its resources. The human species now makes use of most of the world but even the world as a whole is all too obviously finite. In the 1960s, as famines fell over Asia, many felt that Malthus's predictions were coming horribly true.

Two turns of events put Malthusian thinking out of fashion again. One was the Green Revolution of the 1960s and '70s. Within barely a decade, new varieties of wheat and rice put a stop to the Asian famines, and for a few years turned Pakistan and India into grain exporters. The idea seemed to take hold that whatever happened to human numbers, somehow technology would find a way. The Green Revolution was indeed wonderful but, as we will see, this was only one of several extremely damaging misconceptions that have arisen from it. Talk of overpopulation also fell from fashion for political reasons. Numbers were growing most rapidly in poor countries, and people in those countries (and many sympathizers in the West as well) pointed out that total numbers was only half of the equation. The other half was consumption: and per capita consumption – not particularly of food, but of all inputs combined – was far higher in the rich countries than in the poor. In short, poor countries began to see themselves as scapegoats for the world's ills in general. The ecological shortcomings of the world as a whole were being blamed on the large families of the Third World when in truth, most of the world's resources were consumed in the West. In many countries, large families were perceived as a human right, integral to their culture. So it became unfashionable, and then politically incorrect, to talk of 'overpopulation', or to express anxiety about it. Yet several extremely populous countries did bring in methods to control population in the 1960s and '70s (although these were presented as 'family planning'); notably India, which in effect tried to bribe people to have fewer children, and China, which was more coercive, and restricted all but a few favoured minorities to one child per couple. China's policy persists. Their aim was to restrict total numbers to 1 billion, but they have now topped 1.2 billion and surely will reach 1.5 billion before they level out. But at the same time, some other countries, including Malaysia, Tanzania and Romania, became pro-natalist, urging their citizens to have more children. Even

France began to fear that the world might soon have to endure a relative paucity of French people.

But although it seems that Malthus must be right in principle – his idea is common sense, after all – he may well be crucially wrong in practice. For it now seems that animals in general, including human beings, are more subtle than he supposed. Biologists have now shown that many creatures, including many birds and mammals, are able to adjust their birth rate to the circumstances. They do not simply breed to the limits of physical possibility. What applies to animals in general, certainly applies to the human animal. Women in hunter-gatherer societies – people living in what might be called a state of nature – rarely have more than five children in a lifetime. The women are not physically able to bear children until they are about nineteen years old because their diet is not rich or copious enough (their 'nutritional plane' is too low), and for the same kind of reason, and because of prolonged breastfeeding (which induces lactational amenorrhoea, or absence of the menstrual cycle), the interval between births is around four years. Women in hunter-gatherer communities typically have their last child at around thirty-five, so five in a breeding lifetime is the maximum total output. Some of the children die young, mainly through infection and accident, so with five offspring per woman a hunter-gatherer population will grow only slowly, if at all. After at least 40,000 years of residence, the Aborigines of Australia numbered only around a million at the time the Europeans arrived in force in the late eighteenth century.

By the late 1990s it became clear that people in societies worldwide were choosing, of their own volition, to have fewer children. China chose to be draconian, and perhaps it needed to be (China understands its own problems better than any outsider and I would not presume to comment); but the rest of the world was reducing its birth rate too, even without coercive measures. Western Europeans in particular became seriously non-fecund. By the late 1990s the population of Italy was falling. Numbers in West Germany would have gone down too had it not merged with East Germany. Britain would probably be going down now were it not for immigration. In the world as a whole, the families of five, six or eight that had lately been fashionable, in general became less so.

Remarkably small differences in average family size make a huge difference to total population growth, and hence to the prospects of humankind. Since women are half of humanity (as Mao Tse-tung was wont to say), a population should in theory remain steady if each woman has two children. But because some people die before they can reproduce, and some are infertile, and some prefer for whatever reason to remain childless, a population will not grow if the birth rate falls below about 2.3 offspring per woman. If the birth rate averages, say, 2.7 offspring per woman then the population will grow steadily. A small percentage increase rapidly produces massive population growth: such that a steady 2 per cent growth per year doubles the population in forty years, and a 4 per cent increase (as Kenya was achieving in the 1970s) doubles population in not much more than twenty years. But if two children per woman becomes the norm, then the population will soon stabilize, and then start to fall.

So although the reductions in family size worldwide have not generally been dramatic since the 1970s (except in China, where they have been spectacular), the consequence of that modest contraction is very great indeed. Thus, in the early 1970s, when the annual growth rate stood at 2.1 per cent, the world population seemed on course to double every forty years or so. The consequences of this seemed horrendous. Numbers then stood at around 4 billion, so if things had continued as they were we would have had 8 billion by about 2010, and 16 billion by 2050, 32 billion by 2090, 64 billion by 2130, and about 120 billion by the second half of the next century. Some who believe that technology will always find a way have argued that the world could easily feed 20 billion (although this, as we will see, seems very dubious). A few zealots even argue that we could feed 100 billion-plus; but most feel that such euphoria is ludicrous. Yet that kind of figure would have been reached, at 1970s rates of increase, within the lifetimes of our great-grandchildren. Something, clearly, had to give. Sometime in the next century and a half, and probably sometime in the twenty-first century, the Malthusian chickens would surely come home to roost.

But because people worldwide, for the most part voluntarily, have chosen to have fewer children, the yearly percentage rate of increase has been going steadily down since the 1970s, until by 2000 it was a mere 1.3 per cent. In fact it's not so 'mere'. Total numbers in 2000 were around 6 billion, so an extra 1.3 per cent meant another 77 million people a year – roughly the population of Britain plus that of Australia. Still, United Nations demographers now suggest that if the percentage rate of increase continues to fall as it has been doing these past thirty years then it will reach zero by the middle of the twenty-first century.

In other words, after 10,000 years of almost uninterrupted, exponential growth, the world population will stabilize by around 2050. By that time, human numbers seem likely to have reached somewhere between 9 and 11 billion. Ten billion is a sensible middle-ground estimate.

Now we come to an issue I really don't understand. I have an eerie feeling that I must have missed something: that the rest of the world has realized some truth that for some reason has passed me by. For this simple fact – that the world population will probably stop rising by about 2050; well within the lifetimes of many readers of this book – is the most staggeringly important shift in the world's affairs since the Neolithic Revolution itself. Ever since that ancient time, humanity has had a steadily increasing impact on the world as a whole. Over the past few centuries, as demographers and politicians began truly to think globally, they have simply assumed that the world must produce more and more human food with each passing year. At present, policy-makers at the United Nations Environment Programme (UNEP) and the Food and Agriculture Organization (FAO) are planning for a 2 per cent increase in food output over the next few decades, partly to accommodate rising numbers and partly to make good the present shortfalls.

But after 2050, this steady increase will no longer be necessary. The task of feeding the human species will no longer be open-ended. It will be finite. Agriculture will therefore be finite. For the next few centuries after that, for the first time since the Ice Age ended, humanity as a whole will have a chance to consolidate. In that period of consolidation we can secure the rest of our time on this earth – the following 10,000 years, or 100,000, or a million. The prognosis for humanity has thus shifted absolutely within the past thirty years, as its percentage growth rate has slowed; and since the fate of all our fellow creatures depends absolutely upon the numbers and the voracity of humanity, their destiny has changed too. When I first became involved in discussions of food and conservation in the 1960s, doom and gloom, though infinitely depressing, seemed inescapable. Serious discussion of the environment seemed to lead inevitably to despair. That was one good reason for avoiding the whole issue, yet any other attitude seemed merely euphoric. Many felt that the only sensible response was to eat, drink and be merry, in the certainty that soon we must all die. But now the picture has changed absolutely. We don't have to die, and neither do our fellow creatures, and suddenly it's worth taking the future seriously. This is a fundamental shift in world ecology and overwhelms by orders of magnitude every other human concern. Yet it seems hardly to have been noticed. I have read remarkably little about it in the newspapers. I have spotted no programmes on TV. Insofar as politicians take an interest, it has been to express concern at the pending lack of a youthful labour force, to send to war and to provide the future generation with their pensions. Have I really missed something? Or does this demonstrate yet again that the people who now presume to lead the world have misconstrued the nature of it?

In theory, if the average birth rate per woman worldwide remains at less than about 2.3, then, in time, the world population will start to fall again. For various reasons, the fall will not be immediate or dramatic. For one thing, the rise and fall of populations does not depend entirely on birth rate, but on birth rate relative to rate of death. At present, the life expectancy in 'developed' countries is around seventy-five, and in 'developing' countries around sixty-three. But people are living longer, such that septuagenarians in rich countries commonly find themselves looking after their ageing parents. This means, for the time being, that the death rate (as a proportion of the whole) is going down, in parallel to the decline in birth rate. The most significant exception to this general trend is brought about by HIV/Aids, which is now the fourth most common cause of death worldwide. It has killed 20 million worldwide since the 1970s. Forty million now have the virus, of whom 70 per cent are in sub-Saharan Africa. There, at present, it is the leading cause of death. However horrible it is, though, this epidemic will surely be temporary.

A population seems bound to fall if women have fewer than two children each; yet it would continue to rise even if women had less than one child each if everybody lived for ever, meaning that nobody ever died. Clearly, individual human beings never will live for ever (I don't think many people would want to), but centenarians are already becoming almost commonplace and some sober observers now suggest that human beings in the future, even without any dramatic interventions at the genetic level, might routinely live to 120 or so. A dramatic increase in life expectancy over the next few centuries is at least plausible and could in principle upset present demographic predictions, even if the birth rate does continue to fall. However, reasonable guesses suggest that human numbers are liable to rise for the next fifty years to reach about 10 billion and (largely because of increasing longevity) are liable to stay at such a level for the following few centuries, and perhaps for as long as a thousand years. But with present projections, world population would then begin to fall. How low the numbers will become in the distant future will depend on the policies and whims of our descendants. In a couple of thousand years human numbers could be down to 2 or 3 billion, the kind of numbers that prevailed in the nineteenth century.

The period of 500 to 1,000 years in which world population hovers at around 10 billion has been called 'the demographic winter'. That suggests a pessimistic attitude and indeed, it may not be easy to feed that many for so long. Yet surely the challenge can be met; and if we can get through the next 500 to 1,000 years without ecological collapse or major conflict, then life after that should become steadily easier. That, at least, is a reasonable projection.

As the birth rate goes down and life expectancy increases, we will see a dramatic shift in the age structure of populations worldwide. Indeed, this is already apparent. Worldwide in 2000, 69 million people were eighty or over, including 11.5 million in China (which considering that country's traumas since the 1920s is even more remarkable). By 2050, worldwide, there will be 379 million people aged eighty or older; far more than the present population of the United States. This has all kinds of implications. In traditional societies, and indeed up until the mid twentieth century, youngsters and workers greatly outnumbered pensioners, who typically retired at around sixty-five and died five years later at the biblical three-score years and ten. I certainly remember when septuagenarians were considered old. Now, people who retire at sixty-five typically expect to draw their pension for another twenty years. If people continue to retire at sixty-five, and then to add decades to their lives, then those who are not working (the pensioners and the children) will begin seriously to outnumber those who are. By 2050, one in three Europeans will be over sixty; they will outnumber the children (aged 0–14) by 2.6 to 1. Something will have to give. Britain is already somewhat frantically rethinking its traditional policies of retirement, and many who were looking forward to comfortable pensions are now reassessing. At the time of writing (the summer of 2002) this is commonly ascribed to the shaky stock market, but the underlying cause is one of biology: the demographic shift.

Things will get a great deal worse before they get better. It will be a pity if governments respond (as they are all too prone to do) by becoming pro-natalist again and encouraging people to breed, basically to bolster the workforce. That will merely postpone present problems, and when the world is finally forced to face them its plight will be a great deal worse. One short-term solution is for countries with dwindling birth rates to welcome more people from countries who are still breeding.

In the longer term (if we want to avoid simple Malthusian collapse) we will have radically to rethink our attitude to work and retirement; and among other things this implies the creation of jobs that older people are actually able to do, and find agreeable, so they still make some contribution to society. It doesn't take a lot of imagination to see how this is pertinent to farming and why the present overwhelming desire to reduce the agricultural workforce may be absolutely at variance with future needs.

Much more to the point, though, is that the task of feeding people in the future no longer seems open-ended. Clearly (as the FAO emphasizes) we need to double present food output in the next half century both to keep track of rising numbers and to make good present shortfalls; for at present an estimated 800 million people worldwide, about one in eight, are undernourished. Distribution, rather than overall deficiency, is the reason; but even so, more food would in principle be desirable. But we will not have to double output again in the latter half of the twenty-first century, and then again in the first half of the twenty-second, and then again, for ever and ever or until Malthusian collapse ensues, as seemed until recently to be necessary. Once the final doubling is achieved, that should be it. After that we merely have to find ways of maintaining that output without too much strain on the environment, so that we can go on doing so at least for the following thousand years, the period of the demographic winter. Our descendants should also find time to focus more and more, as we have not these past few decades, not just on ways of producing more food but of producing it by more agreeable means. Clearly, though, a huge burden has been lifted from humanity and from the world as a whole. The task for the future is big, but now we can see that it is finite.

This puts all agricultural policies into a new context, and a new light. The prime strategy of the past few decades has been to increase productivity with maximum cash efficiency. The environment has suffered and the labour force has been reduced hand over fist, allegedly in the cause of efficiency in general but in truth in the interests of cash efficiency. When the task is perceived to be finite, then the frenetic emphasis on ever-rising output no longer becomes sensible. On the other hand, when once we perceive that we – humanity – should still be on this earth in 1,000 years' time, or 10,000, or a million, then 'sustainability' becomes paramount: at least it does if we give a damn about the lives of those who come after us. Frenetic reduction of agricultural labour begins to seem even less appropriate than it does now. When there are more people on the land, they can take better care of it: and future people, including a majority no longer in the flush of youth, will want jobs of an agreeable kind.

A few codicils seem in order. First, some still argue that human numbers simply should not be contained; that nature should simply run its course. Some zealots from various religions argue that each new person is a new soul, for the glory of God; and that anyway, God will find a way to help us out of our troubles. Some extreme technophiles continue to argue that science can solve anything. People who welcome restraint are sometimes accused of being anti-humanity; that we should not presume to pre-empt the births of people who might otherwise have lived. In truth, to advocate the containment of numbers is very much to be pro-human. If human numbers are restricted to the 10 billion who now seem inevitable, and then are allowed to fall, then our species, *Homo sapiens*, could still be here in 100,000 years, or a million, or more. Within that time, many hundreds of billions of people will have had the opportunity to enjoy the earth. But if we simply let things rip, then the human party could be all but over within the next few centuries.

There is a grand historical and moral lesson, too, in the rise and possible stabilization and fall of the human population. In particular, it was widely argued in the 1960s and '70s that the only antidote to the high birth rate was a high death rate. I remember otherwise reasonable people arguing albeit with regret that welfare measures and famine relief were simply prolonging lives that were doomed. It would be better, they suggested with a shake of the head, to let them die. It has often been argued, too, that wars and epidemics are a necessary check on human numbers.

Yet the events of the past few decades show the precise opposite. It's clear in particular that a high death rate does not counteract a high birth rate. It is a prime *cause* of a high birth rate. When people fear that their children are liable to die, they have more. In poor countries, too (and in a state of nature), fertility is limited by lactational amenorrhea: when they are breastfeeding, women on a low nutritional plane are not able to conceive even if they want to, and weaning in traditional societies is long delayed. If the infants die, lactation stops and the women become pregnant again. Clearly, societies in which birth rate and infant death rate are very high are not 'natural'; this is not what happens in hunter-gatherer societies, which continue to live in a state of nature. More broadly, human beings reproductively speaking are K-strategists, like orang-utans, elephants and eagles: they have very few offspring, in some cases only at long intervals, but they expect most of their children to live. When the death rate is high and the birth rate is raised to track the death rate, we become like r-strategists: like mice, say, which have a great many offspring and expect most of them to die. High birth rate and high infant mortality represent humanity in biological disarray. This is in no way defensible. After war and famine, too, populations quickly spring back – even though 'quickly' is a relative term. The Second World War prompted an immediate 'baby boom' in many countries, though the population of Ireland is only now springing back to what it was before the Potato Famine of the late 1840s. The recovery in Ireland's case may seem to have taken a long time. But that is because we have a poor sense of time. In biological terms, a century and a half is too small to register.

The evidence shows, as a corollary, that when people really do expect their children to live, then they *voluntarily* have smaller families. This is a prime reason for the extremely low birth rates in much of present-day Western Europe. Of course, we are commonly told that women in many poor countries want to have huge numbers of children, and there are surveys to show that some women in Kenya (where the birth rate has been extremely high) have tended to perceive that a family of eight was altogether too modest. On the other hand, it's clear that many women who have a great many children would rather not. Malcolm Potts is a British family-planning specialist who spent much of the 1970s talking to women in Third World villages. Birth rates were still booming but, said Dr Potts, the women generally told him that all they lacked was contraception: not knowledge of contraception, which they already had in abundance, but the hardware. Clearly, too, women in many societies gain status primarily or even exclusively by becoming mothers, and the more children they have the more they are esteemed. Evidence worldwide shows that as attitudes to women change, and as women have more opportunity to find fulfilment outside the home, they almost invariably choose to have fewer children. Contrariwise, it seems that extreme pro-natalism often results from religious or political zealotry: as Dr Potts has commented, the women in China who are now being told to have only one child are the daughters of women who were exhorted by Mao to have as many as possible. It's clear, too, that in many traditional economies people feel they need children for security in their old age, and the more they have, the more secure they feel. Pensions make an enormous difference to attitudes.

In short, contrary to the common view of previous times, it's now clear that the only measures that can constrain the growth of populations in the long term are the ones that are humane – the things that are independently desirable: the reduction of infant mortality, the liberation of women, the provision of pensions (or other forms of security) in old age. Contrariwise, the measures that were traditionally thought necessary to curb human numbers – high infant mortality, famine, war, epidemic – have no long-term effect and in fact encourage high birth rates. Thus it emerges that cruelty is not only bad in itself, but is counter-productive. Simple humanity does what is needed.

The broader lesson still is that human beings are sensible. It is sensible to limit family sizes for all kinds of reasons: and the evidence worldwide is that most people, most of the time, are only too happy to do so if only they have the opportunity. Societies in which women habitually have huge families are always in various ways odd: either under extreme biological or economic pressure (like the pioneer families in newly occupied continents who needed more hands to work their virgin farms and to fill the wide open spaces) or else harangued by religious leaders, politicians or mothers-in-law. But when people in general and women in particular are able to take their own affairs in hand they do not breed for the sake of it. When people feel self-confident they will override their traditional leaders. The birth rate has been lowest of all of late in Catholic Italy.

All this is eminently encouraging: that humane measures work, and cruelty does not; that people left to themselves are perfectly capable of managing their own affairs. The general demographic trend is encouraging, too.

Even so, the pending population of 10 billion is extremely large: nearly twice as many as we have now. There will probably be that kind of number for 1,000 years. Can we really feed that number for so long? The task may now seem finite, which is a very good start, but is it too big nonetheless? Will we collapse in any case, with the winning post in sight?

Can We Really Feed So Many?

Hundreds of reports, essays and books address this issue but I know none that are satisfactory. Broadly speaking, those of technophilic mien tend to euphoria, suggesting that if we merely scale up the most productive forms of present-day food production – create more greenhouses, say, where yields can beggar belief – we could easily feed 20 billion people, or 100 billion. Those of ecological bent are wont to point out that when all pertinent factors are considered we will be hard-pressed even to sustain the 10 billion that will soon be with us. Ecologists, too, know that it's hard to predict even the fate of a small nature reserve, where everything seems to be under control; let alone the whole world, where nothing is. Those of no particular emotional bent finish up sitting on the fence.

Literally millions of statistics illuminate the picture, but all can be deceptive. For instance, the total amount of fresh water in the world is obviously pertinent. If there was none, we could not farm at all, and if every country had a rainfall like New Zealand's there would be, as they say in those islands, no worries. What matters at least as much is where the water is – and in fact, much is at high latitudes, while the crops are mostly at low latitudes. The way the water is cycled, too, matters at least as much as the total amount. Computer-controlled drip irrigation doles out the water drop by drop to the individual plants which need it most at that moment. In high-tech greenhouses, all the water that evaporates is condensed and may be run through again. Contrast this with the enormous profligacy of the Amazon or the Ganges.

Of course, the world's great rivers do provide irrigation which, for example, provides water for 40 per cent of all the wheat in the Third World. But irrigation raises problems of its own. Most rivers, and hence most irrigation water, comes from mountains which now, in the words of United Nations ecologists, are 'among the most volatile and threatened environments on earth'. Any use of rivers raises political issues, as illustrated not least by the Jordan. Big irrigation schemes typically involve dams, and as the World Commission on Dams put the matter in its report of 2002 (*Dams and Development*, Earthscan, 2002): 'By 2000, the world had built 45,000 large dams to irrigate a third of all crops [and] generate a fifth of all power, control floods in wet times and store water in dry times. Yet, in the [twentieth] century, large dams also disrupted the ecology of half the world's rivers and displaced over 40 million people from their homes.'

Yet the water that is pumped from underground perhaps raises the most pressing difficulties of all. This is a finite resource, for underground water is generally fossil water, the remnant of some ancient lake, and when it's gone it's gone. As it is drawn to the surface it brings salts with it so that the land it irrigates becomes salinated, or in other ways toxic.

In the developing world, according to FAO figures, about 20 per cent of all the irrigated land is now damaged by salinity or by waterlogging. Each year throughout the 1980s an estimated 10 million hectares of irrigated land were abandoned. South Australia is among the regions severely damaged; the saline soils have reached the margins of Adelaide. In such ways the different factors interact; the more we try to develop one part of the farming system (for example, the water supply) the more we may compromise another (for example, the land).

Then again – a key theme of this book – the present must be balanced against the future. Intensive farming, producing fabulous crops, may gobble up non-renewable resources, including ancient, non-renewable underground water. Yet intensive production need not compromise the future. Notably, intensive organic farmers contrive to increase the fertility and depth of their soils year by year. But the enormous productivity of intensive units is invariably deceptive, because they concentrate resources from a much wider area. What matters is not the space occupied on the ground but the total 'ecological footprint'. Thus, greenhouses are fed by fertilizer factories, and typically require fuel energy produced in refineries, and need elaborate drainage once they grow to truly commercial proportions, and so on. Modern industrialists put great faith in hyper-intensive livestock units. Million-strong pig units have already grown up in the US and it has been seriously mooted in high places that all the pigs in Europe should be produced in one great centre in Poland. In such set-ups, the output per unit area is stupendous. But the beleaguered pigs of Poland would still be consuming grain from 100,000 hectares, just as they do now. In terms of output, there is no gain at all (although there would be a massive pollution problem, which increases the ecological impact yet again).

For all these reasons and more, it just isn't possible to gauge the future prospects of the world by totting up the yields of the most productive units, and then scaling up. Even if this were sensible, the problems of scale-up itself would be horrendous. Statistics would surely show, after all, that if we put the wheatfields of the Ukraine under glass we could feed the whole world five times over. But such an endeavour would be a nightmare. It would be foolish to moot it at all were it not that some of the schemes of recent decades (such as the proposed realignment of the Volga, and the actual straightening of the Rhine, and the Aswan Dam, and the million-head pig unit, and so on and so on) have in their ways been just as fatuous, and yet have been put into practice or seriously discussed. In reality, in the future as at present, we will have to rely for our basic food on arable crops – notably cereals – grown on the field scale in the great outdoors.

Euphoria, in short, is not justified. But the same kinds of considerations apply the other way round as well. Yes, pollution and soil erosion are horrendous. But they are not quite inevitable or unstoppable. It is just as foolish to base predictions on the worst of modern-day examples, as it is to base them on the best. All in all, it is hard to improve on common sense. At least, it's a good idea to gather all the statistics that are possible, and subject them to the most rigorous analysis.

A middle, common-sense course between euphoria and despair suggests that we should indeed be able to feed the 10 billion people who will be with us within fifty years, and for the thousand years after that, and to do so to the highest standards. If total numbers fall in the fourth millennium and beyond we should be able to go on feeding everyone who is ever liable to be born until the world itself gives up. But common sense and present data also suggest that this will not be easy. We have to treat the problem as a serious exercise in applied biology. If agriculture continues to be treated simply as another adventure in politics and commerce then we surely will be in trouble.

The most basic statistics (supplied not least by UNEP and the FAO, and summarized in *Global Environment Outlook 3*, Earthscan Publications, London, 2002) tell us that the total area of land in the world is roughly around 12 billion hectares. This means we now have 2 hectares each, which is around 5 acres, which seems plenty.

However, much of that land is under ice, some of it is high, steep mountain and in the end only about 1.3 billion hectares is now used as arable land. This gives us 0.22 hectares per person: about half an acre. In 1960 there were only half as many people as there are now, and the world's arable area was not much different. So each person then had 0.47 hectares. But yield overall has kept pace with numbers – and indeed, the average intake now, worldwide, is higher than in 1960.

Is our allotted half-acre enough? It certainly should be. Half an acre of intensively managed market gardening could feed dozens of people; and although we should beware euphoria, that is a good fall-back position. The most intensive food production is horticulture, meaning gardening with or without glass. Inputs are expensive (either labour-intensive or very high tech or both) so that in the present world, horticulture is geared to the most expensive cash crops: tomatoes, aubergines, cut flowers. Tomatoes and aubergines are good for 'micronutrients' (vitamins, minerals) but are not significant sources of energy or protein, the nutrients we need in bulk. But some high-energy crops are grown by horticulture. Avocados (rich in fat) are raised under polythene (for example in Spain). Many traditional farmers (for example in Africa and New Guinea) grow staple pulses in small, intensive gardens. Mushrooms and other fungi could do more to feed us than they do, and are conventionally raised in sheds. So despite the caveats outlined above (notably the matter of ecological footprint) we could, in future, make more use of intensive horticulture for the purposes of basic nutrition than we do now. In the main, though, as now, the bulk must come from cereals, raised outdoors on arable fields. Is 0.22 hectares of arable field enough?

Again, the answer should be yes; comfortably. Arable farmers in England now average around 8 tonnes of wheat per hectare, and the most productive achieve about twice this. A tonne is a metric ton: 1,000 kilograms (and it happens, by good fortune, to be almost the same as an imperial ton). A good rule of thumb says that one tonne of wheat should provide enough energy to keep three people going for a year. (One gram of wholemeal flour provides about 3.3 calories, which means that 1 kilogram (1,000 grams) provides 3,300 calories, which is more than enough for an average person for a day. Thus 1 tonne (1,000 kilos) would keep a person going for more than 1,000 days, which is about three years; or three people for one year.) So if the world's arable fields yielded like those of East Anglia, the 0.22 hectares that now serves one person could support at least five. Again, there seems to be plenty of leeway.

Statistics relating to wheat are particularly pertinent. The world grows hundreds of different crops but three of them are in a league by themselves: wheat, rice and maize. Wheat is a crop of temperate lands, with moderate rainfall. Rice essentially is tropical, and the commonest varieties like plenty of water. Maize also likes hot countries but prefers drier climates, though it is nothing like so drought-resistant as sorghum. India and China grow wheat to the north and rice to the south, and China in particular is now a massive producer of maize.

The world as a whole produces about 600 million tonnes per year of each of these three great crops. About a third of the wheat, and two-thirds of the maize, though only a small proportion of the rice, are fed to livestock. Whether eaten directly (as cereal) or indirectly (after conversion into meat), these three between them provide humanity with half our calories and two-thirds of our protein. I find this statistic astonishing but again, simple calculations make the point. Six hundred million tonnes per cereal means 1,800 million tonnes between them, which is 1.8 billion tonnes. Since there are 6 billion of us, this means there are 0.3 tonnes for each of us. This, as shown above, is precisely equivalent to the amount needed to sustain a human being for a year (wheat, rice and maize are roughly equivalent nutritionally). In practice, though, the big three are required to provide only half our energy, and we can apparently afford to feed a large proportion of them to animals. Clearly, the amount produced is the right kind of magnitude, but seems to leave plenty of leeway.

Every other kind of crop – even the pulses; even the major tubers, like the potato – is minor by comparison with the big three cereals. Even beef, for all its massive impact on world agronomy, diet, culture and economy, is small beer compared with wheat, rice and maize. In short, if we get these three right, then we have virtually solved our food problems. Such considerations also put genetic engineering into perspective. Genetic engineering has so far contributed nothing of unequivocal value to the development of wheat, rice and maize, and is not likely to do so in the next half-century when human numbers are continuing to rise. Despite the hype from politicians and industrialists and the immense economic and political implications, genetic engineering is, from the point of view of feeding people, peripheral.

Again, then, these crudest of crude but still salutary statistics suggest that we have plenty of leeway. The present yield of the three crops that are far and away the most important seems to be twice what we would need if the chips were really down (provided the produce was fairly distributed). There seems to be plenty of room for expansion, too. Several independent studies suggest that the present area under cultivation could be more than doubled, to 3 billion hectares or more. Then (it seems) even with present yields, with perfect distribution we could feed ourselves four times over. The greater potential, though, is surely to increase yields. While English farmers expect 8 tonnes of wheat per hectare (and many achieve far more), many African farmers are pleased to produce 1 tonne per hectare. If Africa could achieve the best English standards, the world could export food to the entire solar system.

Here again, though, reality strikes. Firstly, the only practical way to expand current farmland would be to encroach more and more on to the forest, with all the horrendous aesthetic, social, and ecological complications that this would imply. Among other things, deforestation leads to flooding. In Pakistan, mountain floods in 1992 destroyed 4,600 hectares of crops, and flash floods in the Swat Valley in 1995 washed away more than 1,200 hectares of agricultural land, with twenty-six watermills for good measure. In China, landslides cause around $15 billion's worth of damage each year and kill about 150 people. Some schemes to 'reclaim' desert are perhaps worthwhile but again, the ecological consequences of large-scale irrigation, including the diversion of major rivers, are impossible to predict and (as much experience already suggests) are at least as likely to be disastrous as to be helpful. On the whole, the best land is already cultivated. Land claimed in the future will be increasingly marginal. It is better to try to feed ourselves from the 1.3 billion hectares we have now. The cities are eating into it, though – and they tend to occupy the land that is best for farming, in the valleys and near rivers. One more reason for restricting urban growth by providing jobs on the land.

In the end, the yield of any one kind of crop plant is limited by its genetics, and the most that any crop can attain in the best possible conditions is called its 'yield potential'. Yield potential is generally realized only by enthusiasts; for example the pumpkins that challenge the strongest wheelbarrows are raised by amateurs in what are essentially intensive-care units. In practice, commercial growers in the great outdoors or even in the greenhouse aspire only to achieve some reasonable proportion of the yield potential. Two kinds of factor determine what proportion is achieved. First, and importantly, the intention matters: what is the farmer actually trying to achieve, and why? Second, in all circumstances farmers encounter 'limiting factors' – particular environmental constraints that click in long before the theoretical, genetically defined 'yield potential' is realized.

In most of Britain, broadly speaking, farmers have to try for maximum yields. The milieu is competitive and commercial, and land is expensive. Only with high yields can they pay their way. In turn, the market (which in recent years has largely been propped up by the taxpayers) is geared to take as much as the farmer can grow, whether the produce is needed or not. In much of Africa, the priorities are quite different. Land can be very cheap (when people have access to it at all). The greatest bugbear is climate. Any one year can be totally different from the one before, and some are devastating. The farmer must aim not for huge surpluses (for which there would be no market), but for security. It's better to grow crops that will reliably yield 1 tonne per hectare in the worst years, than to aim for a 10-tonne crop that will fail altogether in all but the best conditions. I am told by people who know Africa well that many foreign advisers have failed completely to see the distinction, and created much havoc along the way.

The physical, limiting factors differ from place to place. In Britain, yield and reliability of maize is limited by lack of sunshine. For English fruit, late frosts are a problem. Over much of the world, the limiting factors are water and the soil itself: its fertility, and indeed whether there is any soil at all. Again, it is easy to be overwhelmed by statistics, but the most striking is that one-third of all the world's people live in regions that are significantly hampered by 'water stress', up to and including out-and-out drought. On the day I am writing this (5 November 2002) London's *Financial Times* reports that Australia is now suffering one of its worst droughts in a century. Predictions for next year's winter cereal crop is down to 14.8 million tonnes. Last year's was 34.1 million tonnes. This matters to the whole world. Australia is among the top three exporters of wheat, barley and rapeseed (canola). The most certain prediction to emerge from future scenarios for climate change (not least from NASA) is that future climates will be uncertain. Yet, as is all too clear, attempts to alleviate drought by irrigation (which is the most obvious course) bring problems of their own. All in all, production systems that are geared to cope with uncertainty (like much of Africa's) rather than to maximum yields will be at a premium. The notion that the world in general can edge ever closer to the yield potentials of our crops – or even towards the kind of yields now attained by England's wheat farmers – is too sanguine by half.

We might pile statistics on statistics, but the general picture is clear. Taking all into account it is indeed hard to improve on common sense. Common sense says that if we try to farm as well as possible, sensitive to the details of each terrain, not risking too much, not putting too many eggs in one basket, making the best possible use of science, then we should be fine. The broadest statistics suggest that there is still scope for us to do very well indeed. But there is no case for euphoria, or excessive bullishness, or risk, or putting faith in any one particular technology – whether civil engineering or biotech – no matter how commercially attractive it may seem in the short term. Traditional farmers the world over know this. They have known it for centuries. They are worth listening to.

The Joker in the Pack: Climate

Then of course there is climate. The idea that the world may be getting hotter has been around since the 1980s. Warming is caused by an increase in the atmosphere of gases such as carbon dioxide and methane, which inhibit the flow of radiant heat – infra-red – away from the earth's surface. Since this is the way the glass in a greenhouse works, carbon dioxide and the rest are called 'greenhouse gases'. Humans are causing carbon dioxide to rise by burning fossil fuels and felling forests (which take up carbon dioxide as they photosynthesize), while the world's ever-increasing herds of cattle belch out vast draughts of methane as they ruminate. The greenhouse effect now seems irrefutable, accepted in principle even by the recalcitrant governments of the US and Australia.

But although the world will in general grow warmer, it will not do so evenly. Ocean currents carry heat energy from place to place, and general (but uneven) warming could reverse their flow. For example, if the Gulf Stream that now brings warmth to Britain from the tropics stopped running, then Britain could become as cold as Labrador. It is on the same latitude, after all.

In short, although global warming now seems as near to being a fact as science can provide, only two detailed predictions emerge from it. The first is that the world's sea-levels will rise, as the last of the ice left over from the last ice age melts and runs off the lands of the extreme north and south – notably from Greenland and Antarctica. Low-lying countries, like Bangladesh and many an island, seem bound to flood or disappear, along with coastal plains worldwide (which often support the most productive agriculture). Many a great city, from Amsterdam to New York, will be in trouble. The second firm prediction is that the weather will become more unpredictable, and more extreme. We have seen the trend already, of course: more hurricanes on America's east coast; extensive floods in Britain, all exactly as predicted – or at least not *exactly*, because exactitude is not possible; but in general as predicted. The effects on the world's crops can be devastating. Most directly, Hurricane Mitch in 1998 wiped out 70 per cent of the staple food crops in Nicaragua and Honduras: maize, beans, potatoes, plantains. With the crops went the topsoil, which clogged the streams and rivers that irrigate and drive the generators for electricity. Cash crops went too: 80 per cent of the bananas, coffee and tobacco. Various international research centres have helped effect some restoration, but clearly the world as a whole cannot absorb too many such shocks.

Drought is the other side of the same coin. In general, worldwide, land that is now suitable for wheat may have to give way to maize, which needs less water; and land now down to maize may become suitable only for sorghum or even for millet, the most drought-tolerant of all the cereals. The Gobi is currently marching south by 50 km per year and is rapidly closing on Beijing. In 2000 I travelled through the maize fields that lie between Beijing and the Great Wall. The crop should have been two metres tall – 'The corn as high as an elephant's eye' in the words of *Oklahoma*. But it was not. Except where it grew more or less on the banks of the few remaining streams, it stood in crisp pathetic wisps, barely waist-high. The farmers knew it was a waste of time to plant it. But what else could they do? Farmers plant crops, and that presumably is what they will carry on doing until they are bankrupt. Again according to the FAO, an estimated 250 million people have been directly affected by the spread of deserts in recent years, and another billion – one-sixth of the world population – are at risk. Is Australia's present drought part of this grand trend?

Yet the effects on the world's crops could in general be more profound and subtle than this. In particular, crops that grow in temperate latitudes far from the equator tend to gear their growth both to changing day-length (they know when it's spring and when it's autumn) and to temperature (they can tell a cold spring from a warm one). High temperatures at high latitudes could be very confusing indeed, and disrupt their cycles of growth. New varieties could be bred, adapted to the new conditions; but it takes at least a decade to develop a new variety and to multiply the seed to the point where it can provide a harvest. Events may move too quickly for the breeders. Genetic engineering offers no instant solution, despite what the biotechnologists often imply. The introduction of new genes – assuming you know what genes to introduce – is only the start of the breeding programme. The failure of crops in China will have worldwide consequences. As Napoleon said, 'When the lion of China wakes, the world trembles.' It's waking now. If the Canadian wheat crop fails, confused by unprecedented climate, the whole world's food prospects will be turned on their head.

So how should the world respond to the prospect of changing climate? Some policy-makers have refused to accept that the greenhouse effect is real at all, but even the most reluctant now acknowledge that something is afoot. The protocol agreed at Kyoto in 1997 was intended to reduce carbon dioxide emissions by 55 per cent in the period from 2008 to 2012. This would not solve the problems, but was seen as a useful first step. However, the protocol is not yet ratified by all the significant powers, and notably not by the United States. President Bush's stance – that the USA is not prepared to do anything to inhibit its own industrial growth – should surely be seen as anarchy on a global scale, far more frightening than the terrorism that has launched a thousand warplanes. His notion – that it is better to adapt to change than to prevent it – is whistling in the dark. It may be possible to adapt to some extent. But it may not. If the prairie dries out like the Sahara (and nothing can be discounted) then it will at least be interesting to see what form this adaptation would take. The American dust-bowl of the 1930s shows the kind of thing that can happen. The steadily increasing force and number of tropical storms, and their inexorable spread northwards, ought to give pause for thought. If President Bush was simply inviting suicide on behalf of his own electorate it would be reprehensible enough; but of course, the atmosphere knows no national boundaries and if America's climate really does spin out of control, then the whole world is in trouble.

Now Australia, too, has taken a Bush-like line. Its prime minister recently announced that Australia preferred to work its way around future changes in climate, than to take steps to ameliorate them. If I were Australian I would be terrified. The climate of that flat and ancient land swings most athletically of all the continents. Most of the centre is desert.

The monsoon lands of the Northern Territory can be swamp for a decade, and as arid as the Gobi for the next. Agriculture has been established in Australia only through heroic inputs of money and effort from Europe, with many a horrendous failure along the way. The Aborigines did not establish regular farming in Australia in all their 40,000 years of occupation not because they did not know how to do it (for a long time during the Ice Age Australia was joined to New Guinea, and the people of New Guinea are accomplished horticulturalists), but because they reckoned it was not worthwhile. It was better to manage large areas in a loose-reined way than to settle and farm in one place. The agriculture and horticulture of modern Australia, spectacular though they can be with all that rolling wheat and fruit and cattle and sheep, are precarious. The climate is too changeable for comfort as it is, but if I were Australian I would prefer the devil I know. The idea that such a country can simply absorb further climatic instability, like a boxer riding a knock-out punch, is a gamble I certainly would not like to take – or to be taken by an elected government on my behalf. Perhaps the present drought will bring a change of heart.

Whatever the future holds, the key endeavour must be agriculture, just as it always has been. What agriculture is, and how it came into being, and the many ways in which it can be practised, is the subject of the next chapter.

2

Farming: What It Is, What It's For, How It Became and How It Works

Without farming, humanity would simply be subject to the ecological law which says that 'big, fierce animals are rare'. Specifically, so history suggests, the world population would probably be a mere one-thousandth of what seems probable by 2050: 10 million, as opposed to 10,000 million. So it's obvious what farming is for. It's for feeding people.

Is it, though? If some hyper-intelligent and dispassionate Martian, some cosmic wizard versed in biology, were to assess the problems of feeding the human species, and then looked at agriculture as it is practised, he would surely be puzzled. His knowledge of the earth's ecology and of human physiology would tell him that we, human beings, could easily produce enough to feed everybody well; and indeed could go on doing so for many thousands of years to come, and effectively indefinitely. Our Martian, a bit of a time-lord, would take it to be self-evident that we ought to be thinking about the next few thousand years; that we should not be living entirely for the present. Yet he would perceive, for who could not, that while some are starving, others are grossly overfed; that many endure specific diseases of deficiency (like the blindness caused by lack of vitamin A, which affects around 40 million children worldwide), while others suffer diseases caused by specific excess (for example, diabetes). He would also see that millions of people each year are poisoned in one way or another by their food, and quite a few of them die; yet, he would observe, most of this could easily be avoided. Many domestic animals die too, long before they are due to be tidily slaughtered, often in horrible circumstances. Why should this be so? He would perceive, too, that if we continue to handle our own food supply as we do now, our chances of surviving even the next few hundred years in a tolerable state (alive and well and in tolerable democracies) are somewhat up in the air; and that's even before we take probable climate change into account.

Our hypothetical Martian would also be aware that human beings are rational creatures; not quite in the same league as himself, obviously, but brighter than, say, algae. Indeed, modern human beings pride themselves on their rationality. Given that we practise agriculture in the way that we do, it's obvious that feeding people is not our priority. If it were then we, rational beings that we aspire and profess ourselves to be, would surely do it differently. On the back of an envelope he would work out what it is we would need to do if feeding ourselves was really our prime aim, now and in the long term. An envelope would be enough: conceptually the problems are easy. Then he would ask himself why, if the task is so obvious, and so eminently doable, and we are supposedly rational creatures, we choose to adopt a completely different course that in the short term has such obvious deficiencies and on the slightly longer term is liable to prove disastrous. This would take longer, and I will attempt to give my own answer in Part III. First let's look at what we need to do.

The Tasks of Farming

In practice, the grand aim of 'feeding people' has several component tasks. First, and most obviously, we have to produce *enough*, and make sure it gets to the right places. Sufficiency – yield, productivity – is the prime desideratum.

Then, of course, the food has to be of the right kind. The basic, staple components of diet are energy (amounting to an average of around 2,500 calories per person per day), with a proportion of protein (equivalent to around 5 per cent of the total). But human beings also need an array of essential fats, vitamins, minerals, and (it now seems) a newly defined and largely neglected category of 'functional foods', which in status are somewhere between vitamins and tonics. In other words, the food has to meet nutritional requirements.

Mere nutrition isn't enough, however, taken alone. The psychology of eating is immensely important. Like all animals, human beings have a keen sense of what can be counted as food, and what cannot; and to a much greater extent than other animals, our tastes differ from society to society and from age to age, obviously depending largely on experience and upbringing but also on genetic predisposition. Many, for example, cannot tolerate lactose, the characteristic sugar in milk, while others revel in it. People who are starving have been known to refuse food that simply did not conform to their idea of what food ought to be. In better and more normal times, food has endless social connotations: sometimes a pizza on the hoof is appropriate, sometimes a 28-course banquet, as favoured in diplomatic circles in China. We build our lives around the nuances of food. In short, gastronomy is supremely important, as well as mere nutrition.

With sufficiency, nutrition and gastronomy in place, we still have to worry about safety. The most delectable food, carelessly prepared, can be lethal. Priceless and delicious oysters have sent many a reveller to an early grave. Pollutants are of two main kinds: infective agents, including worms, protozoans (such as parasitic amoebae), bacteria and viruses; and toxins, which may be produced by bacteria (as in botulism) or fungi (mycotoxins are an enormous source of morbidity and mortality worldwide), or various human-generated pollutants including organic compounds of heavy metals. Pesticides and other agrochemicals are also undesirable, of course, although direct evidence of specific harm from contaminated food is not so thick on the ground as commonly expected. (Direct damage from, say, being sprayed with dieldrin or some such, inadvertently or merely carelessly, is a different matter.) Still, the generalization remains: the food should not only be sufficient, nutritious and delectable, but also safe.

There is one more key desideratum, immediately spotted by the time-lord Martian. However we produce food, we should be able to go on doing it. Our descendants will still be here in 1,000 years, or 10,000, or a million – or at least they could be. There is no obvious biological reason why not. If we spare a thought for them at all, then we ought to be thinking not just of the next generation, or until the next election as politicians do, but of the next 10,000 or million years. We don't need to think about it all at once. But each generation needs to ensure that it leaves this world as safe, tidy, diverse and potentially productive as it came into it – or preferably better. To the list of sufficiency, nutrition, gastronomy and safety, we must add sustainability.

So far so good: yet all this, as it stands, is a little too human-oriented (anthropocentric) and a little too materialistic. Human beings are flesh-and-blood animals, but we are spiritual beings too. To function fully and to be content we need to be well fed; but we need more than that. Besides, we do not occupy this world alone. There are other species, too

– perhaps hundreds of millions of different kinds, even though we have so far made the acquaintance of, and put names to, less than two million. In anthropocentric vein, we might acknowledge that those other creatures contribute hugely to our own contentment, if only because we enjoy looking at them. In more altruistic vein, we might properly acknowledge that those other creatures are important in their own right. We have no right to obliterate them, or to compromise them more than is absolutely necessary.

Thus, a series of important codicils should be added to the basic policy of human food production. Food should be produced kindly. We like to eat meat and in practice it makes good biological sense to raise some livestock. But having opted to take over the lives of domestic animals, we should then feel duty-bound to keep them (and at the end to slaughter them) as kindly as possible. Animal welfare matters, in short; and should be built into the grand strategy.

Wild creatures matter, too, and the landscapes in which they live. However we farm, we should do it in ways that do not destroy our fellow species, or their habitats, and where possible should encourage them to flourish. Farming, in short, should be wildlife-friendly.

Finally, of course, farming affects humanity itself in many different ways. Agriculture is huge, in all ways: the greatest by far of all human endeavours. Structurally, in human affairs, it is like the neck of an hourglass: all history, all culture, all technology, feed into it; and in turn, it feeds into everything else that we do. Indeed, agriculture at any one time reflects, and contributes to, the entire *Zeitgeist*. In short, the social and political connotations of farming are crucial too. In particular, farming has been at the heart of all the world's rural communities for many thousands of years. Through most of that time it has been the principal employer. In the West at least it seems largely to have abandoned this role, or at least to be careless of it; and where the West goes today, the rest of the world tends to follow, partly of its own volition but largely because it tends to find itself with very little alternative. The social shift out of farming this past hundred years has been among the most dramatic and significant of all human history; noticed less by modern people than it should be because most modern people, particularly in the West, are urban, and on the whole, urban people get to write the commentaries.

Agriculture, though, surely should employ people, and sustain rural communities, as it has always done. The United Nations demographers tell us that by 2050 there will be more people living in cities than currently live on the whole of the earth. Rightly, this is perceived to be a problem. As physical structures, many cities have already outgrown their strength. In Bombay, Mexico City, and a hundred other conurbations worldwide there isn't enough water and there aren't enough drains. More and more major cities are surrounded by shanty towns, which are commonly bulldozed, *faute de mieux*. The scent that greets you as you step off many a plane in tropical countries is hard to pin down at first, until you realize it's the sweet smell of human faeces. In the city streets, there is nothing useful for people to do. As the cities grow, and the engineers and the architects do battle with the arithmetically impossible, and the politicians wring their hands, those same politicians contrive with the help of other experts of a different stamp to reduce agricultural labour by as much and as rapidly as can be managed. It doesn't seem to occur to them first, that agriculture in general benefits from more people, as cities, after a certain point, generally do not; or that the countryside could and should and traditionally did provide the principal solution to urban squalor and all the horrors that go with it.

For those farmers that are left there should be justice, just as there should be for all human beings, though farmers, worldwide, have drawn some of the shortest straws these past few decades of anyone in the world. To complete the list of agricultural desiderata we must add employment, and producer-friendliness. I have put this last on the list not to downplay its importance, but for emphasis.

The first set of considerations – those that have to do with the food itself – are biological in nature. On the one hand they have to do with the ecology of the planet, and on the other with the physiology of human beings and of the animals and plants that we raise for food. The second set of considerations – to do with wildlife, livestock, and the well-being of producers and rural communities – are of an aesthetic, moral and social nature. The biological considerations have to do with material reality: what is possible; how to get the best out of living systems without wrecking them. Morality and aesthetics (I define aesthetics broadly, to encompass what many would call spiritual values) are matters of human value.

To understand why present-day farming does not meet humanity's needs as a sensible observer would surely suppose it should, we have to look at the historical, economic and political trends and ideas that have driven it off course. But first we need to look at agriculture itself: what it is, how it works, and the meaning of good husbandry. It seems worthwhile to begin at the beginning, and to ask how farming came into existence in the first place. How did we make the transition from versatile, knockabout, predatory ape, to farmer? In recent years, theory has shifted; and the differences between the old and the new accounts are salutary. The following account of traditional thinking has a touch of parody, in the interests of brevity. But only a touch.

How Farming Became: The Traditional View

To begin with, traditional scholars took it to be self-evident that agriculture is good. Common sense says that it is better to have food on tap than to track it down and seek it out afresh each day as hunter-gatherers seemed obliged to do. The birth of agriculture therefore represented progress, and westerners are brought up to believe that progress is a real thing and is self-evidently good; indeed that progress is effectively synonymous with destiny. Farming increases the amount of human food that a given area of land can supply. That is what it's for. Therefore farming allowed human numbers to rise, which seemed biologically desirable – and also, by increasing, to fulfil God's injunction to Adam and to the sons of Noah, to 'go forth and multiply'. Indeed it provided surpluses of food and so, as a bonus, freed more and more people from the task of supplying food, so they could specialize in other things. It also enabled people to stay in the same place for long periods – perhaps indefinitely – and so encouraged them to build. Put large numbers, specialization, and permanent residence together and you can build cities. The Latin for an inhabitant of a city is *civis*, root of the word 'civilization'. Thus farming gave rise to civilization: monasteries, universities, scholarship, science, arts, philosophy. Obviously, then, on all fronts, farming was, and is, good.

Traditional scholars were happy to suppose, too, that farming was consciously invented. How else do human beings make progress? Some Stone Age Galileo, seeing further than the rest, must have reasoned it out – for traditional scholars also took it largely to be self-evident that human progress is rooted in reason, and is nudged ahead by great men with broad foreheads; people very like themselves, in fact.

The archaeological evidence seemed to bear out such an interpretation, and scholars of all kinds, for the best of reasons, like to stick as closely as possible to the evidence. Before about 10,000 years ago, the archaeological record shows no traces whatever of cultivation or domestication. Then, about 10,000 years ago, the time of the New Stone Age, or Neolithic, traces of wheat and barley appear in the Middle East, notably in human sites at Jericho in Palestine and Čatal Hüyük in Turkey, that are different from those that grow wild. In particular, the grains are bigger and fatter. Clearly they have been selected, and cosseted in fields: in other words, they have been cultivated. In later sites come the first traces of domestic animals. Their remains tend to be smaller than those of their wild counterparts – perhaps because they have been underfed, and perhaps because they have been selected for ease of handling. Both are signs of domestication. In addition, domesticated animals tend to be killed while young, and all tend to be roughly the same age at slaughter. Those caught wild are typically of all ages.

More and more sites of early agriculture from later dates have turned up all over the world: in the Indus valley in Pakistan; in India; in China; and in various sites in North, Central and South America. Always the pattern seems much the same. Over a long, early period, there is no sign of cultivation at all. Then the first traces appear; and then the people in each spot seem to take to farming with a will, and it quickly becomes the norm.

So the overall picture seemed rather straightforward. Some genius in the Middle East around 10,000 years ago discovered the art of farming. Like Archimedes discovering the means to measure density, he (surely it was a he) was overjoyed. His peers were equally pleased. They knew a good thing when they saw one and perhaps they sensed in their bones, like the Renaissance scholars and financiers of late-fifteenth-century Florence, that they and humanity in general were on the verge of something big.

There were only two obvious problems with such an account. One was to explain why farming, when once discovered, apparently spread rather slowly to other places; for farming seems to have originated at different sites at intervals of centuries or even of thousands of years.

The other was to explain how agriculture spread at all. There seemed to be three plausible scenarios. Perhaps people stayed where they were, and the word spread from tribe to tribe by a kind of Chinese whispers, or a sound wave travelling through a cymbal. Perhaps conquering people – the successful farmers themselves – spread out and took their new ways with them. Or perhaps (so some historians have often felt) human evolution should be seen as a slow fulfilment of destiny. Perhaps each group of people is destined to discover agriculture at some time, but some groups take longer than others. It seemed fitting to Western scholars that the people of the Middle East were first off the mark, since Western civilization clearly owes much of its cultural heritage to the Middle East, and in particular to the stories of the Bible. Obviously the Middle Easterners would be the most advanced people. By the same token, non-Europeans would obviously be slower to climb the ladder to modernity. The Aboriginal people of Australia did not farm, simply because they had not yet reached the appropriate stage of development.

This is a parody of the traditional view of farming's origins – but only slightly so. All the elements in this story are in the old accounts: the idea that farming is a conscious invention; that it self-evidently represents 'progress'; that such progress must have been recognized as such, and welcomed; that it began in one place at one time (in the Middle East, with the Neolithic Revolution), and spread from that single origin; that there is, in the birth of farming, a sense of 'Man' fulfilling his 'destiny'. The story is wonderfully coherent, and it flatters all the conceits of the modern West: that whatever we do is done by conscious choice; that whatever we choose is self-evidently superior; that change over time is progress, and our own progress has been perfectly rational; and that whatever we (westerners) do represents the destiny of humankind as a whole. Taken all in all, at least to a Western scholar and those who wish to be flattered, such an account seems irresistible. The underlying psychology of it persists. Western leaders still speak and act as if they believed that it is the destiny of westerners to make progress, and to lead the world as a whole along the same path, and that the progress itself is rooted in rationality, and is self-evidently good.

The traditional account is also, so more and more scholars have suggested, almost total nonsense.

The Modern View: How Agriculture Really Became

The traditional view of farming's origins depended on a series of assumptions which, seen from the point of view of modern westerners, seemed almost too obvious to be worth questioning. It seemed obvious that people who farmed would be better off than their hunter-gatherer predecessors. But in the short term this turns out to be nothing like as true as it seems, and sometimes, in some ways, the new farmers were clearly much worse off. Traditional scholars took it to be self-evident, too, that the trappings of civilization – including division of labour, trade to increase the range of available materials, and then the building of cities and the development of art – all depended on a solid, reliable food supply that could be provided only by formal agriculture. Certainly, a reliable food supply is vital (Napoleon said that an army marches on its stomach, and this is just as true of societies as a whole). Yet more and more archaeological and modern anthropological evidence suggested that people can and do specialize, and trade over spectacular distances, and develop intricate religions and fine art, and even build respectable settlements that could well qualify as cities, without being formal farmers. Such observations raise a different kind of puzzle, and a deeper one, for modern westerners tend to assume that whatever people do, they do for a good, rational reason. But if people became worse off by farming, and could enjoy the advantages of civilized living without it, why embark on it at all?

All in all it seems that the most significant ecological, economic and social shift in all human history was made without any deliberate policy at all, and indeed was in many ways perverse. This is both shocking and salutary. It illustrates that we human beings are not in control of our own affairs to anything like the extent that we like to think. Actually, of course, some cultures never did assume that human beings are in charge of their own affairs. Many, including the old Greeks and the modern Muslims, emphasize the precise opposite. But it's a powerful theme of modern Western culture that we, human beings, are very much in charge. The rise of agriculture, with very little conscious input at all, and in the face of privation, deals a powerful blow to that most fundamental of Western premisses.

The evidence that the first farmers were not better off than their hunter-gatherer forebears is of three kinds. First, it seems that the members of early farming communities were often sick, in ways that non-farmers generally are not. For example, Theya Molleson at London's Natural History Museum found a very strange pattern of arthritis in the bones of ancient Egyptians, unlike any seen elsewhere. Most obvious were peculiar deformities of the lower back and feet. How on earth had these come about? She recalled Egyptian murals that showed people grinding corn with a saddle-quern. This consisted of two stones, one on top of the other, with the corn crushed to meal or to flour in between. Two workers (commonly construed to be slaves) knelt at opposite sides to push the top stone forwards and backwards. It is a monstrous occupation. The back is under tremendous strain. The toes are bent under the feet to provide extra purchase. The result is as Dr Molleson reports. Among the old Egyptians, too, other studies typically show rows of abscesses – perforations of the bone around the roots of the teeth. These, it is supposed, were caused by eating grain with too much sand in it. There are many other studies showing general feebleness or specific pathologies among early farming peoples.

In contrast, studies of modern hunter-gatherers show that they often have an easy life, and although we cannot assume that our hunter-gatherer ancestors of the Pleistocene lived exactly as their modern counterparts do, there is no obvious reason to suppose they would have been worse off than their modern counterparts. After all, before there were farmers, hunter-gatherers had the pick of the land. Now, they are forced to the sidelines.

The outstanding study was by Richard Lee and Irven de Vore in the 1960s, of the !Kung Bushmen of the Kalahari. Common sense and traditional scholarship supposed that the Bushmen, living in some of the harshest terrain on earth, must spend all their waking hours desperately trying to gather enough vegetation, and risking their necks to eke out their tough and dreary provender with a little stringy game. In practice, Lee found that the men hunted for only about six hours a week, and spent most of the rest of the time sitting under the few shade trees, telling stories. These trees notably included the mongongo, source of so many fruit and nuts that they supplied the !Kung with half their calories, yet still there were so many left behind that many simply rotted where they fell. The women and children gathered enough vegetation to provide the basic staple diet with just two or three days' effort per week. The diet included about 105 different kinds of plant – far more than ours is likely to do.

Even so, wouldn't the Bushmen have taken up farming if they had thought of it? Well, it's clear now that they often did think of it. They were perfectly capable of herding goats, for example. Archaeological studies have now shown that at several times in the past, often for long periods, the Bushmen grew crops. From the seventh to the eleventh centuries AD come traces of sorghum, millet, melons and cowpeas. But farming, it seemed, they could take or leave. Sometimes it suited them, and sometimes it did not. The nineteenth-century scholarly conceit that once the hunter-gatherers sniffed the possibilities of farming they were bound to leap at it, emphatically was not borne out. Other peoples, both in Africa and Australia, have shown a similar reluctance to hang around watching cows when other distractions were on offer, including the thrill of the chase.

The third source of evidence – or at least, some people including me suggest that it should be treated as evidence – is the Bible. Primarily, of course, the Bible is a work of theology. But like any great work of literature (like the novels of Charles Dickens or George Eliot, for example) it also in passing provides a very fine first-hand account of life as it was lived by another people in another age. The first story, in Genesis, is that of Adam and Eve in the Garden of Eden. Eden was Paradise, and Adam and Eve, the first two human beings on earth, had the freedom of it. At least, there was one restriction (Genesis 2: 17): 'But of the tree of the knowledge of good and evil, thou shalt not eat of it'. But, tempted by the serpent ('more subtil than any beast of the field which the Lord God had made' – Genesis 3: 1), they did eat of it: Eve first, and then Adam. So God expelled them. The story is typically explained as a morality tale: the hubristic (as the Greeks would say) flouting of God's will; the loss of innocence, as the world's first people dared to acquire knowledge that had been forbidden to them. But the story can also be read quite literally. Life in the Garden is a folk memory of ancient days of hunting and gathering, when people could simply take what they wanted from a wild environment that was endlessly various and generous. The expulsion is the birth of agriculture. No longer can they simply take what they want. Whatever they want, once they are farmers, they have to work for.

The text reinforces this interpretation in various ways. First, there is the curse that resounded in Adam's and Eve's ears as God threw them out (Genesis 3: 19): 'In the sweat of thy face shalt thou eat bread, till thou return unto the ground'. Surely people made bread before they were farmers, grinding and baking (and perhaps fermenting) the wild grains that they gathered. Querns have been dated from 18,000 years ago, and people seem to have been controlling fire for at least a million years and could have been accomplished cooks long before they were farmers. But in the early days, so the Eden story implies, the grain for the flour was there for the taking. Now, and forever more, it would require sweat.

In addition, Jurin Zarins at the University of Missouri has sought to find the location of Eden; and his conclusions are very intriguing indeed. The Garden of Eden appears to have been in an area that is now several kilometres offshore in the Persian Gulf. How could people have lived at the bottom of the sea?

The answer is that around 8,000 years ago much of the Persian Gulf was not beneath the sea. The Ice Age had not yet run its course. Huge volumes of water were still entrapped in glaciers on the northern continents; so much, in fact, that at the height of the Ice Age, the sea level was 200 metres lower than today. By 8,000 years ago the ice was beginning to melt, but still the oceans were depleted.

When it was dry land, the Persian Gulf could have been wonderful. It would surely have abounded in herbs, trees, fish, molluscs, resident and migratory birds, probably turtles, fallow deer and gazelles. But then the ice on the northern continents melted, and the sea rose, and the people were driven inland, up on to the high, dry hills that they had formerly considered hostile. They found grasses of the genera now known as *Aegilops, Hordeum* and the weedy *Avena*: the forerunners of wheat, barley and oats. With their paradisiacal hunting ground gone, and forced into a small space, the people were now obliged to cultivate the grasses that they once would simply have picked – or perhaps shunned, in favour of more interesting fare. Now, truly, they were eating bread in the sweat of their faces; and we, their cultural if not their literal descendants, have been sweating ever since.

There is one obvious snag in this scenario, however, at least as presented so far. It is certainly plausible that people lived in what is now the Gulf, and were forced inland by the rising seas. It is clear that after about 10,000 years ago, they were cultivating grain inland. But it is difficult to imagine that they discovered the arts of cultivation just at the time they needed to; that as soon as they found it necessary to subsist on upland grasses, they worked out the means of growing them.

The probable answer to this is again commonsensical, and again is suggested by observations of modern hunter-gatherer people: for tens of thousand of years before the Neolithic Revolution, people were already, to a significant extent, managing the local vegetation and wildlife. Adam and Eve, meaning the people of the Persian Gulf, weren't just hunters and gatherers. They were also gardeners and game managers.

It's a mistake to think of farming as one single activity, which you either do or you don't. It has many different components; and the ones that may seem the most fundamental were sometimes, probably, the last to be adopted, and the ones that may seem the most peripheral may often have been the first in place. Any one of the component activities can make a big difference, even in the absence of the others. Thus, farming can develop piecemeal, step by step. Thus, too, farming defined broadly ceases to be unique to human beings. Many other animals clearly practise elements of it.

Thus, one of the key elements of modern agriculture is pest control. At its most fundamental, this simply means keeping rivals at bay, of one's own and of other species. Many creatures guard their favourite prey species. Ants drive other ants, and other insects, away from trees such as acacia where they forage for their own prey. Many reef fish stand guard over their favoured patches of algae. In the forests of South America, spider monkeys drive woolly monkeys away from whatever trees are in fruit; and so on. All are practising pest control. Weed control is a related function. Sheep increase their own pasture as they graze, by eliminating rival plants from the pasture's margins. Pruning is related, too: removing clutter from favoured plants. Woolly monkeys in captivity have been observed routinely breaking dead branches from the trees they intend to occupy.

Animals in general tend to be specialists, each with just a small repertoire of tricks. Human beings are typically versatile (and so are apes); and we can imagine early human beings, though primarily hunter-gatherers, adopting all or several of the devices of cultivation that are practised by other animals. Of course they would have protected their favoured prey. They would surely, from early times, have practised weeding: pulling up and generally attacking plants that did not please them, to make way for those that did. It's a small step to deliberate planting. They would have known that seeds grow into plants: they could see this happening. Besides, especially in tropical conditions, a stick thrust into the ground would often make a new tree. Furthermore, even such ministrations as these change the prey species genetically. A plant that has human beings to look after it does not need thorns to protect it – and since thorns are expensive to grow (they require metabolic energy) the thornless types seem liable to prevail in circumstances where thorns are not needed at all. If people deliberately favour the fruit trees with the sweetest fruit, or the plants with the largest tubers – taking just some, and leaving the rest to grow on – then sweet fruits and large tubers are spread at the expense of their meaner rivals.

All this is speculation, but not of an outlandish kind. It is all commonsensical. It seems, too, that virtually all the hunter-gatherer people who have ever been studied, including the !Kung and many forest peoples, do manage and hence seek to improve their environment in the ways outlined. Rhys Jones, anthropologist at the Australian National

University, Canberra, points out that although the Aborigines have never apparently practised formal agriculture, they are excellent game managers nonetheless. In particular, they set fire to the bush in a very controlled manner, burning off the rank vegetation, allowing green pasture to come through – and thus attracting prey animals, which they then capture. They also use fire to corral animals. In fact, says Rhys Jones, they are 'firestick farmers'. They do not farm Australia's uncertain land in a formal, European sense, not because they don't know how to do it but because, without the kinds of inputs that Europeans have brought to bear in such abundance, European-style farming is supremely inappropriate in a land of such extremes and uncertainty (as many a European-Australian farmer has discovered). Besides, farming gained its foothold in Australia through sheep; and sheep were a Eurasian import.

Formal domestication of animals shows up much later in the archaeological record than formal cultivation of plants. Yet it may well be comparably ancient, and achieved in the same kind of stepwise fashion. Some biologists now speculate that the human association with dogs may extend back for 100,000 years. In this, canine intelligence plays a part, as well as human. We can readily see how each species, our own ancestors and dogs, saw mutual advantage in the association. Dogs are supreme scavengers, and would find rich pickings around the camps of hunters and gatherers. Among other things they eat human faeces and thus, unlikely though it may seem to modern sensibilities, they are agents of hygiene. Dogs are also accomplished hunters – and our ancestors may have cashed in on this, just as the dogs cashed in on human hunting. I have seen this principle in action in Mudu Mulai, South India (Andhra Pradesh) where, early one morning, the local wild dogs, known as dhole, killed a deer. Immediately the villagers went out with sticks, driving the dogs away, claiming the venison for their own. In another village in the same area a wild boar emerged regularly from the forest to feed on the garbage. There is a great deal of overlap in ecology between dogs and pigs. Pigs are clever, too; and in many societies, dogs are food for people at least as much as they are companions.

More broadly, Western historians have tended to suppose that our ancestors, the first livestock farmers, deliberately captured wild animals and heroically domesticated them, overcoming tusk and claw with courage and superior wit. This, to be sure, is how Asian people still capture and subjugate wild elephants. But Stephen Budiansky has argued in *The Covenant of the Wild* (1995) that in practice, many wild creatures would have found it convenient to tag along with ancient peoples, who provided nutritious garbage that suited pigs, dogs and fowl, but also (cf. the Australian Aborigines) may have provided protected grazing. The grazers, like sheep and cattle, may not have deliberately sought the company of people as pigs or dogs may do but, says Budiansky, if human beings did indeed provide protected grazing, then the grazing animals that took advantage of what was on offer would thrive. In short, natural selection would favour the grazers that put up with people, at the expense of those which (like most deer and antelope) prefer to keep their distance. The title of Budiansky's book captures the idea beautifully: *The Covenant of the Wild*. Modern reality surely vindicates his thesis. Domestic cattle and sheep are the most common large mammals. Wild species of cattle and sheep have all become rare and some types (like the European aurochs, the ancestor of modern European cattle) have become extinct. Natural selection is not concerned with the dignity of creatures, or of human beings. Genetic proliferation is the name of the game. And the animals that threw their lot in with us have proliferated while their proud, pristine cousins have faded.

Put such thoughts together and it becomes easy to suppose that our ancestors were managing their environment to a significant degree for tens of thousands of years before the Neolithic Revolution. Perhaps the process began hundreds of thousands, or millions of years ago. Some palaeoanthropologists, including Professor Clive Gamble of Southampton University, argue that human beings radically changed direction around 40,000 years ago, in the Upper Palaeolithic (later Old Stone Age). Then, he says, the tools that had been much the same for the previous million years – generally rather cumbersome and limited in range – became more subtly shaped, lighter and more various. Perhaps it was then that people switched from being mere inhabitants of their various landscapes, accepting their environment as it was, as lions do, to being managers. Whenever the shift took place, whether gradually or in bursts, even small manipulation of the environment could allow the human population to rise significantly above what it would be if people simply took what unimproved nature put on offer. Yet the kinds of changes outlined here, including the steadily deepening symbiosis with dogs and pigs, would not show up in the archaeological record. Not until domestication was well advanced – the grains significantly bigger, the livestock noticeably smaller – would the symbiotic associations be detectable. But the ecological significance would be enormous. By manipulating their environment, cryptically but decisively, human beings would become more common, and more widespread; and since humans are ecologically significant creatures, as hunters and as manipulators, their increase and spread significantly affected other creatures.

Such a scenario, far more realistic than the traditional notion of the Stone Age Galileo, puts a quite new spin on the origins of farming – and explains much that is otherwise mysterious.

Most relevant to the present discussion, this view explains how it was that the people who were driven from the paradise of the Gulf to the unforgiving uplands, were able so quickly to cultivate wheat and barley. In principle, at least, they already knew how to do it. Their ancestors had known for many thousands of years. It also becomes easy to explain how agriculture apparently spread from location to location, even crossing the Atlantic; or (the alternative explanation) how it was that so many different people learned independently to cultivate. The answer is that neither of those unlikely scenarios is actually the case. It was traditionally believed that the first human beings in America were the 'Clovis' people who arrived around 11,000 years ago, having walked across the land-bridge between Alaska and Siberia, which formed as the sea level fell during the Ice Age. Now there is evidence that people first reached America more than 20,000 years ago. In either case, I suggest, the newcomers already had the elements of farming when they made the crossing – and the Clovis, arriving a mere 11,000 years ago, certainly did.

The Clovis and their immediate descendants moved in a wave right down to the southernmost point of South America, Tierra del Fuego, and as the wave advanced, so the native mammals – particularly the big ones – disappeared. The casualties included giant species of elephant (mastodons and mammoths), peccaries, giant ground sloths, giant armadillos known as glyptodonts, and an array of camel-like creatures known as litopterns, which lived in South America. Many giant predators that had preyed upon these weighty herbivores similarly disappeared, including sabre-toothed cats, the giant dire wolf, and a huge, long-legged, short-faced running bear, built like a Rottweiler but half as big again as a grizzly, which apparently hunted in packs; surely among the most fearsome predators of all time. Some suggest that a change of climate killed these beasts, but many feel that the Pleistocene hunters did for them – for if it was climate, why weren't the small mammals affected at least as much? I suggest that if the Clovis people had also learnt the arts of cultivation, then this would increase their prowess as hunters. The principle is like that of suburban cats who are well fed on Whiskas and then play havoc with small birds. Normally, if a hunter kills too many prey, then its own population must go down. But if its diet is supplemented, then it remains fit and strong even when the prey is reduced, and its own numbers stay high. Thus we can envisage that well-established Clovis people, though perfectly well able to survive without mammoths to prey on, nonetheless picked them off at will: and since mammoths were surely prestige prey, the rarer they got the more they would be targeted. Thus the background of environmental management reverses the normal logistics of predator–prey relationships, and greatly tips the balance in favour of the predator. The large mammals that survived in North America were those that by chance had special survival strategies. The big-horn sheep lived high in the mountains, in the Rockies. The moose lived deep in the forest, and to the north. The buffalo, or bison, undertook long, fast, and unpredictable migrations. None were easy to pin down – although some of the people we used to call 'Red Indians', presumably the direct descendants of the Clovis, based their entire economy on the buffalo, for many centuries.

Similar logic, I suggest, could have accounted for the demise of the Neanderthals in Europe, 30,000 years ago. We need not envisage that our own European ancestors – the Cro-Magnons – hunted the Neanderthals to extinction (although they might well have done; genocide is common among human beings). All we need envisage is that the Cro-Magnons, from 40,000 years ago onwards, were managing their environment in a way that the Neanderthals were not. So the Cro-Magnons grew steadily more numerous, relative to the Neanderthals. So, too, they eroded the Neanderthals' prey-base – again, largely mammoths. But this affected the Neanderthals, who were pure hunters (and gatherers) far more than the Cro-Magnons, who were manipulating a greater variety of their fellow creatures, in a variety of ways. This, then, became a rivalry of logistics and economics: and the fight, inevitably, went to the more efficient. In the same way, modern systems of agriculture oust ancient ones simply because they allow the people who practise them to multiply faster.

Such a scenario clearly puts a quite new spin on the Neolithic Revolution of 10,000 years ago. It remains as a watershed – clearly, the unequivocally cultivated barley and the nearby signs of early cities indicate a very important change. But they do not represent the beginnings of agriculture.

On a general and pedantic note, we may observe first of all that single archaeological sites found in isolation, like single fossils, very rarely represent the true 'beginning' of anything. It is rare for any artefact to be preserved, to become evidence for archaeologists, and even rarer for any bone to become fossilized. So in general, nothing is likely to appear either in the archaeological or the palaeontological record until it is already well established. So it was never acceptable to argue that farming began with the sites at Jericho and Čatal Hüyük. We can claim only that farming is *at least* as old as those sites.

What such sites do represent is the emergence of large-scale, settled agriculture. In particular, they represent the first, formal, large-scale cultivation of cereals – the beginnings of arable farming. Arable farming, as discussed later, is by far the most important component of modern farming, and is the prime reason why human numbers have risen so spectacularly since the Neolithic. Because the Neolithic (probably) represents the start of arable farming, its historical importance remains as great as ever.

So now we can re-address the question raised earlier: Why, if farming was so much less agreeable than hunting and gathering – more work, producing less-healthy people – did people take to it full time? Why, when they could combine hunting and gathering with a little casual cultivation as the !Kung Bushmen did until so recently, and as people apparently did in the plain that is now the Persian Gulf? Again, there is a general answer, and probably a series of specific answers.

The general answer is that the casual cultivations, combined with hunting and gathering, allowed human numbers to rise. At the same time rising numbers, and the management practices themselves, caused some at least of the local game to die off: spectacularly so in the Americas, where the native animals had always been completely innocent of humans. Thus while human numbers rose, the ability to live purely by hunting was compromised. As people began to manage their environments to greater and greater effect, the more such management became necessary. Thus, once the first hunter-gatherers embarked on environmental management, they began a vicious cycle: the more they manipulated their surroundings, the more they needed to manipulate them. Eventually, at site after site, human numbers grew so great, and the local environment had been so compromised, that the people needed to cultivate large-scale, and full-time, to survive at all. We have no reason to assume that our ancestors wanted to go down this route. We have evidence that they did not. But the shift was forced upon them. The specific circumstances that may have promoted such a shift in any one place are not usually known; but the flooding of the Persian Gulf at the end of the Ice Age, which drove the ancestors of Israel inland, was surely one of them.

One last note. Natural selection was first described by a biologist, Charles Darwin, to explain the mechanism of evolution, by which species change over time and (so Darwin intended) new species emerge. But in truth it is a universal force, which applies to any group of entities or systems that are obliged to compete. For many thousands of years, people with an aptitude for farming have competed with those who prefer to take their chances with the wild. Farming can be grim, and it is relentless, but it works. It enabled the people who practised it to do as God commanded Adam and the sons of Noah, to go forth and multiply. Natural selection says nothing about the happiness of those who are obliged to compete. The hunters and gatherers of the late Palaeolithic may have been very happy, and the first farmers may have been largely miserable, but the farmers prevailed nonetheless. Happiness and material success are not nearly so closely related as modern politicians are wont to suggest.

There is also a very significant logistical difference between hunting and farming. Hunters can soon overplay their hand. If they hunt too much, they kill too much prey, and then they too die off. They could achieve overkill in a few months, if they really went at it. So natural selection favours hunters who show restraint. Lions sleep for twenty hours a day. Kalahari Bushmen sat under the trees and swapped yarns for five days a week. We may envy them their idleness, but to the predator idleness is to a large extent a necessary survival strategy. More work would be self-destructive.

The logistics of farming are different. To be sure, farmers can overdo things too. By classical times, as we will see, farmers had long been having problems with over-grazing, and over-cultivation. But the timescale is much longer. Generally, at least for farmers working fertile land – of which there was no shortage in the pristine world – the harder they worked, the greater the reward. This is not true of hunting and gathering. The difference is crucial. We have been living in the sweat of our faces this past 10,000 years for more reasons than one. Firstly, we needed to sweat if we were to eat at all. Secondly, those who sweated most would be bound to oust those who were content to take life easier. In short, when our ancestors developed farming, they also developed the work ethic. Make of that what you will.

The logistical shift can also usefully be analysed in terms of game theory. Biologists who have studied evolution in action have often asked what qualities natural selection is most likely to favour. To be sure, birds that fly quickly and with little expenditure of energy seem to have superseded those like the ancient *Archaeopteryx* which clearly flapped with a great deal of effort to very little effect. Clever people like us have replaced our smaller-brained ancestors. So natural selection seems, on the whole, to favour various kinds of technical improvement of a measurable kind.

But there is a more general point. Natural selection favours *systems*: modi operandi of particular kinds that deliver the goods. Supremely favoured is what might be called the 'positive-feedback loop': any system in which any entity, by doing what it does, reinforces itself. Life itself is a positive-feedback loop *par excellence*. Living things, by metabolizing, make more of themselves: by dint of what they do they grow bigger, and then they divide, so there are two where before there was one. Not all systems are self-reinforcing. There are negative-feedback loops in nature, too – systems which, by doing what they do, put a stop to themselves. We might reasonably suggest that the key, logistical difference between farming and hunting is that farming is a positive-feedback loop, and hunting is a negative-feedback loop. The more that people farm, the more food they produce, and the more their population grows, and the more therefore they need to farm and are able to farm, since with passing time there are more people to do the work. By contrast, the more that hunters hunt, the more they destroy their own prey-base and bring their own population to a grinding halt.

Positive-feedback loops beat negative-feedback loops. *But* positive-feedback loops last only as long as the raw materials hold out. Farming remains a positive-feedback loop only as long as the farmers take care to conserve the raw materials of it – notably the soil and water. If they destroy the ingredients, then they slump into negative-feedback mode, like hunters, destroying their own base. One way of looking at the dilemmas of the present world – in fact, perhaps the most fundamental way – is to suggest that we are now shifting from positive-feedback mode into negative-feedback mode, as we finally run down the resources on which farming relies. The world is big and generous, and has stood up to our depredations for a long time, but it is not infinite.

Secondly, the favouring of positive-feedback loops over negative-feedback loops has nothing to do with morality. The positive does not prevail over the negative because it is better in any moral or aesthetic or spiritual sense, but because it works. Natural selection does seem to favour some technologies over others in many circumstances (efficiency of flight, versatility of behaviour) but on the whole it is not concerned with particularities. Natural selection simply says, in the words of the Hollywood adage, 'nothing succeeds like success'. 'Positive-feedback loop' is what 'success' means: a system that reinforces itself; a life form that grows; a financier whose investments make him richer; farming – so long as the resources hold out. In Part III I will argue that present-day 'agribusiness', and the use of science to promote agribusiness, is a prime example of a positive-feedback loop. It works, in the short term, in many circumstances. But it works purely for logistical reasons; not because it is innately good in any moral sense, or is in the long term sensible. It simply feels good to the people who are benefiting from it in the here and now.

This, then, so many scholars now think, is how agriculture really began. Our ancestors did not invent agriculture all of a piece, and shout for joy. They put it together little by little over many thousands of years until, eventually, here and there, they found they were committed to it. Natural selection has favoured the shift. But happiness, rationality, conscious decision and destiny have had next to nothing to do with the case.

Clearly, some societies, like the !Kung, have simply flirted with agriculture. But others, including our own cultural forebears, have steadily increased their control and productivity as the centuries and millennia have gone by. So what, in practice, did they do?

The Paths to Modernity

The Neolithic Revolution was not the absolute beginning of farming but it does represent a very significant shift: from the management of the otherwise wild environment, to the cultivation of fields, and hence to the transformation of landscape and, in particular, to the mass and increasingly controlled production of staples. It was the birth of arable farming. The history of farming since that time has largely been a matter of extending the area under cultivation, and making it more hospitable to crops.

Cultivable fields have been acquired partly by ploughing native grassland (notably in the North American prairie) but probably even more by felling forest. In many tropical countries, 'swidden' or 'slash-and-burn' agriculture has been widely practised. People clear a small area of forest with fire and machetes, cultivate it, and then as the fertility runs down and the pests build up, they pack up and start again somewhere else. Provided swidden is not too extensive, and the farmers do not return too quickly to any one spot, this can be continued indefinitely. Fertility is restored to each patch by the natural processes of nitrogen fixation (of which more later) and the incursion of animals which come to forage and leave their dung, or come and die. The protein-rich and therefore nitrogen-rich corpses just of invading insects soon add up to many tons.

But the people of Europe in particular cleared the forest that grew up after the Ice Age on the grand scale and effectively permanently (though nothing on this earth is ever really permanent). Northern Europe, including Britain, was almost all forested until a few thousand years ago, notably in oak, ash, birch and pine. Some of the pristine forest was cleared for timber. Most of it, to make way for farming.

But forests can be deceptive. Much of the forest of northern Europe in particular grew in deep, fertile soil; and when the trees were removed, farmers could revel in its bounty for century after century – with luck, and care, making it deeper and more fertile with each season. But in general, despite appearances, trees require much less nutrition than most crops. After all, most of the vast biomass of a tree at any one time is dead: timber is composed of dead tissue. A great deal of apparently lush forest grows on remarkably thin and unpromising soil. Take the trees away, and the shortcomings of the substrate are revealed. The Pilgrim Fathers cleared away the conifers on Cape Cod and found sand dunes underneath; very pleasant for modern tourists, useless for serious crops. Tropical forest is now being cleared worldwide, at least in the recent past largely to make way for beef. At the time of writing these lines (July 2002), the Brazilian Congress is considering a plan to reduce the Amazon forest by 50 per cent. But experience suggests that the rich forest of the Amazon will not give way to rich pasture. Without the forest cover, the soil will soon deteriorate, slumping first into savannah and then into desert. Mass removal of trees has a huge effect on local climate, too: to some extent drying, as trees constantly emit moisture: but also leading to flood (notably as in Bangladesh and China) since forest acts as a sponge, holding the water of the sudden rains (notably those of the monsoons) and releasing it gradually over the following months. Nonetheless, much of the world's cultivated land, including most of Europe's, has been expropriated from forest; beginning even in the New Stone Age, and continuing into the present day. The world area under cultivation was increased by 50 per cent between the 1930s and 1960s largely by clearing forest in the USSR and North America. But the land, once cleared (or acquired from grassland) may then be too steep, too rocky, too dry or too wet to be easily cultivated.

The most dramatic solution to steepness is the terrace; dwarf walls built along the contours of the hill, transforming it into a series of platforms like a modern auditorium. There are dry terraces in the Americas and Europe, notably the vineyards and groves of olive and cork oak of the Mediterranean, and the intricate horticulture of Malta, where broad beans and brassicas are raised on cliffsides in fields that are sometimes as small as a suburban kitchen. The terraces of Asia, flooded to make paddy fields for rice, are among the most supreme of all artefacts: a stunning exercise in civil engineering, and extraordinarily beautiful. Terracing is an ancient craft: the earliest known date from at least 3,500 years ago. But they have to be built well, and they require diligent maintenance. If productivity starts to fail, and economies to waver, they collapse, as seen in many a ruin worldwide. Nonetheless, terracing has enormously increased the world's areas of cultivable land and provided some of the best growing conditions, with a range of microclimates. Where the hillsides are rocky, the terraces serve a secondary purpose: getting the rocks out of the way, and stacking them neatly. The tiny, ancient fields of much of Britain, as in the English Lakes and the Yorkshire Dales, each surrounded with a dry-stone wall, are a different solution to the same problem.

The ancient cure for aridity, year-round or seasonal, is irrigation. It has transformed much of the world, and hugely extended the range of farming and hence of the human species. Today, 40 per cent of wheat in the Third World is irrigated. But again, nothing comes free. As described in Chapter 1, water brought up from underground brings minerals with it which, over time, may render the soil toxic; and it is also usually a finite resource. The diversion of rivers for irrigation and industry transforms entire landscapes: it has devastated Russia's Lake Baikal, for instance, and all the surrounding ecology. It has sadly compromised the Jordan. The long-term effect of giant dams is literally unknowable (the ramifications are too complex to be modelled in detail) but certainly are not all to the good. The destruction wrought by Egypt's Aswan Dam, notably though by no means entirely on the erstwhile fishing industries around the mouth of the Nile, was to a large extent anticipated. But the dam was built nonetheless.

Large-scale water works, in short, are a very mixed blessing. We may assume that they are generally built with the best of intent. They may do a lot of people a great deal of good; and sometimes may continue doing good for many centuries. For huge numbers of people, they are necessary; at least, once they are built, people rely on them. Sometimes the good that they do can be easily quantified: for instance if a dam delays the headlong rush of water from mountain to sea, using it for some good purpose along the way, then in theory a great many more creatures (including human creatures) might benefit, than are harmed. On the whole, though, at least when measured over time, dams and other irrigation systems often amount to what military strategists and modern ecologists call a zero-sum game. The good that is done to some people in some places over a few years or even centuries, is balanced by the damage done to others, perhaps at some future time. One huge, complicating factor is that human beings – or at least, the kinds of people who become national and world leaders – have a penchant for the grandiose. Ancient potentates liked to build fountains and aqueducts to demonstrate their power over nature and their social beneficence. Modern institutions such as the World Bank have suffered from the same disease. They like to be seen to do something, and the bigger and the grander, the more appealing it seems. I have tried to show throughout this book that agriculture has not been shaped solely, or even primarily, by the human desire and need to produce good food. Everything that human beings do and feel, reflects upon it. Farming mirrors the entire *Zeitgeist*. A taste for civil engineering on the grand scale has been part of the *Zeitgeist* of all great civilizations, and of their grandiose leaders. It is partly how they define their greatness.

Yet excellent irrigation schemes can be on the smallest scale. Small farmers in Africa commonly contour the ground like engravers, creating a small depression around each plant, to capture even the dew. Modern developments include trickle irrigation in which, in its most refined form, each plant is continuously monitored and a computer doles out every millilitre according to need. In hydroponic systems, water is recycled in a closed circuit. None is wasted.

When the land is too wet, it is drained; either that or the water is otherwise controlled, for example to create water-meadows, not least for the growing of cress as was traditional in parts of south-western

England. Too much water is bad in several ways. It shoves the air out of the soil, and so literally drowns the plants. Water is also slow to warm up, and plants in soil that is too wet are slow to 'get away' in the spring. As a sometimes serious side effect, water encourages parasites, such as the hookworms that caused such damage in Asia and the Americas; and also the vectors that carry parasites, like the snails that carry liver fluke and the worms of bilharzia, and the mosquitoes that carry the parasites of malaria. Malaria (traditionally known as 'ague') used to come much further north than it does now; for example into Cyprus and even into Britain. Many historians feel that the development of drainage from the Middle Ages onwards, but especially in the seventeenth, eighteenth and nineteenth centuries, truly brought Europe's agriculture into the modern age. Clearly, some countries need more drainage than others. In early twelfth-century England the Normans built Ely Cathedral in Cambridgeshire on an island, surrounded by the water of the Fens. 'Isle of Ely' still survives as a place name and so, magnificently, does the cathedral; but the fens are largely gone (although some scholars feel they may come back, as the greenhouse effect begins to bite and the waters rise). As with irrigation, however, drainage has its downside, and can be overdone. Part of the downside is seen to the east of Ely where the ancient peat, which persisted as long it was wet, began to oxidize away as soon as it was dry. The other downside is the loss of wetland, which is rich in its own particular wildlife, and beautiful. Precious little of Britain's unique, pristine Fenland remains and what there is can no longer support some of its traditional denizens, like the large copper butterfly. Worldwide, wetlands are under threat, both through drainage and through expropriation for irrigation. The Florida Everglades, a jewel among the world's wildlife reserves, with skimmers and alligators and manatees and ibises and herons, is sadly compromised by the horticulture to the north which is largely dedicated to enormous tomatoes that are grown to decorate hamburgers.

Once land has been freed from trees and rocks, and watered or dried, it needs to be broken up to remove remaining vegetation (loosely classed as 'weeds') and to receive seed. For this it can simply be dug, but on the larger scale it is ploughed: and of all human inventions, the plough is surely the most significant, more even than the printing press. Farming in general enabled human numbers to rise, and made our species dominant. But the plough is the basis of arable farming, and it is with the plough that we have truly put our mark on the world, for good and not so good.

Of key importance in Eurasia and Africa (though not in pre-European America) was the harnessing of animal power. Oxen (castrated male cattle) or perhaps cows were the commonest traction animals of earlier times. They are cheaper than horses, not least because they eat less. But on the whole they are less powerful (they have marvellous shoulders, but lack the great haunches of the horse) and so, for larger tasks, they were deployed in larger and larger teams. Prints from the eighteenth century show scores of oxen in convoys shifting, for example, mature trees for landscaping, or even entire houses. This is not agriculture, of course, but it shows what animal power can do. The power of the horse was not fully realized until the soft horse collar came into use. It was invented by the Chinese – possibly as early as the fifth century AD – and appeared in illustrations in Europe in the tenth century. Before then, harnesses commonly were tied around the horse's neck, so the more the animal pulled, the more it was throttled. The somewhat panic-stricken expression of horses in Greek statues (as in the Elgin Marbles in London's British Museum) reflects in large part their rising discomfort and fear of strangulation. But the stiff but padded horse collar transfers the pressure to their shoulders and then they can press with their great hind feet with impunity. Boxer, in George Orwell's *Animal Farm*, is the epitome of equine might. Oxen never raised this problem. They could be yoked without choking them.

From early times, improved cultivation was complemented by subtlety of cropping, to maintain fertility and to reduce the build-up of disease. We need not suppose that early farmers understood the chemistry of fertility, or the mechanics of disease transmission; but their crafts, presumably evolved through experience, carried them through. Rotation of crops has been with us for a long time, taking many forms. The Romans practised a three-course rotation. The most famous rotation was the Norfolk four-course, developed in England. It probably evolved over several centuries, and reached its zenith in the seventeenth century. Viscount Charles 'Turnip' Townshend (1674–1738) refined and popularized the idea of crop rotation and in particular introduced the turnip as cheap winter feed. The more livestock that can be maintained through the winter, the more there are to make use of next year's pasture.

From the nineteenth century onwards, animal power was increasingly supplemented and then replaced by mechanical power: first steam in the form of the traction engine and then, much more importantly, with the internal combustion engine and the tractor. Even small tractors may be several times more powerful than a horse and on the whole they are a lot less trouble. Above all, they don't need feeding unless they are working. The internal combustion engine did not take over immediately, however, even in the most mechanized societies. Britain had more horses in the 1920s than at any other time, as the human population rose and the work animals increased accordingly (for transport as well as for agriculture), while tractors and motor cars had yet to establish their niche. Picturesque as they are, too, and generally pleasing, horses were a tremendous drain. When they were at their most numerous, up to a third of agricultural land in Britain and America was devoted to their upkeep. There is still a role for traction animals. In many a context they do an excellent job (oxen carrying panniers of grapes in French vineyards, yaks on the Himalayas, water buffalo in wetlands everywhere), and they fit very neatly into many economies, living as they traditionally do on fodder that would otherwise be wasted (meaning they live essentially for free). But modern agriculture, and hence to a large extent the modern human race, depends very heavily on the tractor; and although it is arrogant and often foolish to assume that tractors should replace traditional oxen, it would be childish to suggest that oxen or horses could ever again supplant the machine.

By the eighteenth century the more avant-garde farmers were demanding more versatile machines to reduce effort, hasten cultivation (speed is especially desirable when the weather is uncertain) and reduce the clamorous workforce. Machines on the whole are easier to handle than armies of people. A machine for reaping was an obvious desideratum but it wasn't until 1827 that Patrick Bell of Scotland came up with a horse-drawn version that actually worked. Yet such machines were slow to catch on: deep into the twentieth century over much of the world, including Europe, cereals continued to be cut with sickles and hay with scythes. Still, though, the machines grew inexorably bigger, more powerful and more versatile, and finally evolved into the modern combine harvester, a factory on wheels that does the work of a thousand labourers, propelled by many hundreds of horsepower and smart with on-board computers. This, clearly, is a good thing in some ways and not so good in others. Some of the traditional work on farms was and is ghastly; but much even of the hardest labour is satisfying, and people need to work at something, and although the novels of Thomas Hardy and George Eliot depict much strife in traditional English farming communities, they were fine societies too. It isn't obvious that the cultures of the Mediterranean, India, rural China or Africa will be improved if and when machines take over the countryside as they already have, say, in the east of England and the heart of the United States. (More of this later.)

The new forms of power, too, extended beyond the farm itself, into processing and transport. Notably, by the end of the eighteenth century fast sailing ships generically known as clippers came into play, soon to be supplemented by steam (painted with great relish by J. M. W. Turner); and the big fast ships in particular opened up the arable fields of America to the rest of the world. In the late nineteenth century ships were refrigerated and meat, too, could be shipped en masse from continent to continent. Here were the technical origins of globalization.

From earliest times, too, the crops and livestock themselves were changed by domestication and, increasingly, by deliberate programmes of breeding, so they were able to respond to ever-richer soils and feeds, and a greater range of conditions. Over the past 10,000 years, cultivation and breeding have developed in dialogue. But the improvement of crops and livestock (improvement, that is, according to the criteria of farming) is more conveniently discussed in Chapter 7.

This, then, is how farming arose (or so it now seems); and how it progressed from Cain's first barley fields into modernity. We should look more closely at what modernity entails.

An Outline of World Farming

From its earliest beginnings, agriculture has been fairly clearly divisible into three broad categories: horticulture, pastoral and arable. Those categories are with us still. We should look at them each in turn.

Gardens, Orchards and Plantations: Horticulture

If farming did indeed begin in the way I have described, then horticulture must be the oldest kind of farming. '*Hortus*' is Latin for 'garden'; and 'horticulture' implies that plants are cared for individually, thinned and planted out and pruned.

Horticulture spans the complete spectrum of technical innovation. At its most basic, it is as non-invasive as it is possible to be. The world's first gardeners must simply have tended wild plants where they stood: removing the worst of the pests, breaking off the dead bits, pulling out rival plants (which would then be categorized as 'weeds'). Later they would have planted seeds, tubers and sticks that served as cuttings, but without preparing any special 'bed'. They would simply have stuck the seed, tuber or stick into the virgin ground (in a basic form of 'minimum tillage'). As the efforts became more intensive, so the managed wilderness eased step by step into garden.

But modern horticulture can be among the most high-tech of all agricultural systems: the plants grown in sterile soil (or no soil at all, as in hydroponics) in greenhouses or polythene tunnels where conditions may be controlled to the nth degree: temperature, humidity, day-length, light intensity, what you will. It is in horticulture, too, that genetic manipulation is most advanced – where genetic engineering has had greatest impact (and so produced 'genetically modified' or GM crops). Such technical input partly reflects the value of the crops (which justifies the costs), but is also a matter of botany – members of the Solanaceae family lend themselves particularly well to genetic manipulation, and include tomatoes, aubergines and capsicums.

Horticulture embraces the orchard and the plantation – thus grading into forestry. It also grades into arable, since many traditionally horticultural crops, such as potatoes, cabbage, lettuce, onions and various beans, are nowadays commonly grown on an arable scale. Sometimes on the same farm, the same general kinds of crops may be grown both on the horticultural and on the arable scale; as in Angola where the farmers grow some beans plant by plant, and others by the field (though not very big fields). Such decisions are not made whimsically. Some crops are better suited to the large scale than others, and minute differences can tip the balance. Those who know Africa's traditional farming well invariably emphasize how much the rest of the world can learn from it, and especially from its attention to detail.

The world could do with a great deal more horticulture. In general horticulture focuses on high-value crops that are nutritionally highly desirable (like tomatoes and capsicums) yet are not 'staples': that is, they are not key suppliers of energy and protein, as cereals and pulses are. Yet some staples are grown in ways that must qualify as horticulture – notably the coconut, and indeed the olive. Perhaps, too, as the world grows more crowded, we should begin to grow more and more crops of a staple kind by horticultural techniques, including, say, avocado which has a high content of fat and therefore of energy, and a greater variety of fungi, from mushrooms to truffles.

More to the point, the fruit, vegetables, herbs and spices that are the traditional produce of horticulture are of immense nutritional significance. They may not be staples – not significant sources of energy and protein – but they are key sources of minerals, vitamins and essential fats, and also of texture and flavour. Traditional farms, however specialized they may have been, always had their own gardens; and cities, too, can produce tonnes of vegetables on balconies and roofs or indeed in allotments, if only people put their minds to it – and, notably, if the planners provide the space. I have seen barley growing between the high-rise office blocks in modern Beijing (and a strange if cheering sight it was); but I don't recall allotments in any big city outside Western Europe. They should be a priority. They could improve people's lives in a dozen important ways.

Texture and flavour are immensely important too – not just for obvious aesthetic reasons, but also for world economics and agronomy. People will keep body and soul together if their diet is high in staples (which mostly means cereals and pulses) and the staples are affordable. People often have lived virtually entirely on staples, like many millions of Asians today, or the eighteenth-century rural Scots who often subsisted virtually entirely on oats. But such plain diets are boring, and human beings, like all omnivorous animals, have a predilection for variety. Add plenty of well-chosen fruit and vegetables – and just a little bit of meat or fish – and the high-staple diet becomes acceptable; indeed it becomes the stuff of great cuisine from Italy to China and all places in between. Without the vegetables, the only way to make the staples acceptable is to add more meat – which is what the modern world tends to do. Yet this is undesirable from all points of view (except from that of the meat wholesalers, the makers of hamburgers, and some sections of agriculture).

In truth, although genetic engineering has generally been employed for somewhat dubious purposes it could make many worthwhile contributions, and perhaps especially in horticulture. So far it has mainly been used for commercial trivia – for instance to prolong the shelf-life of tomatoes – for the benefit of the packers and retailers rather than for the consumers or the environment. But genetic engineering could help people raise more good, fresh fruit and vegetables in cities, where most people are already living. It would be possible, for instance (even easy) to grow vitamin-rich spinach as vines up every apartment block. All this and much more is technically possible although it is not, on the whole, the kind of approach that scientists, and the governments and commercial companies that employ them, are currently putting their minds to.

The Pastoralists: Livestock

'Pastoralist' is a lovely word, from the Latin '*pastor*' meaning 'carer' or, more specifically, 'shepherd'. In an age when animals are raised in 'factories' with all possible speed the term no longer seems appropriate. But perhaps it might be again, and meanwhile we can do our best to keep the idea of it alive.

Like horticulture, livestock farming still reflects the complete spectrum of technical invasiveness. At one end, a few hunters still contrive to manage wild game – like some Australian Aborigines, creating green traps with fire. More hands-on are the Sami people of Lapland, herding and controlling reindeer that still rely on wild fodder and migrate as wild creatures do, with the seasons. The keepers of deer parks ('paradise' is a Persian word meaning 'garden' but implying deer park) have taken one further step; they confine the beasts they aspire to hunt. The deer-park principle of semi-captive game is ancient (depicted, for example, both by Uccello and Botticelli in the fifteenth century). Some say the Romans introduced the rabbit to Britain with just such husbandry in mind. The traditional farmers that we all remember and still occasionally encounter, with their hens in the yard, sheep on the hill and cows in the meadow, exercise even more control over their beasts. At the modern end of the spectrum is the factory farm: just one species of livestock raced from conception to grave as fast as food can be shoved into them, and bred for nothing but 'feed efficiency'. Beyond modernity lies the unknown future. We might hope that the future will be more like the past, at least in general structure and appearance: this, at least, is what I am arguing for in this book, in the name both of long-term biological efficiency and of common morality. But if some modern scientists, industrialists, and their supporting politicians have their way, then animals could be reduced to nothing more than flesh, conveniently pre-shaped for processing.

Animals of many different kinds and zoological classes have been raised for food (and for hides and wool and milk and eggs and – as with bears and musk deer – for folk medicines and perfumes). Domestic invertebrates include edible snails, oysters, mussels and prawns of many kinds. Fish farming now, amazingly, provides half the fish we eat, and is of huge importance economically, nutritionally, gastronomically and ecologically (but it has become so big, it needs another book to itself). Amphibians and reptiles are raised for food, including edible frogs, turtles and alligators (although alligators are kept mainly for their skins). Most important, however, are the domestic birds and mammals.

The birds include chickens (which have become the most numerous of all land vertebrates, with the possible exception of mice and rats), ducks, geese, turkeys, guinea fowl, quail, ostriches, emus and pigeons. These at least are the main food birds. Others are domesticated or semi-domesticated for other purposes, like the Chinese fishing cormorants. The mammals include cattle of various species, including the descendants of the extinct aurochs which now form the breeds of Europe, plus the zebu, yak, water buffalo, and anoa; sheep; goats; a few antelope, including eland; deer, including reindeer, but also variations on a theme of red deer (including the wapiti and Père David's); horses and donkeys; pigs; camels; llamas, vicunas, alpacas and guanaco; dogs; rabbits; guinea pigs; and edible dormice. Again, others are domesticated for other purposes, including elephants, but the above are the main food species.

However, the most helpful way to classify livestock is not into birds and mammals, but into herbivores and omnivores.

The herbivores, of course, are the specialist plant eaters; either browsers (eating the leaves of bushes and trees) or grazers (grass-eaters). Grass on the whole is more important – the world contains vast grasslands and grass is among the biggest of crops – but browse is not to be written off. Camels, for example – supremely important in hot deserts – are primarily browsers, not least of acacias. So are goats, of mixed wild herbs and bushes (and they can climb trees, if they are not too steep). Cattle, sheep and horses, though primarily grazers, will browse when appropriate.

The specialist herbivores are able, as human beings and other omnivores broadly speaking are not, to digest cellulose. Cellulose in refined form is the stuff of cotton and more generally forms the tough sheeting that is the chief component of plant cell walls, and as such is the commonest organic material in nature. It is also a carbohydrate, compounded from molecules of sugar, and when broken down it is a fine source of energy. However, no animal produces 'cellulase' enzymes of the kind needed to digest cellulose (occasional rumours that some may do generally turn out to be false). The specialist herbivores therefore employ micro-organisms, alias 'microbes', generally bacteria but also sometimes (as with termites) including protozoans as well, which live in their guts, and which do produce cellulase. In practice, these resident, symbiotic microbes do not break cellulose into sugars but into organic acids. But the animals that harbour the microbes absorb these acids, which typically form their principal source of energy.

Among mammals (and birds), the specialist herbivores include foregut digesters, and hind-gut digesters. The former harbour their symbiotic bacteria in specialized stomachs. Kangaroos do this and so too, which is more relevant here, do the ruminants, whose specialist stomach is called the rumen. The domestic ruminants include cattle, sheep, goats, eland and deer. Hind-gut digesters include horses, donkeys, guinea pigs and rabbits (which are raised in large numbers in, for example, China and Malta).

The various ruminants and the hind-gut digesters approach vegetation in different ways. Thus sheep and cattle tend to digest their food very thoroughly, the physiological equivalent of Britain's erstwhile prime minister William Gladstone, who chewed each mouthful thirty-seven times. On very low-grade fodder they do badly, basically because they spend too much time trying to extract every last dash of nutrient. Horses and asses push the food through quickly, skimming off the top and abandoning the rest. Thus they do well on low-grade fodder, since they don't waste time on it. Overfeed them, and they swell and die (unless of course they are working hard, in which case they need rich pasture and are traditionally fed additional oats). Horses are prodigious defecators and their dung is rich and especially good for crops because so much of the organic nitrogen is left in it; it is also high in fibre, since so much remains undigested. The strategy of red deer, although they are ruminants, tends to be more like that of horses. Goats specialize in eating browse of a dubious nature – tending to be toxic – and they like a very mixed diet; moving from plant to plant so that they don't get poisoned by any one of them.

The general point is, though, that cellulose is potentially a cornucopia. There are billions of tons of it out there in nature. Hedgerows, chemically speaking, are made of sugar. Yet all of this cellulose is useless for creatures like human beings, or lions or wolves, who cannot digest it. Still, though, these omnivores and specialist predators can join the feast, simply by eating the herbivores that can digest cellulose (with the aid of their alimentary microbes). Thus, for us, the specialist herbivores provide the *entrée* to nature's best-stocked larder.

Traditional pastoralists, particularly in the semi-deserts of Africa, combine different classes of livestock to exploit the meagre, difficult and highly seasonal vegetation to the full. Nomadic African tribes typically travel with mixed herds of sheep, goats, camels, donkeys and cattle. Broadly speaking, the acacias are in leaf during the dry season, which is when the camels, which browse on them, lactate; and the grass and other herbs grow in what relatively speaking is the rainy season, when all the other animals do most of their feeding, and can breed and lactate. All the animals are kept mainly for milk, but young sheep and goats are killed for meat in the brief periods between the fading of the grass and herbs and the sprouting of the acacias. Other animals are eaten for meat as they become superannuated. Thus the people subsist either on milk (of various species) or meat, depending on season, but never on both. If milk is available, they don't eat meat. (Such thinking surely lies behind the Jewish kosher directive not to consume milk and meat in the same meal.) When Job's fortunes were finally restored after seemingly endless tribulations, he had 'fourteen thousand sheep, and six thousand camels, and a thousand yoke of oxen, and a thousand she asses' (Job 42: 12). A full house; and exactly the same range of beasts as the present-day tribes of desert Africa, in the same proportions. Job had milk and meat aplenty but to eat them together was in principle uneconomical, and desert people have little margin for error.

These nomads show what wonderful use can be made even of the most unpromising landscape through the judicious use of livestock. Livestock enables people to live vicariously on provender that otherwise would be lost on them. We also see how intricate apparently 'primitive' farming typically is. Modern farming, which typically raises one species at a time on the simplest possible diet, is kiddicraft by comparison.

In traditional settings ruminants graze on areas of land where grass will grow but staple crops will not – as on steep hills and undrained meadows – and on temporary grass ('ley') grown as part of a rotation. They also, importantly, feed on straw. Straw effectively contains nothing but cellulose; but when mixed judiciously with a source of nitrogen (which can be something very crude, like urea) it becomes perfectly acceptable feed for cattle. All cereal crops produce copious quantities of straw. So too do the pulses, and in India their stalks too become cattle food. In the arable farm without livestock, the straw may be wasted, or turned into fibreboard or biofuel or whatever (these are not bad uses, of course).

Many birds and mammals are omnivores, eating both animal and vegetable matter – though not, as a rule, digesting much or any of the cellulose. Among domestic livestock, the outstanding omnivores are pigs and poultry, particularly chickens but also ducks. (Dogs are important meat animals in some countries, and although they are primarily carnivores they too lean strongly to omnivorousness.) Omnivores eat the same kinds of foods as human beings do, in roughly the same ratios (although poultry are efficient grazers, and pigs can be too, particularly traditional outdoor breeds such as the Berkshire). In general, though, pigs and poultry have lower aesthetic standards than human beings, and so make do on leftovers, or surpluses. Poultry traditionally picked up any grain that was spilled in the farmyard (before the mice got to it), and picked the little red worms out of the dung-heap. Ducks come into their own in the Chinese paddy fields, gobbling the snails (a useful medicinal measure, since the snails often carry schistosomiasis, alias bilharzia). Pigs eat anything. Traditionally they were raised on the household leftovers, and/or were turned into the fields after the crops were pulled to fossick for whatever remained, or roamed in the woods for acorns and other tree seeds (I once sat in a traffic jam in Holland as pigs snuffled up beech-mast from the road). Gloucester Old Spots were traditionally bred to feed on windfall apples.

From all this we see the general importance of livestock. First, the ruminants enable us to exploit a huge resource – cellulose – that is otherwise closed to us. Second, by virtue of this, ruminants enable us to exploit territory that would otherwise be unproductive: grasslands in general; cold hills, too steep and windy for crops, which cause no problems to the tougher breeds of sheep; semi-desert and dry and rocky hills, where judiciously mixed herds may thrive on what may look like almost nothing at all. They also eat straw, inevitably produced in superabundance with cereal crops and useless as human food. The omnivores fill in the spaces, making do on surpluses and leftovers that would otherwise be thrown away. Both groups help to exploit forests in various ways, particularly woodland cattle, pigs and fowl. Vegetarians argue that because crop plants provide a great deal more energy and protein per unit area than livestock can do, it would be far more economical, and therefore far better for the planet, if human beings contrived to live entirely on plants. But there is no system of crop production that could not be even more productive if a few appropriate livestock were slotted in.

The key words, though, are 'few' and 'appropriate'. Livestock truly becomes a drain on resources when we focus on just a few species (notably cattle, pigs and chickens) and contrive to raise them in the greatest numbers possible – which is the ambition and the prime strategy of modern agriculture; and is becoming the prime strategy of agriculture worldwide as fast as the creators of policy can make it happen. Then, crops that human beings could perfectly well consume as they are, are grown expressly to feed livestock that is nutritionally superfluous, and indeed nutritionally pernicious. In particular, livestock consumes a third of the world's wheat, two-thirds of the maize, and at least three-quarters of the barley and soya. Thus in today's systems farm animals are nutritionally equivalent to at least an extra 2 billion people; and by 2050, when the world population is 10 billion, and meat consumption per capita is even higher than now, the virtual population of people-plus-livestock will be 14 billion. Thus, by 2050, livestock will be consuming more food than was consumed in 1970 by the entire human population; although, at that time, there were some major famines and many doubted if the world's people could be fed at all. Thus a tremendous asset, the biological versatility of animals, instead becomes a huge additional burden. Thus, too, livestock production is turned, more and more, into an adjunct of arable farming. The whole world is jeopardized, and for no good biological or aesthetic reason at all. There is nothing behind the present zeal for more and more livestock except a belief that nothing can or should be done that is not maximally profitable: that plus a misunderstanding of human nutrition. This is discussed in Chapter 4.

The Biggie: Arable Farming

Horticulture and pastoralism are huge, but it's arable that's the key to human survival. The word derives from an old Latin root meaning 'ear', referring to ears of corn. Indeed, cereals are grown on the arable scale but so too, nowadays, is a vast array of other crops; including many pulses (notably various beans); oil crops such as sunflower and rape; and tubers such as potatoes. The term 'arable' no longer refers to the specific crops grown, but to the manner of growing. Arable farming is the most invasive of all systems. Although minimum or zero tillage has re-emerged in recent years with the use of chemical herbicides (the seeds planted straight into uncultivated ground), arable farmers traditionally begin by ploughing: turning over the soil, and any vegetation with it; then breaking up the entire surface into a fine tilth in a series of further passes with harrows and other crumblers of soil. In other words, they do not minister to the crops plant by plant, as horticulturalists do. They work directly on the earth. Their canvas is the ground itself. They grow crops on the field scale; and although arable fields don't have to be big (there's a lovely late medieval Dutch painting in the Prado that shows a field of barley in what would otherwise be the village green), in the modern world they have been growing bigger and bigger, as the machines of cultivation have grown to the size of fire engines and now, often, are as big as traditional cottages.

Most (all but a minute amount) of the world's staples are grown as arable crops; staples being the principal sources of energy (mainly carbohydrate but also fat) and protein. The big seeds of the grasses known as cereals are by far the most important. As discussed in Chapter 1, the big three (wheat, rice and maize) provide humanity with the equivalent of roughly half our daily calories and two-thirds of our protein, so that all other crops and even the most conspicuous and economically important livestock, including cattle, are minor or even marginal by comparison. As human food, wheat and rice are widely considered the most desirable. Bread wheats have the tough, springy proteins collectively known as gluten, which allow bread to be fermented with yeast and leavened by the expanding bubbles of carbon dioxide, but maize also makes more than tolerable flour and meal and is the traditional source of Mexican tortillas. Rice, in all its variety, is grand when simply boiled; and delectable when tricked out with slivers of onion or fish or whatever is available.

There is also a range of minor cereals – or least, minor by comparison: though each of them serves as a staple for at least some people. Oat has always been present among the wheat, barley and rye, essentially as a weed. But it is nutritious – particularly rich in nutritionally desirable fat – and has long become a significant staple in a variety of difficult conditions, notably Scotland. Barley is fairly salt-tolerant, and although grown mainly for brewing and animal feed, is also the source of barley bread and various porridges, eaten for example in Tibet. Rye withstands harsh conditions, including cold and wet, and has been especially important in Eastern Europe (and for whiskey, while its stiff straw makes serviceable thatch). Maize is generally more tolerant of drought than wheat, but sorghum is even more tolerant and the various millets and the Ethiopian teff are the most drought-resistant of all. In times of prolonged drought – the kinds that suggest serious climate change is afoot – farmers will sometimes shift from wheat to maize, then maize to sorghum and even from sorghum to millet. Beyond millet lie some varieties of groundnuts, which are pulses. They are the end of the line.

Pulses are extremely important too. They have two supreme assets.

First, they are legumes, which means that their roots harbour bacteria (of the genus *Rhizobium*) which convert atmospheric nitrogen into soluble nitrogenous compounds, notably nitrate. This is 'nitrogen fixation'. Thus, when legumes are grown, they can leave the soil more fertile than they entered it. The term 'pulse' refers to legumes that are grown for their seeds – beans, peas, lentils, chickpeas (and lupins, which are grown for livestock). But green legumes are also grown for fodder, and to be ploughed in as 'green manure'; notably clover and alfalfa (aka lucerne).

Second, the seeds of pulses are high in protein and, as described in the next chapter, pulse protein complements cereal protein beautifully.

The principal bean worldwide is soya, which originated in China although the US is now the chief exporter. (The history of agriculture, before the world lost its innocence and discovered patents and property rights, is supremely cosmopolitan. Tit for tat, the Chinese grow enormous quantities of maize and potatoes, which originated in the Americas.) But many other pulses are important too, primarily those of the kidney bean series which include the haricots and the 'navy' beans which are grown for baked beans.

The third main category of arable staples are tubers of various kinds: potatoes, sweet potatoes, yams, cassava. In their various countries they are of huge importance. Entire populations have subsisted virtually solely on the European potato, notably the rural people of early-nineteenth-century Ireland. This of course is extremely undesirable from all points of view, and reflected the extreme poverty of the rural Irish and the cynicism of their English and Irish landlords. But the fact that survival was possible at all shows how astonishingly nutritious potatoes are. They even contain adequate protein (for an adult) and although they are not especially rich in vitamin C (compared, say, to oranges), in practice they often serve as the principal source of it.

Many other kinds of crops, across almost the whole botanical spectrum, serve different peoples in various special ecosystems as staples. Many subsist largely on bananas and plantains, others on coconuts. It seems to me at least possible that in Neolithic times in the Middle East the olive, rich in oil, was at least as important as a staple as cereals were. It is still a major source of energy in the Mediterranean. In the Americas before the Spanish arrived the native maize (which is a grass of the genus *Zea*) was supplemented and often replaced by quinoa, amaranth, and 'wild rice', a wetland grass. Buckwheat has also been important to many people at different times, as have squash seeds. The list goes on. But as arable crops, the cereals dominate, supplemented by the pulses. Wheat, rice and maize are half of agriculture; and of those three, wheat and rice stand supreme.

Put the three great categories of farming together and we have what can properly be called a 'system'.

Farming in the Round: The Idea of the 'System'

Horticulture, pastoral and arable are different kinds of activity that work the land in different ways, provide foods that contribute differently to the human diet and, traditionally, have often been practised by people of different temperaments. Because of the differences there has often been conflict; indeed, the running battle between pastoral and arable in particular has been a powerful leitmotif of literature since the most ancient times. It surfaces first in the story of Cain and Abel, and since Cain and Abel were the sons of Adam and Eve their falling out in Genesis 4: 2–8 takes the conflict of pastoral versus arable back to the very beginnings of human history, at least as traditionally understood. Again the Bible emerges not just as a treatise in morality and theology but as an important work of journalism, the first draft of history, and always finely tuned to the issues of agriculture:

> And Abel was a keeper of sheep, but Cain was a tiller of the ground. And in process of time it came to pass, that Cain brought of the fruit of the ground an offering unto the Lord. And Abel, he also brought of the firstlings of his flock and of the fat thereof. And the Lord had respect unto Abel and to his offering: But unto Cain and to his offering he had not respect. And Cain was very wroth, and his countenance fell . . .
>
> And . . . Cain rose up against Abel his brother, and slew him.

Although the Bible does speak highly of arable farmers here and there, on the whole it seems biased towards pastoralists. In their early form, after all, the keepers of livestock were non-invasive. Grazing animals may transform a landscape over time but in principle the early pastoralists took the environment as they found it. Their animals were metaphors of serenity and beauty, as in Solomon's Song of Songs (4: 1–2): 'thou hast doves' eyes within thy locks; thy hair is as a flock of goats, that appear from mount Gilead. Thy teeth are like a flock of sheep that are even shorn, which came up from the washing . . .' A quaint metaphor, but sheep certainly come out of it well. Much later, it was the shepherds, successors of Abel, who attended the birth of Jesus. No one turned up with a sack of barley.

Arable farming is innately aggressive. The arablist takes the landscape by the scruff, and rebuilds it from scratch. To a desert people, still with a hunter-gatherer background, this surely was an offence. It is arable farming that requires people to eat bread in the sweat of their faces; and arable, perhaps more than anything, that obliterates the hunting grounds for doe and gazelles (which also feature in Solomon's songs). It is also intriguing that vegetarians have often quoted the story of Cain and Abel for their own purposes – seeking to demonstrate how the bloodthirsty meat-eaters have put pressure on gentle herbivores. But their memory is false. *Cain* was the herbivore. His victim was 'the keeper of sheep'. The same conflict crops up in many a western movie – and with the same reversal of roles, since the arablists are generally depicted as gentle toilers, and the cattlemen as the buccaneers. It is the subject of a song in *Oklahoma*: 'Oh, the cowman and the farmer should be friends!'; implying of course that they very definitely were not.

Yet the three approaches, horticulture, pastoral and arable, can and certainly should complement each other. We see this in the traditional mixed farms that people of my generation remember from their childhood, which feature in nursery rhymes. Rural children of my generation were steeped in the details: they must have been deeply shocked by 'Little Boy Blue', for the sheep in the meadow and the cow in the corn was anarchy indeed. In Britain, the mixed farm perhaps reached its zenith in the 1950s; in New England, in the nineteenth century. It persists, still, in much of the world. I saw a form of it in all its glory in Yunnan, in subtropical China, in 2000. Primarily the people were growing rice, which indeed was (and is) the main component by far of their diet. But there were patches of high ground in the paddy fields too, hillocks, where they practised horticulture with yams and cabbage and spinach. There were ducks everywhere, frantically squeezing between the rice-stalks. I don't know if there were carp in the irrigation water, but there could well have been. In the village there were pigs and chickens, largely just knocking about, as if they themselves were villagers (and since they lived on the ground floors of the wooden houses, they effectively were). A Western health inspector with his clipboard and nylon trilby would have closed down the entire province, but I would as soon have taken my chances there as in any Western supermarket. At least in the village, what you see is what you get. The modern supermarket, when you look closely, is all too often a whited sepulchre. Altogether this Yunnan village was an idyll; and not just a nostalgic one, for I found that some of the people at least, upstairs in their own living rooms, had wall-to-wall computers and hi-fi. The village was wonderfully remote but the people were in touch with the whole world, or at least had the technical potential to be so.

In fact, the mixed farm is the key to the future of all humanity. For when crops and livestock are judiciously mixed, agriculture mirrors nature; and nature works. The principle is of course known to every eight-year-old who has ever had lessons in Nature Study. Animals, plants, fungi and all the myriad variety of other organisms complement each other, and feed off each other. Plants create organic material by photosynthesis. Animals eat plants, and return the materials to the soil in their manure, in forms that the plants in turn can feed upon. Fungi, bacteria, and other 'detritivores' mediate the interactions. This simple cycle is elaborated in a myriad ways but this is the essence of it. The key issue is that of ratio: the right proportion of animals to plants. We can see the principle in action in the early-twentieth-century US: yields in the central, essentially arable states began to fall until livestock were reintroduced, to raise the fertility. Traditionally (at least until the end of the eighteenth century), natural philosophers supposed that the whole system worked so beautifully because God had designed it that way. Following Darwin, the moderns invoke ideas of natural selection. The individual creatures are co-adapted to each other. Animals have evolved to eat plants. Plants have evolved to thrive on the excretions of animals. Natural selection seems virtually bound to ensure that the numbers of plants and animals is matched, each to the other. Natural selection also works on whole systems. Any association of creatures that is particularly productive is likely to overwhelm and to oust any system that is less productive. Productive systems may suffer setbacks, as tropical storms cut through the forest and ice ages come and go, but on the whole, natural selection seems bound, in general, to favour greater productivity rather than less. The main caveat is that any one pathogen, parasite or over-zealous predator may rip through the system, briefly flourishing at the expense of the rest, and thereby, at least for a time, reduce the productivity of the whole.

Farming is intended to modify nature. It is designed to bias the outcome: to ensure that any one stretch of landscape produces more organic material that is suitable for human food than it otherwise would. More advanced and intensive farming takes the process one step further: it raises the total productivity of the landscape by raising the overall fertility, while also ensuring that the highest possible proportion of what grows is suitable for human food. Farming, therefore, is very much an artifice. Still, though, it is a modification of nature. If we want to be maximally productive, and to go on being productive – if we don't want simply to run the system down – we should work to nature's rules: march to the drum of biology.

This is what mixed farming does. In general, the farm animals feed on some of the crops, and their manure goes back to the crops. We eat the animals (and most of the crops), which means that the system as a whole would run down if we did not replace the materials that we ourselves expropriate. This is done in various ways. Some traditional systems of agriculture (including traditional Chinese) return human sewage to the fields. There have been many modern variations on this theme, more aesthetically pleasing than the traditional slop-bucket and also safer. Most importantly, however, fertility is traditionally restored by variations on the theme of nitrogen fixation – turning atmospheric nitrogen into nitrogenous compounds that plants can absorb. Traditionally, this was done mainly by growing legumes: pulses, clover, alfalfa and the rest. Now it is done largely and indeed mainly by industrial means, as discussed in Chapter 6. But the central pattern should remain the same – farming essentially as a mirror of nature: livestock and crops complementing each other.

In reality, the interactions are complex, and can become extremely so. Traditionally, the specialist herbivorous livestock (mainly cattle and sheep) fed on grass, some of which was up in the hills or in wet meadows and was not otherwise cultivated (that is, was permanent pasture), and some of which was grown temporarily, as part of a rotation, between crops of cereals and other crops (a ley). The herbivores in turn were further subdivided. Thus Britain (which ecologically is more complicated than most small countries) until well into the twentieth century had at least forty species of sheep in commercial service (that is, not kept merely as rare breeds) which broadly speaking were divided into lowland types, like Dorset and Leicester, and upland types or hill sheep, like Cheviots and Scottish Blackface. The former were (and are) round, often big and generally fairly docile creatures, and the latter are fast on their feet, agile, and able to get out of the way of bad weather. In Yorkshire I have watched hill sheep leap the low stone walls around a holding pen as fast as the farmers put them in, and shepherds catch them with their crooks as in a rugby tackle. It's not elegant but there's no easy way. Farmers commonly cross-bred upland ewes with big lowland rams to produce fast-growing lambs, bigger than their dams, which spent the summer on the hills and then came down to the lowlands to be fattened. There were and are many variations on this theme; for example, the ewes are often cross-bred to begin with, so the resultant lambs are 'three-way crosses'.

Britain also had a wide variety of cattle breeds, which were traditionally divided into specialist dairy animals, and specialist beef animals. The former, typically, were bony-looking creatures with pretty faces and nicely shaped though not vast udders, which produced comfortable quantities of fine milk from traditional pastures that included a variety of grasses and other wild herbs. Their calves were taken away on day one (or day two) and generally raised for veal: not, traditionally, the whited anaemic kind but simply, in effect, juvenile beef. The point is, of course, that cows do not produce milk at all unless they first have a calf, so they are 'put to the bull' (or, more usually, impregnated by artificial insemination, or AI) once per year. Thus for nine months of each year they are pregnant, for ten months they are lactating, and for seven months, both. After their brief adolescence as 'heifers' they are always either pregnant or lactating or both; but despite that, if they are well cared for, they may live for twelve years or so (though the modern average is less than half that). Ayrshires and Jerseys were and are archetypal dairy animals. The specialist beef animals were much more muscular. The cows of beef breeds formed 'suckler herds', typically on the hills; feeding on grass, and raising their own calves. Aberdeen Angus and Hereford are archetypal beef breeds.

Nowadays, in a different economic climate, the commonest cattle are black-and-white Friesians (or Holsteins, which are a variant) which are essentially dual-purpose. They give an enormous amount of milk when fed on custom-bred, protein-rich ryegrass, plus plenty of concentrate (typically based on imported soya) to top up their energy and protein; and they are cross-bred with beef bulls to give big-bodied calves that can be raised for beef on grass (often supplemented with cereal).

In village India and traditional Africa, cattle are multi-purpose. The cows in India have traditionally been sacred under Hindu law (they still cause traffic jams in the cities), which to Western eyes has often seemed perverse. But they feed essentially on rubbish (notably on straw from the crops); yet they produce calves at intervals and these, when castrated, become the oxen that pull the carts and ploughs. Their dung provides fertility, fuel and building material for good measure. You see the pats lovingly spread out along the streets and on low roofs. Thus cattle, in a traditional system, were virtually a free resource on which the whole system ran. Traditional Hindus seem to have had an ambivalent attitude to drinking cows' milk (though they certainly take that of water buffalos); but in traditional Ethiopia, say, a cow might be expected to produce a calf, yet provide some milk for human consumption, and also pull a cart herself. These versatile creatures also have to withstand the heat, and keep their strength up on a poor diet. Western scientists of the more enlightened kind are currently helping to breed truly multi-purpose animals, for half a dozen different tasks.

In Europe, through their many combinations of livestock and crops, interweaving the uplands and the lowlands, the north and the south, summer grazing and winter hay and silage (fermented grass), and with many different rotations, farmers traditionally got the best out of their animals, their crops, and the landscape. The general trick is to let the ruminants feed primarily on grass, typically in areas where other crops cannot be grown (and sometimes on short-term leys grown as part of a rotation); and to raise the omnivores – pigs and chickens – on grain surpluses and leftovers in the form of swill. In most years there were grain surpluses because it always pays to aim for some surplus to insure against shortfall; and in most years, what is aimed for, is what is achieved. Without animals to take up the slack, the surpluses would be wasted – another example of the thrift achieved by keeping them. If the general emphasis is on cereal production, with livestock used primarily as fillers-in, then the ratio of plants to animals produced is obviously high. In fact, the output of traditional mixed farms roughly reflects the ratio of plants to animals found in nature.

There is one further proviso. Agriculture that balances livestock against crops in the way suggested here imitates wild nature; and, since nature clearly manages to persist, farming that mirrors its general structure ought to persist too (and there is plenty of theory to show why it would). However, whereas traditional books of ecology show chains of predators feeding on prey animals feeding on plants, with everything recycled by fungi and bacterial detritivores, modern textbooks recognize one extra element. For now it is abundantly clear that in nature, parasites of the kind known as pathogens – organisms that cause disease – play a huge part. It is commonly assumed that wild nature is healthy: that diseases are caused by overcrowding or poverty or (in folklore) by sin. But now it's known that the affairs of wild creatures, down to and including the details of much of their evolution, are largely driven by their pathogens. In recent years populations of wild dogs in Africa have been brought to the brink by rabies. 'Common' seals in the North Sea have been decimated by phocine morbillivirus, a relative of canine distemper and of human measles. In North America, mule deer are known to carry pathogens that are very bad news for moose: as the former move in, the latter disappear (even though moose are much bigger and stronger). In the 1970s, Britain's elm trees were all but wiped out by Dutch elm disease, caused by a fungus carried by beetles.

Thus, modern diagrams of wild ecosystems show parasites festooning the whole system; a penumbra of invasive bugs and microbes. Husbandry that aspires to mimic nature and yet to improve on it, should contrive to reduce parasite attack to the minimum. The general trick is to break the chains of infection, or prevent them arising in the first place. Parasites tend to be fairly specific, attacking only one class of livestock or crop – and often having more effect on some specific breeds or varieties than on others. The general rule, then, is to avoid keeping too many of the same kinds of creatures together in any one place; certainly for prolonged periods. Doctors have recognized this general principle since earliest times: in the Bible, lepers were isolated; during the Great Plague of seventeenth-century Britain, isolation was the principal defence; the isolation of the sick child is a common theme of Victorian novels (the lovers from neighbouring but stricken houses calling to each other across the void). In agriculture, mixed farms score once again. To a large extent, each kind of creature is surrounded by others of a quite different kind.

If all the world followed these elementary rules – the most basic principles of biology – then we could feed 6 billion people easily. We can reasonably suggest that agriculture that does play to the strengths of the animals, plants and landscape, and is structured to interrupt or pre-empt possible chains of infection, is good husbandry. Farming that deploys good husbandry, and which can therefore feed everybody well, and in principle can do so for ever, might reasonably be called 'enlightened'. It seems enlightened, after all, to want to feed people well; and wicked or at least perverse to farm in ways that detract from this goal. We might also reasonably hope that the people who do most to influence society, politicians and financiers of all kinds, would be keen to develop systems of farming that are designed to feed people. That, one might reasonably suppose, is their job: what they were elected for; what they draw their salaries for. It is obvious, too, that good science can in general help us to achieve anything we put our minds to; so we might reasonably hope to see agricultural science deployed in the interests of good husbandry, and hence of enlightened farming. All in all, it doesn't seem much to ask for: simply a system of farming that is designed to feed people, in a world that is eminently capable of doing this, and contains tens if not hundreds of millions of farmers, men, women and children, old and experienced, and young and energetic, of enormous competence and keen to get on with the job. On the face of things, it really doesn't seem too much to ask at all.

But this is not what is happening. In practice, the world's agriculture is not necessarily arranged along these simple, biology-based, enlightened lines. We do not have systems of farming expressly designed to feed people. Indeed, world agriculture is in general moving further and further away from the basic, common-sense position; and the further it moves, the more the people with the most influence seem to congratulate themselves on their far-sightedness. Neither is science deployed to enhance good husbandry. Increasingly its task is to find short cuts: to override good husbandry; to flout good sense; to help governments and industrial companies that have quite different ambitions to get away with them.

Why Farming Is Not Designed to Feed People

A system of farming that was truly designed to feed people and to go on doing so for the indefinite future, would be founded primarily on mixed farms and local production. In general, each country (or otherwise convenient political or geographical unit) would contrive to be self-reliant in food. Self-reliant does not mean self-sufficient. A self-sufficient country would produce absolutely everything that it needed, and would not trade with outsiders and this, for most countries, would be a nonsense. It would mean that, say, Canada would either have to do without spices, or else grow them at enormous expense in hothouses; while a country like India, which can grow spices of all kinds very beautifully, would be robbed of a valuable export trade. Self-reliance does mean, however, that each country would produce its own basic foods, and be able to get by in a crisis. Strategically, this can be highly desirable. Britain found this in both world wars, when the entire country was under siege. Today, surely, most poor countries would benefit from basic self-reliance, and might well make this their prime goal, even if they also attempt to compete in world markets with rivals that have various kinds of head start.

But no country could achieve self-reliance in the long term unless it strove to provide essentially the same ratio of livestock to crops that is found in nature, for this implies that in general the system is in balance, and is sustainable. Of course, a country could achieve such overall balance without insisting that all its farms were mixed (and in theory it could do this even if none of its farms were mixed). But if the mixed farm was perceived as the norm, then this in general would ensure that each piece of land was in balance; and so the agriculture of each country would be as robust, biologically, as it is possible to be. I would not presume to suggest that every country in the world ought to strive for self-reliance based on the mixed farm, because I would not presume to suggest that anyone, let alone any country, *ought* to do anything. I do want to point out, however, that *if* all the world's countries opted for systems of agriculture that were aimed at national self-reliance, with trade restricted mostly to non-essential delectables, and achieved this primarily via the mixed farm, then world food production as a whole would be biologically robust, and the general standard of nutrition and gastronomy could and should be very high indeed – as high as it is possible to be. It would in fact be relatively easy to feed everybody well. Emphatically not do I suggest that we should turn our back on science, for science is needed more and more to make the systems work in an increasingly crowded world that also aspires to raise its standards of living. But we should certainly turn our back on science that is intended to pull us away from the strategic ideal. I have presented this argument in many venues and have yet to see it seriously challenged. The usual defence of the status quo is to point out that the status quo makes some people rich, so it must be OK.

So why, if we truly think it is desirable to feed people now and for the indefinite future; and why, if in principle this should be easy, don't we contrive to do so? Why, historically, have systems that are expressly designed to feed people been rather rare? Why, in practice, do we strive to create systems of farming that have many different kinds of goal apart from that of feeding people – and why do we deploy science in ways that lead us to these other goals? Why (to be more specific) do we go through the charade of pretending, for example, that GM (genetically modified) crops are vital to feed the world, and that (say) BSE and foot-and-mouth disease are acts of God (yet somehow are the fault of the consumers) when in fact, in a well-designed system, GM crops would merely provide the gilt on the gingerbread, and epidemics of the kind that we are still routinely enduring would be a nonsense? Why in fact do we allow politicians to lecture us, and experts to pay themselves large salaries and play the role of benefactor, when in reality they have created the most dreadful mess, which if anything is getting worse?

I will explore these issues in more detail in Parts II and III. But in general, just to set the scene, there seem to be four main forces that pull the world away from the simple structure of biology-based, enlightened farming.

Ironically, the first problem is one of biology. It simply is the case that some regions are better suited to some kinds of crop or livestock than others. England, which has a mixed ecology and climate, illustrates the point very well. The west is relatively wet; while the east is much drier. Thus it is easy to grow fine, lush grass in the west. It is ideal cattle country and Cheshire, in particular, was the traditional centre of English dairy farming. The east is good for ripening grain and East Anglia is the heart of Britain's arable country. On the larger scale, in the Americas, prairie and pampas are ideal for beef and/or for wheat and maize (corn), which of course accounts for the frequent historical conflict between the two. And why grow fruit in northern Europe when they have so much more sunshine in the south? Why should the British bother with peaches or cherries (let alone grapes)? It seems perverse to mix everything up, when some parts of a country (or the world) seem so obviously to outstrip others, sometimes spectacularly so. In Britain, which historically has always been 'advanced', regionalization of agricultural output was well established by the Middle Ages; and well before that, the Romans imported much of their grain from North Africa.

Then again, we have created a world in which commerce prevails: its mechanisms, and its ideals. I will discuss the reasons and the consequences of this in Chapter 9 but the essence is as follows. First, everything that happens in agriculture, as in the world in general, is given a price. Then the productivity and the efficiency of the whole operation, whatever the operation is, is measured solely in cash terms. Productivity is the total wealth produced; and efficiency is the ratio of the wealth produced to the cost of production.

Productivity and efficiency certainly matter in agriculture – but, traditionally and sensibly, they are measured primarily in biological terms.

If a farmer plants 100 seeds of grain and so produces 100 plants, each of which carries 10 seeds, then he or she has produced a biological mark-up of 10 to 1. If the farmer plants 100 seeds and only 10 germinate, and between them they produce only 100 seeds, then he or she has wasted their time. The Sumerians were famed at one point for their very high ratios of grain harvested to grain sown; far higher than the Romans ever managed. Similarly, pigs are innately more efficient than beef cattle because they produce up to twenty piglets a year (at least in conservative systems; thirty-plus is now the goal), whereas a cow produces only one calf. Cattle become worthwhile because beef cows traditionally feed on grass, which otherwise could not be utilized at all. Small hill ewes are crossed with big lowland rams because in this way the farmer gets a big meaty lamb out of a small ewe, and small ewes eat less than big ewes; and by the same token, twins on the whole are better than singletons (although triplets tend to raise problems with shepherding). In short, productivity and efficiency in agriculture have always been the measure of success – but the criteria, traditionally, were those of biology.

When everything is minutely costed and cash becomes the sole measure, the criteria of excellence are changed, and biology is flouted. The general aim is still to maintain high productivity (turnover is crucial, as any shopkeeper will attest) and also to maintain efficiency. But productivity now must not simply be high, but maximal. Efficiency is seen to be achieved by reducing the total cost of the inputs, and adding value to the product. The greatest input by far in traditional agriculture is labour. Labour must be cut, cut, and cut again. The cutting of costs, and hence the cutting of labour, has been the prime policy of British ministers of agriculture since the Second World War (following a brief period of honeymoon in the immediate post-war years when governments realized that a little home production was no bad thing). Indeed, the cutting of labour has been almost the only coherent policy in Britain this past half-century. I know a farmer in Lincolnshire with 1,000 hectares of cereal, who employs just one labourer. He says he will not feel completely relaxed until his solitary employee is running 2,000 hectares. Within living memory, such areas would have provided good jobs for several hamlets.

Labour can be cut, of course: but only at a price. The point is, however, that the price does not appear on the balance sheet – or at least not initially; not for the first few years or decades. There is time to make a quick fortune before the cracks start showing.

The first requirement, if labour is cut, is to simplify husbandry. One man cannot control 1,000 hectares (or even 10 hectares) if there is any intricacy at all. Basically, it has to be cereals, horizon to horizon: not a mixture of varieties and species, but a monoculture. The farm has more or less to be 'run by numbers': routine applications of all the various chemicals, fertilizers, pesticides and herbicides, according to the calendar rather than to perceptible need. Where dairy farmers were once content with a dozen cows (or had just two or three on a mixed farm), now we find a hundred or more per worker. They are raced into the parlour, and raced out again. There is no question of mixed farming. It's not 'realistic'. When labour is reduced, too, machines must take over. Of course machinery is desirable: much farm work is appallingly hard, and anything that reduces drudgery and physical wear and tear must be desirable. But the role of machines (like that of science, I suggest) should be to assist good husbandry: not to replace it with some unvarying and simplified routine. In the modern farm the tractors and the combines grow bigger and bigger. They also grow more expensive. The more that's invested, the bigger the yields have to be, to justify the costs. Here we have another kind of positive-feedback loop, similar in general form to the vicious cycle that brought arable farming into being in the first place: the more that's done, the more needs to be done.

Thus the cycle that's set in train once labour is reduced leads us inevitably and rapidly towards the industrialization of agriculture: big fields, big machines, monoculture of a standardized product. Once in train, the system must run faster and faster to justify its own costs. If all the world is at the same game, then the individual players must run faster still to compete with all the others, although an army of bureaucrats is on hand to ameliorate the commercial disasters that would otherwise ensue, with cartloads of taxpayers' money to iron out the flaws. All this is in the interests of cash efficiency. The basic principles of biology are not mentioned, and neither are morality, aesthetics, or anything resembling common sense. Those who appeal to such principles are accused of elitism, gratuitous nostalgia, or general muddle-headedness, and held to be irresponsible. 'Unrealistic' is the favoured pejorative. Somehow it's realistic to behave in ways that produce an inferior end result (or at least, significantly below what is possible) in ways that clearly cannot be sustained, but unrealistic to suggest that we should explore alternatives.

Then, thirdly, general politics – strategic, fiscal, social – combine to pull any one country, at any one time, away from the basic tenets of biology.

All governments like to demonstrate that they are able to provide their citizens with basic foods, and preferably at affordable prices. Sometimes, to be sure, governments that are not elected and do not need to curry their people's favour have abandoned this principle. Thus despots have sometimes allowed people to starve as if to punish them; it is not obvious that Stalin was particularly bothered by the famines in the USSR in the 1930s for instance. But on the whole, elected governments, beholden to their electorate (and, to be fair, not generally wicked) have wanted their people to be well fed.

Few governments, however, have had the political nerve or wherewithal seriously to tackle poverty within their own shores. As Jesus put the matter (John 12:8): 'For the poor always ye have with you'. Given that all countries have some poor people, it seems at first sight to be sensible, and basically humanitarian, to keep food prices low. Again, this philosophy has not always prevailed, and Britain's government in the opening decades of the nineteenth century operated Corn Laws that were expressly designed to exclude cheap imports of cereals, and so to keep cereal prices high for the benefit of the landowners – much to the grief of the new industrial workers who had to work long hours just to buy bread. On the whole, though, democratically elected governments favour cheap food (often because they don't know how else to cope with poverty, and see it as a way of papering over the social cracks); and this was Britain's overriding aim through most of the latter part of the twentieth century. To keep prices down, governments have typically encouraged farming methods that reduce and indeed minimize the costs of production. In truth, cheap production does not necessarily lead to cheap prices in the shops – especially in societies like modern Britain, where food is typically processed between farm and supermarket, and value is officiously added to increase the ratio between production costs and selling price. In addition, cut-price husbandry has enormous dangers, as manifest in recent years in the epidemics of BSE and foot-and-mouth disease, both of which resulted entirely from cost-cutting (of which more later).

Thus, fourthly, the perceived need to maximize profits and cut costs drags agriculture away from the enlightened course, rooted in sound biology. Thus for long periods during the past 200 years, British governments have been content to play the world market – pursuing a policy of buying in goods from whoever sold them most cheaply. This was helped, of course, by our own command of a global empire. Britain effectively regarded the wheat of Canada and the sheep of Australia and New Zealand, the tea of India (and even China) and the coffee of Africa as its own: and although Argentina was never part of the British Empire, Britain has long claimed 'close links', based not least on Argentina's beef. Thus it was that on the eve of the First World War Britain grew only 40 per cent of its basic food commodities; and not much more than half at the start of the Second World War.

I am talking specifically of Britain partly because it is most familiar to me; but partly because it illustrates the problems of all countries particularly well – since it is both a small and crowded island, and a world-player. For all countries must perform their own balancing act. On the one hand, self-reliance in food has obvious strategic advantages (made very obvious to Britain in the world wars). On the other hand, it can be far cheaper to buy in whatever is needed from elsewhere – for other countries with different climates and labour forces can sometimes produce the same or equivalent goods for a tenth or a hundredth of the price. Then again – the other side of this same coin – many countries can balance their books by exporting the crops they grow particularly well. Many African countries face just this dilemma. Why grow sorghum for home consumption, worth cents and pennies, when French beans can be sold for a relative fortune? But when everybody is doing this – all Africa, all Asia – the game will change. More to the point, such trading flouts the fundamental rules of biology; and the more the rules are flouted, the more dangerous the game becomes. Whatever the short-term commercial advantages of the global market and the global farm may seem to be, I know which I would choose.

Why Don't We Farm Sustainably?

The hardest thing to get right, of course, is sustainability. A government that sets up a system that does well for a decade, congratulates itself. It has fulfilled the terms of its own office. Well, if we care about the long term, the short term is important too. If the system that works in the short term also lays the foundations for its own continuation, then congratulations are indeed in order. But of course we all know (because it is common sense, and because the evidence is all around us) that it is possible to succeed in the short term by borrowing against the future; by clearing forest, for example, and cashing in for a few brief years on the residual fertility before it all degrades into desert; by mining fossil water; by irrigating, without regard for the slow build-up of saline; by cashing in the organic content of ancient peaty soils; by competing too hard commercially, and destroying other economies, which in the longer term rebounds on the ostensible winners. But on the whole, governments in particular and humanity in general seem to handle the long term poorly. Sustainability is not our skill. Why not?

The basic and crucial fact is that human beings have a very poor sense of time. Perhaps this is partly evolutionary: the here and now is most pressing, and our brains have been geared by natural selection to the next few seasons. Dreamers, with their eyes on distant horizons, tend to lose out to the get-up-and-go pragmatists. But our sense of time must largely be cultural too – for some societies are clearly more aware than others of the deep past, and perhaps of the deep future. Thus in Plato's dialogue *Critias* an old Egyptian tells the eponymous hero that the Greeks are mere parvenus; his own people, he says, have been civilized for 6,000 years. This old sage clearly did have a commendable sense of time. But our own, Western culture seems never to have recovered from the chronology of the Old Testament, which led Bishop Ussher of Armagh to conclude in the mid seventeenth century that God had made the world in 4004 BC. The fact that we still refer to the 'ancient' Egyptians, Assyrians, Greeks, and even Romans, as if they were somehow primeval, suggests that we still retain the bishop's truncated sense of history. The Egyptians, Romans, and the rest were modern people (as the Old Testament makes clear), with bureaucrats and taxes and intricate laws of ownership and civic responsibility – all the paraphernalia of modern life; all rooted in perfectly recognizable farming that was in many ways advanced.

In a culture that has such a poor sense of time it's hardly surprising that economists tend to refer to the next thirty years as 'the long term'. In the starkest contrast, earth scientists invite us to think in hundreds of millions of years; and 10,000 years is a comfortable interval for the ecologist – time for a forest to cover a continent and retreat again; yet even that is only a fraction of humanity's total history so far. If we are serious about our own long-term future then we have to think on the biological/ecological timescale, rather than that of Bishop Ussher. (Of course, I would not presume to say what people *should* do. I am merely pointing out what is necessary, if we take the future seriously.) Yet to the politicians and financiers of recent years such timescales have clearly seemed ridiculous. I can envisage the expression of amused and puzzled incredulity that would suffuse the face of George W. Bush, if ever he were exposed to such wild notions.

Whether or not people are able to think long term, it's clear that they often have not done so: almost always, for obvious and doubtless multifarious reasons, people commonly sacrifice the future for the present. Thus it has been clear that the peat of East Anglia, the centre of Britain's arable farming, has been steadily diminishing this past few centuries as the ground was drained and ploughed for cereals. Once exposed, and dried, the peat oxidizes away. The farmers of past centuries must have noticed this loss. For one thing, many of the roads that were laid out level with medieval fields are now causeways, sometimes several metres above the eroded soil to either side. Perhaps the old farmers just assumed, as many are content to assume today, that their descendants who inherited the depleted landscape would come up with their own solutions. Perhaps they just felt there was nothing they could do to stop the rot, and perhaps to a large extent they were right. Perhaps they simply could not afford to change course. They had to go on producing and producing and producing or go out of business, whatever the long-term consequences. That is certainly the dilemma for many farmers nowadays.

The Romans, supreme civil engineers, often messed things up in other ways. The Roman commentator Lucius Junius Columella (dates unknown, but presumed first century AD) remarked in his celebrated *De Re Rustica* that farms, once beautifully managed by their resident peasants – citizens of Rome – had been incorporated into great estates, or *latifundia*, to be run by slaves. So they were going to rack and ruin – 'handed over to the slaves as if to a hangman'. This illustrates what I believe has become the single greatest flaw in modern agricultural policy: the conceit, among rulers of all persuasions, that they can simply impose their own political philosophy upon the land, and that the land will be infinitely forgiving, and prove infinitely flexible, and continue to yield as before. The obvious parallel is with Stalin's collective farms of the 1930s through to the '50s. Traditional peasant holdings were merged into rural food factories, and political dogma (and the crackpot, politicized science of Stalin's chief adviser Trofim Lysenko) contributed to horrendous failures of crops, and mass starvation. Some observers in Stalin's USSR knew that the science was crackpot and his system would fail; notably the incomparably great and saintly Nikolai Ivanovitch Vavilov, who in the 1930s was one of the world's leading agricultural scientists. Vavilov could have got Russia's new, post-feudal farming off to a flying start. But he was not politically correct, at least in Stalin's view, and was overruled. (In the event, Vavilov was sent to a Siberian labour camp, where he died in the early 1940s. It was either 1942 or '43. Nobody is quite sure. Out of sight, out of mind.)

Mao Tse-tung, from the 1950s onwards, also had a vision of communist agriculture, but in absolute contrast to Stalin's. He did not try as Stalin did to turn peasants into rural factory workers. Instead he sought to build on peasant skills, creating agriculture from the bottom up. For a time, and in some places, I believe he succeeded brilliantly: feeding huge numbers of people from fields that were fertilized with nothing but their own excrement and that of animals; with diets adequately based on rice, and tricked out nutritionally and gastronomically with horticulture, ducks and carp in the paddy fields, and chickens and pigs fed on scraps in the villages. The systems collapsed (I believe) when they became a parody of themselves, not simply pro-peasant but expressly anti-intellectual. Pol Pot's post-Maoist agrarian revolution in Cambodia in the 1970s was a parody in spades which led (as Stalin's industrialized version had done) to famine, and brought rice production to a virtual standstill. Until the wars and the bombing that allowed Pol Pot to take over, Cambodia had had no trouble in feeding itself.

The conflict between productivity and sustainability is of course compounded when the prevailing policy of governments is to bring farmers into commercial, cut-throat competition with each other. The problem is made worse by globalization and the World Trade Organization: nowadays farmers everywhere are increasingly obliged to compete with farmers everywhere else in the world, even though they all operate under vastly different conditions. Then farmers have to produce more and more to stay in business at all. In the modern business ethos this contrived competition is perceived to be a good thing even though a little analysis shows how it squanders both resource and effort, as well as making people miserable for no good purpose. As John Goodman, a small farmer in La Valle, Wisconsin (a rare breed indeed in Wisconsin) told a recent meeting in Oxford, 'I just want to farm well. I don't want to compete with anybody.' What is wrong with that? This is how life was traditionally. People need to eat, and others are happy to farm, and there is no a priori need for the producers to be at each others' throats. In fact, if we apply a little game theory, we can see why cut-throat competition (the fight to the death, as opposed to friendly rivalry between people who fundamentally feel safe and confident) is innately counter-productive. Competition leads to a zero-sum game. One competitor wins, the other loses. It is as the witches say in *Macbeth*, 'When the battle's lost and won'. Constructive cooperation, by contrast, can be a positive non-zero-sum game. By working together (for example swapping machinery or digging common drainage channels) small farmers can produce a better result than any could manage alone, and benefit everybody. Europeans could never have colonized America at all without such cooperation. The gunfights are more dramatic, and have launched a thousand movies; but it's the camaraderie that made the whole nation possible. This kind of point, in some circles, has long reached the status of political correctness. Many argue for more cooperative games in primary schools, for instance. But at the 'serious' level, producers of vital commodities, including food, are now obliged to slug it out, like bull terriers in a dogfight. The more you think about, the more it makes no sense at all. It is organized brutality; and, as in a literal dogfight, only the organizers gain. Yet this is the modern dogma. Competition in all contexts, and to the death, is deemed to be good.

Furthermore, since success is measured in wealth – only the rich can stay in business – the competitors must raise their productivity while keeping their costs as low as possible. There just isn't time or leeway to keep the soil in good heart when the market price of produce barely exceeds the cost of production (and often, these days, is less than the cost of production), while the bills mount up. It is all so obvious. It has all happened before. I do not propose that we should abandon the market economy, or suggest that good, sustainable, realistic farming cannot be accommodated within a capitalist system. But if success is to be measured in wealth, then we have to make sure that the people who do farm sustainably are suitably rewarded. There has to be a premium on sustainability, delivered 'at the farm gate'. The market has to be adjusted to ensure that this happens. In various ways many different groups of people including some modern politicians (notably, to my knowledge, in Europe and Africa) recognize this. But these voices of sanity are not the loudest, any more than Columella's was in first-century Rome.

Overall, the parallels with modern times seem all too obvious. Many, many societies in the past, expert farmers though they obviously were, triumphant though they may have been for centuries at a stretch, got into trouble and disappeared. Sometimes, perhaps, there wasn't much they could do to stop the rot. Sometimes they contributed to their own downfall. This is certainly true of many modern countries where irrigation schemes are coming to the end of their useful life and leaving a legacy of polluted soil.

Sometimes, shockingly, our forebears saw trouble coming, knew the reasons for it, and did nothing. We are witnessing soil erosion worldwide, on a vast scale, and the official policy seems largely to be to keep the fingers crossed. Global warming is a parallel case. It took a long time for successive US governments to admit that it was happening at all or, if it was, that human activity (much of it American) was a contributor. In 2002, five years after the Kyoto protocol, Mr Bush's administration now acknowledges that both are the case, but concludes that it's best for America to continue more or less on its present course. Australia has now taken the same line. The general logic in both cases is that it would, at this stage of world history, be too difficult for major industrial powers like theirs to change course; or at least, too politically inconvenient.

In general the powers that be, such is their psychology, believe that their own political convictions can override good husbandry, and the biological principles in which good husbandry is founded. The Romans thought this. Stalin thought it. Modern governments think it. All had a gung-ho, can-do attitude: boundless self-belief; the conviction that they knew what they were doing. We can see in retrospect that the Romans and Stalin were fooling themselves. We have yet to reap the harvest of present endeavours. But the American philosopher George Santayana famously commented that 'Those who cannot remember the past are condemned to repeat it.' Ironically, one lesson that leaps from history is that we do not learn from it. Each age repeats the mistakes almost as if they felt obliged to do so. Commentators who point out the need to change course, from Columella to Vavilov, are just as routinely ignored.

Farming practice that truly conforms to the demands of biology, and is tempered at least to some extent by morality and aesthetics, can reasonably be called 'good husbandry'. An agricultural strategy based on good husbandry, deeply rooted in biological reality and shaped by morality and aesthetics, can be called enlightened. In truth, the term 'enlightenment' has many contrasting connotations. In eighteenth-century Europe, it tended to be equated with rationality: with the rise of science, and with significant attacks on traditional theology. But Buddhists use the term in a quite different sense: enlightenment for them is a grand grasping of truth, an emotional, and more to the point a spiritual response to the universe we find ourselves in. For me, Jane Austen's expression captures the essence of what I mean in this context: 'Sense and Sensibility'. That's what we need.

Our hypothetical Martian, looking down on our species from a very great height, seeing a middle-sized planet as the earth's own astronauts do, largely covered in water, bright with cloud, and very crowded at least here and there, would take it to be self-evident that in the practice of agriculture, the most important endeavour of humankind, considerations of biology must prevail. Obviously, sensible creatures like us (and he would concede that we have some sense) would not set out to sell ourselves short, or to foul our own small and precious nest. He would also look inside our heads, considerate and knowing being that he is, and perceive that we do indeed have values. We do not regard ourselves purely as flesh on the hoof, waiting to be fed. Morality and aesthetics feature very highly in our thoughts and feelings; and social matters occupy most of our waking time and our dreaming too. He would assume, then, that of course we would organize our agriculture along biological lines, and temper it with morality and aesthetics and social concern. Of course we would farm along enlightened lines, with sense and sensibility. He would be astonished to find that this was not the case. He would shake his head. Being a Martian, a truly rational being, he would not be able to get to the bottom of it.

In the following chapters of this book I do aim to get to the bottom of it. I would tell the Martian, were he here, that he has made one fundamental mistake, albeit one he shares with much of humanity. For although we are rational beings we do not, in the end, run our affairs rationally. And although we are emotional creatures, driven by feelings of aesthetics and morality, and capable of great concern for other people, other forces of various kinds seem to take over nonetheless, and we lead ourselves astray. The Martian's mistake, in short, is to assume that we, thinking beasties, are truly in control of our own affairs, and that whatever we do is bound to reflect our own deep desires and ambitions. But it ain't necessarily so; and in the case of agriculture, spectacularly, it is not.

Of course, if you argue with modern agriculturalists – the few remaining, successful farmers; the agribusiness people; the agricultural scientists and the politicians – they will argue, like Voltaire's Dr Pangloss, that in reality, everything is fine. Actually they will of course acknowledge that life is not perfect. But, they will say, life never can be perfect, and the status quo is as good as it could realistically be. They will point out that today, more people are better fed than previous generations could have thought possible; and in this they are perfectly correct. Henri IV of France famously dreamed of a chicken in every peasant's pot and now chicken is positively commonplace, so cheap that in every high street it's almost given away. At the end of the eighteenth century, when the world population had scarcely reached 1 billion, Malthus predicted that population must soon outstrip supply, and we would surely starve. Now there are 6 billion, and we haven't.

The defenders will also suggest that humanity began to get seriously on top of its own food problems only when agriculture became more industrialized, and as science came on board. Here, though, they have seriously misread history. Large-scale farming does need some industrialization, and we certainly need science; but correlation is not cause. Agricultural science has built on the back of craft, knowledge and skills that have evolved and been devised over many thousands of years, without modern, formal theory. Or at least, science has been useful when it has been used to enhance traditional husbandry. Nowadays it is largely used to override basic husbandry and in that guise it can be extremely destructive, both biologically and socially. Science is vital, in short, but it needs to be reorientated.

Finally, the defenders of the status quo will argue that if we want truly to get on top of our food problems, and continue to do so into the future, then we must carry on as we are now: become even more industrialized, and apply science even more bullishly. This, in effect, is the prevailing policy. It is a profound error. This is the notion of which, above all, the world's leaders must be disabused. Yes, we need science. Yes, the techniques of industry have much to offer. Yes, capitalism broadly conceived can provide the necessary economic framework. But if we want seriously to survive long term, in a tolerable state, then biology and human values must come first. The techniques of industry, and the science that underpins those techniques, must be secondary to those greater goals. The economic structure must be adapted to the realities of biology and the demands of human values, and not the other way around. If we simply industrialize and corporatize as a matter of dogma, and divert science purely for that dogmatic end as now has become the fashion, then we are bound to be in trouble. Biology will not be flouted; and unless we keep morality and aesthetics firmly in view, then they will go by the board. The idea that human values simply install themselves by default is another illusion. In agriculture as in all things, human values must be written specifically into the act.

I don't want to argue that modern agriculture is an unmitigated disaster. Clearly it is in many ways brilliant: in many ways more successful than anything we have had before, or many earlier commentators thought would be possible. Much of what is done now, including much of the science, is necessary, and worth persisting with. But what we have now, nonetheless, has many serious flaws. Clearly, too, it is nothing like as good as it could be. Neither is the status quo sustainable. Any of these reasons taken alone should prompt us to change direction. Take them together, and the case is beyond debate.

Nature is wonderfully flexible, but – unlike human imagination and human conceit – it is not infinitely flexible. It has its own rules. It marches to the drum of biology, and geology, and climate; and none of these is exhaustively comprehensible, or precisely predictable. If the system of farming that any one society devises matches the requirements of biology and the physical environment, then that system of farming, and the society that creates it, will survive. If the farming does not match the underlying realities, then it will collapse. It may take decades or centuries to fail, but fail it will. Since we have such a poor sense of time, a few centuries may seem like for ever. But of course it is not. The farmers of twenty-first century East Anglia are now having to cope with the insouciance of medieval farmers. Modern Australians now face a dozen different disasters (the salination of South Australia is only one), caused by the zeal of the pioneers of the past two centuries (and their casual disregard of Aboriginal know-how). Overall, past mistakes accumulate, and the world as a whole is running out of leeway.

If we, today's people, care about our descendants, then we have to ensure that present-day farming does match up to biological and physical realities. Since farming reflects the entire *Zeitgeist* – economic, political, aesthetic, moral – then, if we want the changes to be robust, we have to adjust the *Zeitgeist*: the whole hierarchy of technique, ambition and attitude. It really is as simple and as stark as that. The present *Zeitgeist* is of boundless confidence, at least in the West; at least matching that of Stalin. We have allowed ourselves to believe that with science and technology, we can put our own stamp on nature. That is a very big mistake.

11

Food: The Future Belongs to the Gourmet

How good one feels when one is full – how satisfied with ourselves and with the world! People who have tried it tell me that a clear conscience makes you very happy and contented; but a full stomach does the business quite as well, and is cheaper.

Jerome K. Jerome, *Three Men in a Boat* (1889)

3

Good Farming, Good Eating and Great Gastronomy: A Lightning Tour of Modern Nutritional Theory and the Wondrous Serendipity

If we want agriculture that is truly enlightened, then we ought to have a good idea of what kind of food we should be trying to produce and why, and most of this chapter is a lightning overview of nutritional theory as now understood. It takes an evolutionary view because nutrition in the end is a matter of biology and, as the great Russian-American geneticist Theodosius Dobzhansky commented in an essay of 1973, 'Nothing in biology makes sense except in the light of evolution'. There is at the end, though, what I believe is the most wondrous serendipity in the whole food story: that good farming, based on enlightened husbandry, provides the kinds of food that are nutritionally perfect, and which also support the world's greatest cuisines. In short, if the world took its lead from good farmers and good cooks, and if science was content to serve those traditional crafts, humanity would have nothing at all to worry about. It's only the economists and politicians who are screwing things up – they, and the scientists who have so complaisantly flocked to their cause.

Nutrition and Taste

Human beings, like all animals, are said to be 'heterotrophs'. That is (unlike plants, which are 'autotrophs'), we need to take in the bulk of our food in the form of organic molecules – ready-made for us by plants or other animals. That food has to supply us both with energy and with 'nutrients' – materials from which to build our flesh. Both energy and nutrients are provided primarily in the form of carbohydrates, fats and proteins. Some of the carbohydrate, fat and protein that we ingest we burn as energy, and some we incorporate into our bodies. But all of it (once it has done service as part of ourselves) tends to be burned sooner or later, and replaced, for self-renewal is the essence of life. We also need to take in a recondite array of organic materials known generically as 'vitamins'. Finally (like plants), we need a fairly long shortlist of minerals.

All this, as heterotrophs, we share with all animals. But animals differ in the ways they acquire the necessary foods. Some, like lions, are carnivores, focusing on meat. Others, like cows, are herbivores, obtaining virtually all they need from plants. Human beings, in common with pigs, rats, foxes, crows and cockroaches, are very definite omnivores. We can in theory get the bulk of our provender either from plants or animals, and in general prefer a judicious mixture of the two. The fact that we are omnivorous makes us (and foxes, and rats) extremely versatile. We can live in a great many habitats. But omnivorousness is risky, too. Everything that grows and a few things that don't (like salt) are potential provender to the omnivore, and the only way that we (or rats or foxes) can work out what it is safe to eat is by watching what others do well on, and by trial and error. The diets of omnivores tend to differ from place to place partly because different things grow in different places, and partly because different populations make different choices from among the things that are available – and all kinds of factors influence those choices.

So how, in general, do human beings satisfy their nutritional needs?

Energy: The Race from Starvation to Obesity

For ordinary, day-by-day, second-by-second purposes, animals like us burn carbohydrates, which we store in our muscles in the form of glycogen. When the glycogen store runs low, we start to burn body fat (and convert some of the burned fat into carbohydrate, to top up the glycogen stores). When (and in general only when) the fat store is depleted, we burn protein. This all makes perfect sense. Carbohydrate is easiest to acquire and is chemically the simplest, so it's a perfect short-term energy store, easy to create and easy to break down. Energy is traditionally measured in calories; and carbohydrates in pure form supply roughly 4 calories per gram, or 400 per 100 grams, or 4,000 per kilogram.

Fat is more complex, and more biochemistry is needed to make it in the first place. Once made, however, fat is a wonderful, weight-efficient, long-term energy store, supplying 9 calories per gram in pure form, or

9,000 per kilogram – enough to last most adults three days, unless they are taking very heavy exercise. Protein is even more complex than fat. A lot of metabolic effort goes into making it. Plants store protein to some extent in their seeds, to nourish the embryo. But animals do not create proteins purely for storage. All our body protein is at work at any one time, in the form of enzymes, or of hormones, or the antibodies that are the front-line troops of the immune system, or – in greatest amounts – as key components of all body cells and most conspicuously in muscle fibres. Red meat is muscle; and muscle is largely protein. To burn protein just to provide energy, then, is literally self-sacrificial. It involves burning functional tissue. The body does this only *in extremis*, when glycogen and all spare fat are spent. A body that is burning protein just to stoke its own fires is a body in trouble.

The hierarchy of preference, and the discrepancy in energy content between glycogen and fat, leads to great consternation among slimmers; or at least among those who attempt to lose weight simply by starving. For the first few days they lose weight hand over fist, and rejoice: the heavyweights calculate that they will be down to a comfortable welter, or even achieve supermodel status, within a month or two. Then, despite the continuing privation, body weight seems to stabilize. The body has got wise. It begins by burning all its glycogen stores to provide the energy that has gone missing from the diet. Since glycogen provides only 4 calories per gram, it takes about 500 grams to provide a day's energy: half a kilo; more than a pound. What's deceptive, though, is that glycogen is accompanied by large amounts of water, so as glycogen is burnt, water is shed. The water outweighs the glycogen by about three to one. So at first, at least four pounds of body weight (including water) may apparently be lost in each day's total fast.

But soon the body becomes alarmed, and mobilizes its fat stores. Only now has 'slimming' really begun. Pure fat provides more than twice as much energy per gram as pure carbohydrate (glycogen), and in the body it is accompanied by much less water. To provide 2,000 calories, the body needs to burn not much more than 200 grams (less than half a pound) of fat; which is about 300 grams (roughly 12 ounces) of adipose tissue (meaning fat plus water). Thus the rate of weight loss slows abruptly. Furthermore, when the body feels it is being starved it may invoke other mechanisms to reduce energy use: for example, reducing the amount of energy lost as body heat. Thus, after a few days' fast, the body goes into high-economy mode. To compound all the problems, the body uses some of its fat to replenish its glycogen stores, which also enables the body to take on more water, to accompany the glycogen. Thus the initial weight loss might even be reversed. Meanwhile, appetite inexorably increases. The slimmer, both desperate and disappointed, gives up. Fasts, more often than not, are followed by binges. The body bides its time, pursues a devious course, but in the end it overcomes the will.

In truth, as all modern nutritionists emphasize, the only way to keep weight within bounds is to match daily intake to daily needs. That's physics. Needs can be increased by exercise. Only if exercise is extreme does its impact seem dramatic. Most of us could get by on about 2,500 calories per day, for instance, but riders in the Tour de France need around 7,000 calories. Yet even modest exercise can make the difference between aesthetic acceptability and obesity. Furthermore, increased exercise does not significantly increase appetite, as is sometimes claimed. Exercise in general helps to regulate appetite. Exercise will increase appetite if it is heavy and habitual (as in traditional lumberjacks, for instance, who probably got through 4,000 or so calories a day and ate accordingly), though not necessarily within the first few days. But for most of us, a little exercise may if anything tend to reduce intake, not least because it alleviates the boredom that can lead to overeating. Bodies are built for exercise; and for people who are resolutely sedentary, eating may be the only exercise they get.

But in the modern world, it is not so easy to gear intake to real physiological requirements. We see this most easily by considering the human species as a whole: asking again, as in Chapter 1, what kind of animals we really are. Of course, different individuals need different amounts of food (where 'need' steers a course between obvious deprivation and obvious excess). Adults in general need more than children because they are bigger, although children may need more per unit of body weight because they are growing. Men generally need more than women both because they tend to be bigger, and because they have a higher proportion of muscle in their bodies, and a lower proportion of fat, and muscle tissue burns more energy per unit weight (is more 'metabolically active') than fat tissue does. Young people need more than old people, because they are more active (although older people in rich societies go on eating anyway, and get fat). All these generalizations are obvious – they all derive directly from the laws of physics – but all are hedged round with 'in general' because there are many variables to confuse the issue. In particular, some people are innately 'thrifty'. Every spare calorie is assiduously stored away, ultimately in the form of fat. Others utilize their food with remarkable prodigality. In them, all spare calories are immediately burned off in the form of body heat. All of us have special tissue ('brown fat', located largely in the back) whose specific task is to burn surplus energy (by means of metabolic pathways known as 'futile cycles'). Some of us have more active brown fat than others.

So, physically, human beings are obviously extremely various; yet their energy requirements are remarkably similar. In general, if adult people have less than 1,500 calories per day over a long period, they will suffer. They will be noticeably thin (and small in general, if the deprivation began before their bones stopped growing) and common sense suggests they are liable to be more prone to infection, since, when active, the immune system requires a great deal of energy. If daily intake falls to much below 800 calories then, in general, death from starvation is well on the cards. On the other hand, if people who are not extremely active (like lumberjacks or professional cyclists) have more than 3,000 calories per day, they are prone to obesity. So the physiological requirement for most of us lies within a very narrow band and if we venture far out of it, either way, we are liable to suffer.

This raises a huge problem for the food industry. As societies grow richer, so they can and do increase their consumption of real estate, motor cars, fossil fuel, electronics, clothes, or what you will. Builders and manufacturers, and taxi drivers, lawyers, plastic surgeons and waiters, see their markets expanding. But rich people can eat only a little more than poor people – at least, if they want to stay healthy. The market for food seems far less 'elastic'. Farmers have limited scope for expansion; and for farmers, surplus can be almost as disastrous as crop failure.

Billions of people are still horribly poor but the total buying power of humanity as a whole has vastly increased in the past half-century. This expansion has presented the food industry with huge opportunities – yet its fundamental problem remains: there is a physiological limit on the amount that people can sensibly eat, and the ceiling is remarkably low. So the food industry has adopted three main strategies. First, it has raised the cash value (or at least the sale price) of food in general by increased packaging, and by offering new alleged advantages: goods out of season and from all parts of the world – and, above all, 'convenience' and 'fast' food. Second, it has focused on foods with an addictive quality (see Chapter 4): rich in sugar, fat and salt. Research has shown, too, that people can take in huge amounts of sugar (at least 2,000 calories per day), for example in the form of coke, without significantly depressing appetite; and so a consumer's total intake can be increased too, well beyond the desirable physiological ceiling.

Finally, and in the end most significantly, the output of staple crops (like wheat, maize, barley and soya) has been boosted artificially by feeding them to livestock, and then selling meat, eggs and dairy produce rather than the original grain (this is discussed further in Chapter 4). Thus the ostensibly non-elastic food market can be expanded remarkably – especially if, as is now the trend, most of the animal is then turned into dog food, and people eat only the fillets and ribs. Diets that are high in meat (and in sugar) are highly concentrated: they supply far more energy, weight for weight, than traditional diets bulked out with fibrous and watery grain and vegetables. Thus it is easy for anyone on a modern, industrialized diet to drift well outside the narrow band of physiological comfort.

The overall change both in food and in wealth has been rapid. Americans who grew up in the 1950s and '60s in an apparent cornucopia were the children of parents who had endured the lean and often desperate years of the Depression. The parents of modern Chinese children lived through the Cultural Revolution, in the wake of Mao. Many modern people have hardly known what it is like to live in a society where people eat traditional diets in normal amounts, and at regular meals. Many have gone straight from the constant threat of hunger to a state of easy excess, when they are positively assailed with provender of unprecedented richness. That obesity is the result is not at all surprising. The diets and the customs that once effectively prevented it have largely been forgotten, first due to deprivation, and then to the modern industrialized food industry that rushed to fill the gap.

Obesity is defined by 'body-mass index', which is the body weight in kilograms divided by the square of the height in metres. People whose body-mass index works out at twenty-five to thirty are technically 'overweight', while a figure of thirty or more signifies 'clinically obese'. The WHO's new International Obesity Taskforce now reports that clinical obesity has increased by 50 per cent between 1995 to 2000, from an estimated 200 million adults worldwide to 300 million. Roughly 18 per cent of the total world's population are now overweight (including clinically obese), which is roughly the same proportion as are undernourished. Obesity traditionally signified wealth. Now, in at least equal measure, it is the stigma of the poor. In 'developing' countries, obesity and frank under-nutrition commonly sit side by side. Ten per cent of Chinese were judged overweight or obese in 1996 and by 1999 it was 15 per cent. The children are particularly afflicted – and this is not helped by China's present restriction on childbirth; it is common in modern Beijing to see two svelte parents with one rotund child, on whom they lavish all their new-found, relative affluence. In Brazil and Colombia, traditionally poor and often horribly so even now, 40 per cent are overweight or obese: figures that compare with Europe. There is some evidence that people who in past ages lived on very meagre diets are physiologically adapted to be 'thrifty', and are particularly prone to fat when their diet becomes richer. Women, who are physiologically geared up to pregnancy and lactation, are in general thriftier then men. Thus one UN study has found that 44 per cent of black women in the Cape of South Africa are clinically obese. But the people of West Samoa and Tonga hold the world record: 65 per cent of men and 77 per cent of women are obese. Ethnically, these are the same smooth, brown and beautiful people whom Paul Gauguin painted in Tahiti, not much more than a hundred years ago. What would he make of them now?

Yet obesity is not the end of the matter. People are taller, too, with more massive bones, and children mature earlier. More importantly, obesity is associated with a range of degenerative disorders – the 'diseases of affluence' are now rife among the world's poor. Obesity itself does not cause these disorders, or at least is only weakly causative, but it is a marker of excess intake. The diseases include Type 2 diabetes (the kind that comes on in middle age), coronary heart disease, and various digestive disorders. The WHO estimates that by 2025 more than 300 million people worldwide will be diabetic – a staggering 5 per cent of the present world population, roughly equivalent to the present population of the United States.

Over-nutrition, and all that follows from it, should be seen as a disaster. Human beings are wonderful creatures, supremely adapted to life on this planet and beautiful to look upon: 'What a piece of work is a man!' said Hamlet (meaning 'humanity'), and he was right. On aesthetic grounds alone, the newly invented body shape is a sad departure. Well-rounded people can be beautiful too, of course, as portrayed not least by Rubens and Renoir. The present vogue for stick-thin supermodels is a commercial aberration (although thinness is now associated with wealth as fat once was, and is becoming a status symbol). But the obesity that's now becoming common worldwide is nothing like conventional chubbiness.

The human frame has been overridden, as if the genes that lay down its basic specifications have given up: flesh is packed in wherever it will fit, typically in great drooping masses on the thighs, like landslips. More and more Americans travel everywhere on little electric carts because they are too heavy to walk. Death comes early and horribly, with amputated limbs and blindness brought on by diabetes. And what is it all supposed to be *for*? Drunks may die early, but some of them at least have some riotous times along the way. Before the days of antibiotics syphilitics traditionally died early too, but they too presumably had some fun. But to endure such morbidity for burgers and coke? It makes no sense at all.

Carbohydrates: Fibre and the Big Change of Mind

Carbohydrates caused enormous controversies throughout the twentieth century, and particularly in the second half. But theory seems to have been reasonably settled over the past twenty years or so.

Basically, carhohydrates are the sugars and their compounds. Some sugars that occur in nature are very simple, like glucose and fructose. Sucrose, the stuff we stir in our tea that comes from grain and beet, is slightly more complex: each sucrose molecule is compounded from one sucrose and one fructose. Starch (the standard storage carbohydrate of plants) and glycogen (the standard storage carbohydrate of animals like us) are both compounded from glucose, though arranged somewhat differently. Cellulose, the tough material of plant cell walls, is also compounded from glucose molecules, with yet another arrangement. Traditionally, it is assumed (and is more or less true) that human beings can digest carbohydrates in the form of sugars, starch and glycogen, but cannot digest cellulose. Nutritionists traditionally refer to digestible carbohydrates in pure form (like table sugar) as 'refined' carbohydrates; while plant material that contains sugars and starch plus the cell walls, is said to be 'unrefined'. One hundred per cent wholemeal flour is unrefined carbohydrate, but 70 or 80 per cent white (or off-white) flour that has had most of the bran removed, is partially refined.

The controversies have been of several kinds. First, some nutritionists in the mid century, notably John Yudkin who was one of Britain's first professors of nutrition, felt that refined carbohydrate itself, notably sugar, was in various ways damaging to health if taken in large amounts – which means in the amounts that are typical of Western diets. The title of John Yudkin's 1960s book, *Pure White and Deadly*, summarizes the general drift. From the 1970s, however, this notion was given a new spin; or rather, two different spins.

First, many nutritionists now agree with the general notion that too much sugar is bad. But they do not commonly suggest that it is innately toxic. The main trouble is that people tend to eat sugar for the taste (we quickly acquire a 'sweet tooth'); and it is possible to take in huge amounts of sugar without depressing appetite. Extreme obesity is likely to ensue; and so, too, the disorders that are associated with extreme obesity, including diabetes.

Second, and more far-reachingly, the 1970s saw a volte-face in the orthodox attitude to non-digestible carbohydrate. Thus, before the 1970s, the materials of which plant cell walls are composed were commonly and somewhat dismissively known as 'roughage', or sometimes just 'bulk'. Roughage was assumed to hasten the flow of food through the gut, and school matrons and old-fashioned general practitioners recommended it to ease constipation. This was one of several reasons why small children were urged to 'Eat up your greens!' But since plant cell walls were assumed to be indigestible, and effectively chemically inert, few orthodox nutritionists thought that 'roughage' had more than the crudest physical effects. Most (as I know from talking to nutritionists in the 1960s) did not take it particularly seriously.

The change of mood came when a number of clinicians and scientists noticed that various diseases of the modern world, including diabetes, coronary heart diseases, gallstones and various cancers, including cancer of the colon, were associated with a diet that is conspicuously low in unrefined carbohydrate – in other words, low in roughage. In particular, the British physicians Denis Burkitt, Neil Painter, G. D. Campbell, T. L. Cleave and Hugh Trowell argued the perils of what they called 'saccharine disease', where 'saccharine' refers to sugar, not to the sweetener that serves as a sugar substitute. (T. L. Cleave, G. D. Campbell and N. S. Painter, *Diabetes, Coronary Thrombosis and the Saccharine Disease*, 2nd edn, John Wright & Sons, Bristol, 1969; D. P. Burkitt and H. C. Trowell, eds, *Refined Carbohydrate and Disease*, Academic Press, London, 1975.) Of course, these 'diseases of affluence' were also associated with a diet high in fat: high fat and low roughage tend, in practice, to go together. Of course, too, correlation does not mean cause: the fact that diets low in unrefined carbohydrate are associated with diseases of affluence does not mean that the diet is the cause of those diseases. High fat could be the cause – or, in principle, some extraneous factor that had nothing to do with diet at all could be the root cause. Less fibre and more fat in the diet are also associated with watching more television, for instance. But common sense should be allowed to play a part in scientific thinking: and if clinical conditions that seem to be of dietary origin (for example including obesity) are also associated with a change in diet, then it is sensible at least in the first instance to look for the causes within the diet itself.

Various other developments in the 1970s strengthened the idea that roughage does not serve merely to scour the insides. Dr Hugh Trowell set the tone by pointing out that the term 'roughage' was in various ways misleading, and suggested 'dietary fibre' instead. Now everyone speaks of 'fibre' (and 'roughage' has become more or less obsolete). Secondly, it became increasingly clear that plant cell walls, which are by far the main source of dietary fibre, do not consist solely of cellulose. They also contain a range of other carbohydrate compounds, including pectin and a range of hemicelluloses. Cellulose is present in virtually all plant cell walls, and indeed is the commonest organic polymer in nature. But it is not necessarily the principal component of any one cell wall.

It became clear, too, that dietary fibre as a whole – cellulose plus hemicelluloses and the rest – was not so chemically inert as had been supposed. For one thing, it became clear that the many large molecules of which fibre is composed adsorb many of the other nutrients on to their surface, and so influence both the speed and the manner of their digestion and absorption into the body. Furthermore, it became clear that fibre does not rush unchanged through the human gut like a pan-scourer. Most of the fibre does survive the depredations of the stomach and small intestine, but then it confronts the bacteria that are the permanent residents of the colon; and although those bacteria do not digest the fibre as thoroughly as those of the cow's rumen, they do break it down significantly.

It was also clear by the '70s that the colon is not the inert conduit that had traditionally been assumed. It absorbs a great deal of material (besides absorbing water, which had long been recognized). In particular, the colon significantly influences the physiology of bile. Bile is produced in the liver, stored in the gall bladder, and flows down the bile duct into the small intestine where it aids in the breakdown and hence the digestion of fat. The bile itself contains various 'bile salts' which are derivatives of cholesterol. Gallstones form in the gall bladder if the salts crystallize out. After the bile has done its work in the small intestine, it carries on down to the colon where most of the bile salts are reabsorbed. They circulate back to the liver and are then reused. However, the bile salts may be changed chemically in the colon by the bacteria that live there; and the bacteria themselves, and the effects they have on the bile, are in turn influenced by the presence or absence of dietary fibre in the colon. Hence fibre in the colon (and the nature of that fibre) affects the chemistry of bile. It turns out that the bile salts that are produced and reabsorbed when fibre is present in the colon in good amounts are less likely to precipitate in the gall bladder and so form gallstones. But the bile salts produced in the absence of fibre in the colon do tend to crystallize out in the gall bladder, and so form stones.

This by itself does not prove beyond all possible doubt that a high-fibre diet prevents gallstones; but it does provide a very good reason for thinking this might be the case, and in the absence of absolute, godlike certainty, it seems prudent to act on what seems likely. Furthermore, Ken Heaton's work on bile showed perhaps more clearly than any up to that time that the role of fibre in the body's physiology may go far beyond that of mere abrasion – and in fact, that abrasion has virtually nothing to do with the case. By such insights, the grand hypothesis – that the amount of fibre in the diet influences our chances of heart disease, gallstones, diabetes and the rest – became more and more plausible. Plausibility is a dangerous principle in science. Ideas may be plausible and yet untrue (that the earth is flat seemed plausible for a long time), and they may seem implausible and yet be true. (It is astonishing, and yet is virtually certain, that the continents do wander about the surface of the globe.) In practice, though, plausibility is necessary. Once doctors and scientists saw that fibre *could* influence the body in various recondite ways they felt free at least to take the grand hypothesis seriously. Now, dietary fibre is very much a part of nutritional science and nutritional lore; and nutritionists are actively probing the pros and cons of particular fibres. In general, cereal fibre seems particularly effective and among cereals, oat fibre seems outstanding.

Finally, and most obviously, nutritionists began to point out that diets high in fibre tended to be more 'dilute' in terms of energy per unit weight or per unit volume than low-fibre diets. To some extent (though by no means entirely), appetite is regulated by bulk. Fibre not only bulks out food without providing extra calories, it also creates an overall architecture that makes food more difficult to eat and slows the rate of eating. It is hard to eat too many calories if you subsist on whole apples, or whole potatoes, for example. Their fibre content is not impressive, but the fibre holds water and the overall structures take some eating. Again, we see that poor people on marginal diets and rich people on concentrated diets approach dietary fibre from different angles. Women in Africa attempting to suckle their babies while subsisting on a diet of maize porridge need less of it: the diet is too bulky, relative to its energy contribution, to give them the calories they need. A bit more fat in their diet can make a big difference. But people in the West, eating burgers (with white bread) and coke need more fibre. The slices of tomato and onion and the wisp of lettuce that typically accompany the burger are symbolic merely. They make no serious contribution to the daily fibre intake.

Now, in the early twenty-first century, advice on carbohydrate has two main strands, both very different from the advice of the 1960s. Then, high-carbohydrate diets were not generally recommended (although I remember the great British middle-distance runner, Gordon Pirie, advocating 'a nice thick slice of bread and honey' before attempting the world record over three miles). Now, competitors in all sports are advised for example to eat plenty of pasta and bananas before and during the events, both of which are rich in starch (and generally low in fibre). Pete Sampras gets through sackfuls. In affluent countries where diets tend to be too concentrated, everyone is advised to eat a high-fibre diet. This means a diet high in vegetable material in general: fruit, leaves, roots and seeds.

Fats

Dietary fat also caused great controversy throughout the twentieth century, again on various fronts, which reflect the innate heterogeneity of fats, the different roles they play in the body, and the contrasting needs of different people. Thus, fat in general is of supreme importance as a store of energy. Take in too much energy and fat accumulates, and this of course is obesity. Secondly, one of the forms that fat takes is cholesterol. Cholesterol is an essential player in the body's physiology, intimately associated with the metabolism of steroids which, among other things, are the key components of a wide range of hormones. (Hormones are compounded either of steroids or of proteins.) Cholesterol is ferried around the blood in various molecular forms, both as 'high-density lipoprotein' (HDL) and as 'low-density lipoprotein' (LDL); and as everyone now knows, high concentrations of LDLs are associated with atheroma (blockage of the arteries by fatty deposits) which, in particular, predisposes to coronary heart disease (the heart muscle itself is supplied by the coronary arteries, and heart attack occurs when these are blocked). Finally, as emphasized in particular by the British nutritionist Michael Crawford, there are various peculiar forms of fat which in small quantities play essential roles in the metabolism and in the body structure, and so are known as 'essential fats'. Some of these the body can make for itself, when plied with suitable precursors. But these precursors have to be supplied in the food itself, and so they essentially have the status of vitamins. In general, these essential fats are found in leaves, seeds, and the particular parts of animals that tend these days to be thrown away, such as the liver.

Many learned committees have advised on fat intake over the past thirty years but none has significantly improved on the advice of Britain's Royal College of Physicians, which they issued in 1976. The physicians pointed out that in most traditional diets, particularly in low latitudes, as in the Mediterranean, India, China and South-East Asia, only about 20 per cent of the calories are supplied by fat. Most of the energy in those traditional, low-latitude diets is supplied by unrefined carbohydrate – notably cereal (usually wheat or rice). By contrast, diets in affluent societies, and particularly in higher latitudes (northern Europe and North America), typically supply 40 per cent or even more of the calories in the form of fat. Livestock makes all the difference. The high-fat diets are high in meat and dairy products. In the low-fat diets, meat serves primarily as a garnish, a source of flavour. A typical southern Chinese meal is mostly rice and cabbage, with tiny slivers of pork or duck. A typical Western meal (particularly in the 1960s) was built around a hunk of meat. When I had an expense account in the late 1960s, many of my more fashionable lunching companions felt they were being virtuous and even austere if they confined themselves to a nice juicy steak and a token salad. But good beef is 'marbled', the fat coursing through it in white veins; and although a steak looks lean, a juicy one, product of the modern farm, is very fatty indeed by the standards of wild nature.

In tribal societies, coronary heart disease is rare or unknown. But by the 1970s it had reached epidemic proportions in Western Europe and North America. Many societies were growing richer at that time, at least in the sense that they had more disposable cash. As they grew richer so they tended to shift from a low-fat diet (meaning a low-meat diet) to one high in meat, and therefore in fat. As they shifted, so their incidence of coronary heart disease tended to rise from low or effectively zero, up to Western levels. Thus in Italy, there was still a clear division between the south, which retained its traditional low-meat diet, and the north, which already had a high-meat diet more typical of northern Europe. The difference is reflected in the sauces that north Europeans put on spaghetti: Bolognese, which is meaty, reflecting the north Italian diet; and Neapolitan, which is mostly tomato, reflecting the more traditional (and less affluent) south. The northern Italians suffered far more coronary heart disease than the southerners, but as the southerners grew richer, its incidence among them increased. Japanese people on traditional diets in Japan had a very low incidence of coronary heart disease but their relatives in California were already achieving (if that is the appropriate verb) Western levels. Clearly, in such cases, the difference between populations with a high incidence of coronary heart disease and those with a low incidence was not primarily genetic. Environment was making the difference. Diet was not the only difference between 1970s California and 1970s Japan, and correlation is not cause. But diet clearly emerged as a strong causative candidate.

But late twentieth-century nutritionists distinguished sharply between fats that are saturated, and those that are unsaturated. Chemically speaking, 'saturated' means that the fat contains all the hydrogen that can possibly be attached to the carbon atoms in the fat. Physically speaking, saturated fats tend to be hard – suet is the archetype – and typical of terrestrial mammals, such as cattle. Unsaturated fats could, theoretically, contain more hydrogen than they in fact do. They may be polyunsaturated, meaning in general that they could contain two or more hydrogen atoms per molecule than they in fact do; or monounsaturated, meaning that one more hydrogen atom would saturate them. Unsaturated fats typically are oils, as found in oilseeds and fish. Most edible oils are polyunsaturated, like sunflower and safflower. Olive oil is monounsaturated. Coconut oil is saturated, but it has small individual molecules and, nutritionally speaking, behaves more like an unsaturated oil than like mammalian fat. Nutritionally speaking, saturated fats are bad news; and unsaturated fats are much better, or even positively good. A high intake of saturated fats raises the level of blood LDLs, which are the kind that predispose a person to coronary heart disease and hence to heart attack (and other diseases); but a high intake of unsaturated fats lowers blood LDLs, and raises HDLs. Eskimos have a high-fat diet but most of the fat is unsaturated fish oil, and they have a very low incidence of coronary heart disease.

The Royal College of Physicians in the 1970s acknowledged that the case against high-fat diets, and particularly against diets high in saturated fat, was not cast-iron. People with a pernickity turn of mind could still argue that the chains of reasoning are too long: there is no absolute proof that a high intake of saturated fat leads to a high blood-level of LDL which in turn increases the likelihood of coronary heart disease. The association is statistical, too, rather than absolute. Always we are talking about an increased *chance* of heart attack; not the absolute certainty. Thus Japanese people in California generally suffer far more heart attacks than those in Japanese villages. But even on 1970s Californian diets, there will be some individuals of all races who do not suffer heart attacks. Always, in a discussion on the relationship between fat intake and heart attack (or smoking and heart attack) someone will cite some old gaffer in their village who lived to be ninety while smoking forty a day and living exclusively on lard. Lucky chap, is all one can say. *In general*, people on diets high in saturated fat have a far higher chance of coronary heart disease; and in societies where the diet is very low in fat coronary heart disease is virtually unknown, and people die with arteries clean as flutes.

So the Royal College of Physicians made common-sense recommendations based on what they called 'the balance of probabilities'. They recommended a general reduction in fat intake for north Europeans and Americans, from around 40 per cent of total calories to 30 per cent (suggesting 35 per cent as a reasonable first target); and a general shift from saturated (generally animal) fats to unsaturated. The physicians recommended sunflower oil; but modern opinion seems to have shifted towards monounsaturated olive oil. Add in Michael Crawford's concerns with regard to essential fats, and we find a general recommendation towards a low(ish) fat diet, high in vegetation and relatively low in meat, dairy and eggs, but also very varied. The plants should include seeds and leaves, and the livestock should include at least some of the bits that commonly, these days, are thrown away; such as the liver.

The last of the big three nutrients is protein: also the source of great controversy, and volte-face, in the twentieth century.

Protein

Modern nutrition is a twentieth-century science, and from the lofty vantage point of the early twenty-first century we can watch the early nutritionists feeling their way. The story is wonderful, full of dedication, ingenuity and insight, hacked out by scientists burning midnight oil and by doctors who often chose to live in the most difficult circumstances, in the poorest villages of Africa, Asia and South America. It is not fashionable to present the history of science as heroism, but it often has been heroic nonetheless.

Occasionally, however, the pioneers of nutritional theory went astray. In the 1930s, and for a few decades afterwards, they seemed to misconstrue protein. The mistake was made for commendable reasons, and the analysis was not grossly inaccurate. Yet it led to a huge and potentially extremely damaging perversion of food policy.

Proteins are the body's executives and general functionaries. They form much of the stuff of flesh. They also provide the raw materials for a large number of hormones; haemoglobin, which carries oxygen in the blood; antibodies, which are the body's defence against infection; and enzymes, which are biological catalysts, which conduct the whole metabolism. Chemically speaking, proteins are compounded from carbon, hydrogen, oxygen and nitrogen, with a small but highly significant content of sulphur (highly significant because the position of the sulphur atoms largely determines the overall shape of the protein molecule; and its qualities depend on its overall shape). More specifically, proteins are compounded from chains of amino acids.

Twenty-something amino acids are known in nature, of which about twenty are found in the human body. About a dozen of these latter are conventionally said to be 'non-essential', which does not mean that the body does not need them. It simply means that they do not need to be present in the food, because the body is able to make these 'non-essential' types from other amino acids that are present. But eight are said to be 'essential' because the body cannot manufacture them from other amino acids. These have to be present in food. Ideally, dietary proteins should contain all the essential amino acids, in the ratios in which the body requires them. If any one essential amino acid is present in less than the ideal quantity, then the quality and the dietary value of the whole protein are compromised. The essential amino acid that is least well represented becomes the 'limiting factor', dragging down the quality of the whole. Animal proteins in general (not always, but usually) contain all the essential amino acids that humans require, in ideal quantities – as they seem more or less bound to, since animal flesh is very like ours. But plant proteins are often (relatively) deficient in one or more essential amino acids. Notably, cereal proteins tend to be low in the amino acid known as lysine; and pulse proteins, as in beans, tend to be low in tryptophan. Traditionally, then, nutritionists have declared that animal proteins are 'first-class'; whereas most plant proteins have been dubbed 'second-class'.

In the 1930s the belief developed that people in general need a massive intake of protein, and that this protein had to be of the very highest quality. In other words, the belief developed that people need to eat animal protein. By the early 1960s, when I first started taking a serious interest in nutrition, it was commonly argued in effect that the more animal protein we ate, the better. (The dangers of animal fat, which are always liable to accompany animal protein, had not yet become fully apparent.) Correspondingly, nutritionists of the mid twentieth century tended to argue that plant proteins, with the possible exception of soya, were inadequate both in quantity and in quality. By the early 1960s potatoes, and even cereals, were virtually being written off as 'empty starch' – at least in popular articles, of the kind on which people at large typically rely.

For various reasons, this emphasis on protein in general, and on animal protein in particular, was pernicious. Yet as Professor Donald S. McLaren, then at the American University of Beirut, outlined in the mid 1970s ('The Great Protein Fiasco', *The Lancet*, 1974, vol. 2, p. 93), this huge and damaging mistake arose from fine observations that were made for the most humane reasons – but with a measure of bad luck thrown in. For McLaren pointed out that malnutrition was widely studied in the 1920s and '30s, and that it presented a confusing picture. It can manifest in many forms. Poor diets may make people thin (although thinness may also be normal and at times of course may simply be fashionable). Or malnutrition can make people sluggish (although sluggishness too can have a hundred other causes, ranging from infection to temperament). Or it can lead to a vague puffiness, or an increased susceptibility to infection. All in all, it can be extraordinarily difficult to tell, when faced with a thin, sick child, whether he or she is thin for purely genetic reasons (what used to be called an 'ectomorphic' body type), or is wasted by infection, or whether the infection and the thinness are both rooted in poor diet – and if so, where the dietary deficiency lies.

Serious light began to dawn in 1932. Dr Cicely Williams, who was the first woman medical officer in Africa's Gold Coast (now Ghana) described a 'deficiency disease of infants' in which, she said, 'some amino acids or protein deficiency cannot be excluded'. This, of course, is a supremely guarded statement, as preliminary observations in science should always be. It makes no claims; it merely raises possibilities that should be looked at. The deficiency disease in question was kwashiorkor, a well-recognized syndrome in which the hair and skin are dry, the hair is reddened, and the belly is swollen not by obesity but by oedema. Biochemists later demonstrated distinct alterations in blood chemistry, which gave laboratory support to the more subjective clinical findings.

Kwashiorkor was distressingly common in Africa throughout the twentieth century; and although the precise mechanism remains uncertain, lack of protein is definitely implicated. Dr Williams' observation was a great advance. But fate took over. As Professor McLaren recalls, 'During the late 1930s and 1940s, when international meetings were virtually non-existent and travel was limited, discussion of the nature of malnutrition in children was carried on through the correspondence columns of journals.' Significantly, this correspondence was 'mainly between workers in different parts of Africa', where many people rely upon cassava. Cassava, a starchy root, just happens to be the only one of the world's major staples that really is deficient in protein.

The Second World War interrupted further research. When the war was finally over, the newly created World Health Organization and the Food and Agricultural Organization of the United Nations were anxious, in a new and optimistic world, to get on with the task of feeding the world's hungry. They built on pre-war observations, which were all that was available. In 1953, in an influential report entitled *Kwashiorkor in Africa*, they made the much-quoted statement that kwashiorkor was 'the most serious and widespread nutritional disorder known to medical and nutritional science'. Yet, says McLaren, this statement was made 'without reference to the rest of the world and other forms of malnutrition'. The die was cast, however. Kwashiorkor was perceived to be the greatest problem; and the single greatest cause was deficiency of protein.

The general idea of protein deficiency was in the air at the time that Cicely Williams first wrote about it. John Steinbeck seemed to have just such a notion in mind when he wrote his first major novel, *Tortilla Flat*, published in 1935. Steinbeck tells us it is 'the story of Danny and of Danny's friends and of Danny's house', and is set near Monterey, 'that old city on the coast of California'. Danny is a paisano, meaning he has 'a mixture of Spanish, Indian, Mexican and assorted Caucasian bloods. His ancestors have lived in California for one or two hundred years.'

One of Danny's friends is Teresina, who has an extraordinary number of children, among whom is Alfredo. And:

At about this time in California it became the stylish thing for school nurses to visit the classes and to catechize the children on intimate details of their home life. In the first grade, Alfredo was called to the principal's office, for it was thought that he looked thin.

The visiting nurse, trained in child psychology, said kindly: 'Freddie, do you get enough to eat?'

'Sure', said Alfredo.

'Well, now. Tell me what you have had for breakfast.'

'Tortillas and beans', said Alfredo.

'[Do] you eat at noon?'

'Sure. I bring some beans wrapped up in a tortilla.'

Actual alarm showed in the nurse's eyes, but she controlled herself. 'At night, what do you have to eat?'

'Tortillas and beans.'

Her psychology deserted her. 'Do you mean to stand there and tell me you eat nothing but tortillas and beans?'

Alfredo was astonished. 'Jesus Christ', he said, 'what more do you want?'

Perhaps the school nurse was alarmed by the lack of fruit and vegetables; but meat and milk were surely uppermost in her mind. At that time, few seriously believed that people and especially children could do without them. Twenty years later, by the 1950s, this supposition had become official theory. The production of protein, in the form of meat and dairy products, became a mission and then, almost indistinguishably, a bandwagon. The US, with all its cash, space, grass and surplus cereals, was perfectly positioned to oblige the world at large. As always, it was generous; and as always, the boundaries between altruism and self-interest were blurred. Crudely speaking, the US unloaded its dairy surpluses on to Third World countries; and when those surpluses dried up, as they had by 1964, they were replaced by what McLaren called 'protein-rich food mixtures', the manufacture and distribution of which were widely acknowledged as necessary succour for people in poor countries. To be sure, malnutrition was still widespread. Yet it continues to astonish me that so few people in high places seemed to ask how it was that the populations of Africa and Asia had thrived for so many thousands of years before the Western world opted to shower them with food supplements. In *The Hungry Planet* (Macmillan, London, 1967) and later in *Too Many* (Macmillan, London, 1969) the geographer Professor Georg Borgstrom of Michigan State University argued that a large proportion of the human race must already be doomed since they clearly did not have enough protein, which mainly meant meat. Yet if the people of Africa and Asia had really been so deficient in their traditional diets, they would not be here at all.

But some scientists, including Professor McLaren, doubted even from the early 1950s whether the emphasis on protein and more protein was really appropriate. As the 1970s progressed, it became clear that it was not. The Indian nutritionist P. V. Sukhatme was prominent among those who stressed the concept of 'protein sparing' (see for example *British Journal of Nutrition*, 1970, vol. 24, pp. 477–87; and J. C. Waterlow and P. R. Payne, 'The protein gap', *Nature*, 1975, vol. 258, pp. 113–17). Professor Sukhatme found that malnourished, ostensibly protein-deficient children in India improved not only when given protein-rich foods, but also when given more 'ordinary' local foods, such as cereal and beans. Of course, cereals and beans do not simply provide energy, as sugar does. They are also sources of protein. But they are not officiously 'high-protein' foods of the kind that had been considered necessary.

So it turned out that malnourished children, even those with kwashiorkor, were not usually short of protein in particular, but of food in general. More specifically, the body needs both energy, to keep it going, and protein, from which to construct and reconstruct its own fabric. But, as outlined above, the body has its priorities. If it is short of energy it will first of all metabolize glycogen in the muscles and liver; and as that runs low, it mobilizes its stores of fat. But if it is still short of energy after that, then it burns protein. If you give protein to children who do not have enough energy foods, then their bodies burn the protein. Then they suffer protein deficiency and they may (although not necessarily) begin to show the specific signs of kwashiorkor. Conversely, high-energy foods will 'spare' the body protein and even a modest intake will be seen to be adequate.

In general, protein nutrition has two separate aspects. First, the body must receive a basic, absolute quantity of protein that ideally should be eaten every day, because surpluses of protein (unlike surpluses of sugar and fat) are not stored. On any day that the body misses out on protein, it has to draw upon body tissue, notably muscle, to make good the shortfall. The immune system, which is heavily protein-dependent (antibodies are protein) is presumably compromised as well.

Proportion matters, too – the ratio of protein ingested relative to the total energy. If the body receives too little energy, then it will burn some of its protein to make good the shortfall, so even if it takes in an amount that should be adequate, it will still experience deficiency. On the other hand, if the body is plied with sufficient protein and also with excess energy, then it will put on fat. Hence a person could get enough protein (in absolute terms) even by eating a low-protein food such as cassava, but in doing so he or she would take so much surplus energy on board that their bodies would soon become spherical.

But thanks to scientists like McLaren, Sukhatme and others, official nutritional authorities reduced recommended intakes of protein about threefold between the late 1940s and the mid 1970s; and recommendations remain at the more modest level. Thus in 1948 the National Research Council of America (NRCA) recommended that children aged one year should receive 3.3 grams of protein per kilogram of body weight per day. In 1957 the FAO had brought this down to 2.0 grams per kilogram for one-year-olds, and by 1971 made a further, even more dramatic reduction, to 1.2 grams per kilogram. Adults, metabolizing more slowly and no longer piling on muscle, require only half as much protein per unit of body weight as children do.

But, during all that time, the recommended energy intake remained steady (as it has ever since). After all, energy requirements are easier to calculate, and to get right. Thus the NRCA recommended 100 calories per kilogram body weight for one-year-olds in 1948, while in 1971 the FAO were recommending 105 calories per kilogram.

Since recommended protein has gone down (dramatically) while recommended energy has remained steady, the ratio of protein to energy that human beings are thought to need has also dropped. According to the 1948 figures, a child would need to obtain 15 per cent of his or her total diet in the form of protein. According to the 1971 (and present-day) figures, 5 to 6.5 per cent is adequate; and for adults, at least when they are not pregnant or lactating, about 3 per cent is enough.

This change of mind has profound implications. Even when you make suitable adjustments for the quality of protein, all the world's great staple crops, including wheat, rice, millet and even potatoes, emerge as perfectly adequate sources of protein. If you eat enough of any of them to satisfy your daily energy requirements (which, after all, is what staples are supposed to do) then you will automatically satisfy your protein requirements as well. You do not need to eat excessive amounts of those staples, which would make you fat; or to supplement them with protein-rich supplements – of which meat is the prime exemplar. To be sure, an exclusive diet of wheat or rice would lead to serious deficiencies of vitamins and minerals (although an heroic Danish physiologist did once contrive to live for a year on an exclusive diet of potatoes; and people in parts of Western Europe through much of the nineteenth century also subsisted virtually entirely on potatoes, as depicted not least by Vincent Van Gogh). Nonetheless, a diet adequate in staples should not be deficient either in energy or in protein; and once these 'macronutrients' are taken care of, the rest should be relatively easy.

Even the issue of protein quality is easily resolved. Traditionally the proteins from cereals were called 'second-class' because they contain relatively low quantities of some of the essential amino acids, notably lysine. However, once the overall requirement for protein is seen to be fairly modest, then the shortfall in quality ceases to matter. There is not as much lysine in cereal protein as is ideal, but there is enough nonetheless. Besides, nature has laid on a wonderful serendipity. The proteins of pulses – beans, peas, lentils – are relatively rich in lysine. So when cereals and pulses are eaten together, their combined proteins have an excellent spectrum of amino acids: together, they become 'first-class'. We find (should it be a surprise?) that cereal–pulse combinations are prominent in all the world's great traditional cuisines. The Indians eat chapattis or rice with chickpeas or dhal, made from a variety of beans and lentils. The Chinese and Japanese traditionally eat bean curd (tofu) with rice. Rice and peas is a staple of the West Indies. The English are not great pulse eaters (apart from mushy peas) but even they eat baked beans on toast (although the beans are generally American). The Mexicans, of course, eat tortillas (made from maize) with frijoles (beans). If Steinbeck's nurse had known what became apparent forty years later, Alfredo's diet would not have bothered her at all (give or take a few vitamins). As Alfredo put it himself, 'What more do you want?'

But in the days of the protein myth, roughly from the 1930s through to the '70s, nutritionists were in effect telling humanity that we could not survive, or at least live to our full potential, unless we had access to high-protein foods, which in particular meant meat. Farmers were very happy to oblige. Commercially this was very good for them.

This over-emphasis on meat, made respectable first by nutritionists and then by sociologists, has been the most pernicious of all trends in modern agriculture: damaging to humanity in many different ways –

nutritional, economic, cultural; damaging to the beleaguered beasts who have supplied the feast; and very damaging indeed to the earth as a whole, and to the future prospects of all the creatures, including us, who live on it. The drive to increase livestock is of such crucial importance that it must have a chapter to itself. In the rest of this chapter, we should look at the final component of diet, the mixed bag of micronutrients: the minerals and vitamins. Here, too, I believe, modern biological theory – meaning modern evolutionary theory – is bringing about, or should bring about, a radical shift in thinking.

Vitamins, Minerals, 'Functional Foods', and the Concept of Pharmacological Impoverishment

From early times medicine men, apothecaries and physicians have known that people need various odd ingredients in their food of a somewhat medicinal nature. By the seventeenth century the Royal Navy's advisers knew full well that fresh fruit can protect against scurvy; and the Navy's particular preference for limes led the Americans for evermore to refer to the Brits as 'limeys'. By the late nineteenth century physiologists were beginning to identify a range of ingredients which, when deficient in the diet, led to specific 'deficiency diseases'. These ingredients were a mixed bag, chemically and in their effects, but they were lumped together under the general heading of 'vitamins' and, for convenience, known by letters. The agent that prevents scurvy is of course vitamin C, alias ascorbic acid.

It has long been clear, too, that all animals and plants require a list of minerals, as outlined above. The minerals that human beings are known to require now embraces a fair slice of the periodic table (about a third, by my calculations) and the list continues to grow: the subtleties of nutrition are far from probed. From time to time, different minerals become fashionable, and supplements containing them are billed as panaceas.

In recent decades, too, nutritionists have identified a range of recondite ingredients that they have labelled 'functional foods', or 'nutraceuticals'. The general notion is that although functional foods are not quite as vital as vitamins are – you don't become conspicuously ill for lack of them, at least not in the short term – they are nonetheless beneficial. A prime example is, or are, the sterols found naturally in plants. It transpires that these, taken regularly, lower blood cholesterol, and in particular reduce the harmful LDLs. Accordingly, various commercial companies have taken to adding them to margarine. (I eat such margarine myself. I have no particular prejudice against big commercial companies per se, but merely object to many of the things they do, and to the uncritical support that so many modern governments give to them. Insofar as big companies do good things, they are worth supporting; and sterols in margarine, at least *faute de mieux*, seem on balance to be a good thing.) Functional foods seem to occupy a niche somewhere between vitamins (lack of which is obviously damaging and perhaps lethal) and tonics. You don't die from lack of tonics; but even if you are reasonably healthy, a tonic will make you feel even better (or that at least is the idea). Western medicine in general treats people who are ill, with curatives; but much traditional medicine treats people who are well, with tonics. The ideal, one feels, would surely combine the two.

The vitamins are a very mixed bag. They are chemically very distinct one from another, and they perform many different tasks within the body. So too are the essential minerals: both metals and non-metals are included and some (like selenium) are of the kind that most people who are not students of chemistry have barely heard of. The 'functional foods' are at least as heterogeneous as vitamins are, in their chemistry and their roles – although what most of the 'functional foods' do and how they act is not yet clear. These are early days.

Such untidiness sits badly within the canon of science. Why should bodies require such a peculiar list of materials, doing so many different things? Scientists hate implausibility, too, as already mentioned; and the fact that so many disparate claims are made for so many different functional foods has made the whole subject seem somewhat suspect. The whole area needs a unifying theory for science to get its teeth into: some grand explanation of why it is that nature should have produced creatures (including us) that in addition to the carbohydrates, fats and proteins that are the stuff of energy and flesh, should also require what seems like a totally arbitrary pharmacopoeia of so many recondite elements and compounds. What's going on?

I suggest that we have simply to refer to Dobzhansky's comment that 'Nothing in biology makes sense except in the light of evolution'. Basic evolutionary theory seems to tell us everything we need to know.

Consider the realities of human evolution. A key component of every organism's environment is other creatures: creatures which, like us, have a living to make and are in effect our competitors. From long before the time that they were recognizably human (or even recognizably mammalian), our ancestors grew up in the presence of plants. Of course, they depended on those plants for their own survival. But also, the plants fought back against the humans and other predators who wanted to eat them: and they did this by evolving, in their turn, a range of thorns and tough integuments – and chemicals of all kinds, including tannins, phenols, sterols, esters, essential oils and so on and so on. Plants (together with bacteria and fungi) are the world's most proficient chemists.

All of these materials were part of our ancestors' environment. So what did our ancestors do about it? Answer: they evolved mechanisms to cope with the toxins and other extraneous agents. Among other things, they evolved a range of enzymes that would detoxify and otherwise disarm whatever noxious chemicals nature threw at them. At least: natural selection favoured those individuals who possessed such enzymes, which amounts to the same thing. In other words, our ancestors adapted to the presence of those toxins and other extraneous materials that plants (and fungi and bacteria) produce.

But this is where the opportunism of natural selection clicked in again. Once a body has evolved a mechanism to detoxify or otherwise deal with some extraneous material, why not press that detoxifying mechanism into some other service? The best-known example of such a progression is provided by atmospheric oxygen. Atmospheric oxygen is produced almost entirely by photosynthesis, as carried out by plants and bacteria. But photosynthesis is a highly sophisticated process, which clearly did not evolve on earth for many millions of years (probably about half a billion years) after life itself first appeared. So before photosynthesis evolved, there was no oxygen in the atmosphere, and all organisms respired anaerobically. Indeed, to these organisms, oxygen gas was highly toxic – because it is so chemically active. Thus organisms that suddenly found themselves in the presence of oxygen had to evolve ways of dealing with it. Some did this by confronting the spare oxygen with sugars, which the oxygen then harmlessly oxidized away to form carbon dioxide and water. These sugars initially were essentially sacrificial, intended to detract the oxygen from the fabric of the organism's own body; like the hamper thrown from the back of the troika to detract the pursuing wolves.

But evolution then went further. As sugars are oxidized, so energy is released. Some organisms evolved means to harness that energy – and so aerobic respiration was born, which in fact is far more efficient than anaerobic respiration. Thus oxygen, the most lethal of toxins, was tamed, and became an asset. Indeed it became so much of an asset that the earth was soon dominated by creatures that practise aerobic respiration, for whom oxygen has become vital. Through the opportunism of evolution, the former toxin became the essential requirement; for while our most ancient ancestors were poisoned by oxygen, we would die if deprived of it for more than a few minutes.

I suggest that an analogous process explains our dependence on vitamins, and our slightly lesser dependence on the newly identified functional foods and tonics. All these are materials produced by other creatures, notably by plants, bacteria and fungi, that our ancestors had to evolve ways of coping with. They did evolve such means; and then pressed the mechanisms that coped with those materials into further service. Thus we have become dependent on materials that once, in the extreme past, might well have poisoned our ancestors. There is little direct evidence for any of this, basically because the idea is new and nobody has yet looked for any. But the example of atmospheric oxygen, which biologists accept universally and for which there is abundant evidence, shows that such an explanation is certainly plausible. Plausibility is not the end of the story, but in practice it is the start.

It is also the case that modern people eat a far smaller number of plants than our hunter-gatherer ancestors did. Anthropologists have discovered time and time again that the hunter-gatherer people who remain (or remained until recently) typically make or made regular use of eighty or more, and sometimes a hundred or more, wild plants. They also tended to eat, say, the roots, leaves and flowers of a plant where we might eat only the fruits and seeds. Thus there seems to be tremendous variety in a modern supermarket; but the plethora of goods on sale are all variations on a few simple themes. The twenty kinds of pasta, for example, are all fashioned from the flour of the same durum wheat. Furthermore, modern cultivated plants have been selected and bred for their rate of growth, their appearance, texture, and sometimes for their flavour. Some, too, like tomatoes, potatoes, cassava and parsnips, have had the toxicity of their ancestors bred out of them. In short, modern diets based on modern crops are far more bland, biochemically speaking, than the diets of our hunter-gatherer ancestors. We should not forget the hunting side of this either: our hunter-gatherer ancestors also ate a greater variety of animals – and they ate the whole animal, brains and guts and all; not just the bland lean muscle, the red meat, that is favoured today.

If it is the case that our ancestors adapted to many of the extraneous materials found in wild plants and animals (and fungi and bacteria); and if it is the case as surely it must be that the human lineage has adapted to the presence of those materials, and indeed to some extent become dependent upon their presence; then surely we should think ourselves deprived. Since the missing chemicals have a pharmacological effect (over and above any effect they might have as nutrients) then we might say that most of us, today, are 'pharmacologically impoverished'. I have written about the putative phenomenon of pharmacological impoverishment and some people in influential places have taken it seriously. I think it needs looking into. It seems to explain, at a stroke, why we need vitamins and such a weird array of nutrients; and it brings the whole otherwise arcane business of 'functional foods' into the arena of plausibility. For example, we need plant sterols to keep our blood cholesterol down for the simple reason that our ancestors were constantly exposed to high doses of these materials, and cholesterol metabolism is adapted to their presence.

Actually (although this is a different story), I believe we should extend the principle of pharmacological impoverishment to include psychotropic agents – the kind that manifest as 'drugs', like marijuana, cocaine, mescaline, and the opiates including heroin; and, of course, alcohol, nicotine and caffeine. We have in general a puritanical attitude to these agents. The legal case against them is not based simply on the perceived harm that they may do. It is rooted, more deeply, in the belief that our bodies and minds *ought* to be free of all such materials. The moral puritanism is reinforced by a kind of scientific puritanism: the deep (and necessary) belief of scientists that all explanations and descriptions of nature should be as simple as possible – according to the principle known as 'Occam's razor'. Doctors, traditionally, tend to be both moral puritans and scientists; and putting the two together they have in general concluded that our bodies (and minds!) are more or less bound to function most efficiently when their chemical intake is as bland as possible; when, indeed, it is more or less confined to those materials that are recognized as bona fide nutrients – proteins, essential fats, unrefined carbohydrates and the approved list of vitamins. The idea that our bodies and minds might actually function *better* in the presence of a weird assemblage of apparently arbitrary materials seems to run counter to those most fundamental premisses.

Yet evolutionary theory explains why this might be so. Our bodies feel deprived of plant sterols, and since our modern diets are so biochemically innocent, sterols must be added. Additives in margarine are not ideal; but for the time being, in the absence of a wild diet, they are perhaps the best we can do. Our minds, perhaps, feel and are deprived of the stimulation and the relaxation that our ancestors once derived from the berries and mushrooms around them. I am not a 'druggie' myself, incidentally, apart from a perpetual intake of tea and coffee. I need the caffeine buzz but otherwise have no stake in the drug culture. But I do feel that the Western world's 'war on drugs', which it has so obviously lost, probably does far more harm than good; and the deep reason is that it is misguided. It is rooted in the belief that it is good, in all senses, to be as biochemically innocent as possible, and that may simply be wrong. It runs against the tide of our evolutionary history. Like most government policies in the modern world, in all spheres, it is rooted in bad biology.

So what can we conclude from all this?

Good Husbandry, Good Nutrition and Great Cooking

The first and most obvious conclusion is that modern nutritional theory is complex. The body needs an extraordinary number of different ingredients, not all of which are yet known. The ingredients cannot simply be considered in isolation: the ratio of one to another may matter (for example, the ratio of fat, carbohydrate and protein, one to another) and some ingredients may affect the metabolism of others (for example, some materials in plants inhibit the uptake of calcium). Different people, and the same person at different times of their lives, differ in their requirements for different foods.

To make matters worse, there are many matters of key importance that cannot be tested exhaustively. For example, pharmacologists who seek to understand the effect of some new drug rely heavily on the 'double-blind trial'. Two sets of patients matched as far as possible for age, sex and other relevant parameters are given either the drug in question, or a placebo, and neither the physician nor the patient knows which is being administered (which is why the trials are said to be 'double-blind'); and as a final refinement, in the kinds of trials known as 'crossover', the patients who had been receiving the placebo are then given the real drug, and those on the real drug are given the placebo, to see if it makes any difference; and all the effects are assessed by neutral observers who do not know who has been receiving what.

In principle, it would be good to carry out a double-blind trial to pin down once and for all the relationship between fat intake and heart disease. But this would require at least two vast sets of subjects, whose diets were controlled over decades. It would take too long, cost too much, and of course be ethically and politically unacceptable. So in seeking to establish the relationship between fat intake and heart attack, scientists and doctors must rely on evidence that is all, in one way or another, obviously flawed. They do experiments on animals – which can be controlled, but animals are not humans, so the relevance of the findings are always in doubt; and of course, animal experiments raise their own ethical problems. They can observe human beings who for some reason or another prefer or are forced to follow restricted diets: patients with particular diseases, vegetarians, people in wartime, monks, prisoners. But such groups almost invariably are in some other way 'selected', and it is difficult to find matching groups over time. Or they can compare societies who seem genetically and in other ways similar but have different diets (southern Italians versus northern, for instance). They can also see what happens when people change diets (like the Californian Japanese), but of course there are many differences besides diet between Japan and California that might also be confusing the picture.

All in all, then, the great generalizations of nutritional science have usually to be inferred from data that are very diverse, but are rarely ideal. Usually, too, we are talking about likelihood, rather than certainty. It simply is not true that *everyone* who eats vast amounts of fat will get a heart attack, and the people who break the rules and stay healthy strike us more forcefully than the tables of statistics which suggest that they ought to be dead. In the end, as the Royal College of Physicians said in their excellent report of 1976, you just have to weigh 'the balance of probabilities'.

The innate complexity and the inevitable uncertainty give endless opportunity for commentators effectively to say what they like. Thus a few physicians still choose to argue that a high-fat diet does not predispose to heart attack. You can bend the statistics in ways that suggest this (at least at first glance) and of course, the long, double-blind trial that could do most to pin the matter down cannot be done. Endless magazine articles pick up on particular, arcane nutrients that swim in and out of fashion: zinc, selenium, potassium, whatever. There is huge uncertainty in the field of 'functional foods'. This is a new subject, which in many ways overlaps the claims of many forms of traditional medicine; some of which surely is excellent, and very much to the point, although (so far) seriously unexplored by Western science.

It is also sad but true that more and more nutritional research these days is financed by the food industry. This is encouraged by governments, such as those of Britain and the US, who seek to reduce public spending, and also believe as a matter of political principle that societies can be run most efficiently by private companies, and for both reasons are happy to allow private companies to take over large areas of research, and encourage scientists, even in the ancient universities, to seek commercial funding. It surely is rare for private companies to publish mendacious results, and most scientists in universities who accept the commercial shilling stress their own independence. Yet it is widely acknowledged that scientists who published research which put their sponsors in a bad light would not find it easy to renew their grants. Perhaps there is nothing bad to be said. But then again, many nutritional disorders of many kinds worldwide have reached epidemic proportions, and the companies that one way or another provide much of that food must surely be implicated somewhere.

Then of course there is advertising. Advertisers are not allowed to lie, but they do put their own spin on things. A small example, but it makes the point, is that in Britain at present semi-skimmed milk is sold as a 'low-fat' product on the grounds that it contains only 2 per cent fat. So it does. But 95 per cent of milk is water. Only 5 per cent is nutrient: fat, carbohydrate (mainly the milk sugar, lactose) and protein. So 2 per cent fat represents 40 per cent of the bit that actually matters. But that is 40 per cent by weight. Since fat provides twice the calories, gram for gram, as either protein or carbohydrate, the fat in semi-skimmed milk in fact provides around 60 per cent of total calories. When nutritionists speak of a 'high-fat' food or diet they are referring not to the weight of fat, but to the proportion of total calories that fat provides. Thirty per cent or less is the recommended proportion. So on this basis, semi-skimmed milk is a very high-fat product. Actually, a pint a day of semi-skimmed milk on cornflakes and tea is by no means excessive: the proportion of total daily calories provided by the milk is low, and milk has many other nutritional assets including provision of calcium. Still, though, the idea that semi-skimmed milk is in itself a low-fat food is not true; not, at least, in the sense in which nutritionists usually use that expression. My point here is not that the advertisers of semi-skimmed milk are bad, or that they set out deliberately to deceive. Still, though, it is very hard for people at large who do not have ready access to nutritional textbooks to make sense of what's being said.

Yet, through the murk, a shortlist of simple principles have emerged this past few decades. Surely by now they are familiar to most people; yet they are worth running through, because so much rests on them.

First, and most obviously, it is not wise to eat too much. Just a few calories more per day than are actually burned off in exercise will cause fat to accumulate, unless you are one of those fortunate people with an alert body-stat, and active brown fat, who immediately turns excess into body heat. Almost all adult human beings who are not taking part in the Tour de France need only around 1,500 to 3,000 calories per day. The gap between deprivation and obesity is remarkably narrow, and many people worldwide have raced from one to the other within a generation or less. Obesity per se may be only slightly life-shortening, but it is associated with other conditions that definitely are life-threatening, of which the most obvious is diabetes.

Second, the lion's share of calories in the diet should be provided by unrefined carbohydrate, which in practice means the fibrous staple foods of the world: cereals, pulses, tubers. Ideally, fat should provide only between 20 and 30 per cent of daily calories.

Furthermore, the fat should as far as possible be unsaturated: either polyunsaturated, or monounsaturated. This means in general, plant oils (such as sunflower and olive oil) rather than hard animal fats. But fat is complicated stuff, and in general it seems wise to take varied sources. Fish oils in general are much recommended.

The total intake of protein need not be vast, however: nothing like what was typically recommended in the 1950s and '60s. This means that in practice we can get most of our protein (all of it, actually) from plants (notably cereals and pulses). Meat and other animal products are always useful as guarantors of protein quality, and they also provide other essentials such as zinc. But meat does not need to be eaten in vast amounts. It need not and should not be the centrepiece of the meal, as has lately been fashionable in north European and North American diets.

Finally, the easiest way to solve the micronutrient problem – that vast and as yet untracked array of vitamins, minerals and functional foods – is to eat as wide a variety of foods as possible. Try many cereals and pulses, the biggest range of vegetables and fruit you can lay hands on – and different varieties, particularly old-fashioned varieties, closer to the wild state; herbs, spices, fungi, and fermented foods, like miso and pickles. These come with no special recommendations (at least, not from me) but it makes good evolutionary sense to suggest that human beings must be adapted to biochemical variety in general, and the new science of functional foods supports this.

At present, plant breeders worldwide are striving to provide a range of staple foods that are enriched with particular nutrients, including yellow rice (enriched with vitamin A by genetic engineering) and vitamin-A-rich maize and sweet potato (produced by conventional breeding). Some experts argue that at least in the short term, the easiest way to correct deficiencies of vitamin A and other micronutrients is through such enriched staples, and since some of those advocates are clearly good people and well informed, I would not want to take issue with them. In general, however, such enrichment is not necessary if people have access to horticulture, especially if the horticulture is eked out with, say, chickens and ducks and a few pigs. In general, too, traditional agriculture includes horticulture, and small livestock. The present deficiencies result from systems of farming, and economies, that are in some way compromised – commonly because the people have lost their autonomy, no longer farm for subsistence, and are now growing cash crops, in monocultures, for export. In the Second World War the British government ordered that bread should be reinforced with calcium. But this was because the country was under siege – normal agriculture had been interrupted. When the war ended and normal service was resumed, this nutritional reinforcement was withdrawn. In other words, such officious enhancement of staples is necessary only as an emergency measure, for countries under siege. To propose that people in poor countries should build their diets around such staples is to acknowledge that for them, a siege economy is the normal condition. So perhaps vitamin-A-rich staples (and other reinforcements) do have a role. But in the long term, and indeed as soon as possible, it surely would be better by far to get every country's agriculture back on to a stable basis, properly mixed, as has been generally the tradition throughout the world.

How does modern nutritional advice fit in with the demands of good, benign, sustainable husbandry – that is, with 'enlightened' agriculture? Like a hand in a glove, is the answer. When the land is treated well, it produces variety. A variety of plants and animals, working in harmony, is needed to bring out the best in it. In general, plants would prevail, with staples dominating and a patch of vegetables at every opportunity. But there would always be some animals: sheep on the hills, cows in the meadow, pigs and chickens in the yards, ducks and carp in the paddy fields.

How, then, would this combination of sound nutrition and enlightened farming match the demands of gastronomy? Absolutely perfectly, is the answer. Every country – and when tradition prevailed, every region and every village in every country – had its traditional cooking. Some was clearly better than others, and few would dispute that among the supreme masters are Italy, France, India, China, and the Arab-influenced countries of the Middle East and North Africa. I would put in a *sotto voce* bid for Britain for reasons of chauvinism and nostalgia, and others would surely bid for Turkey, others again for Japan, some for Mexico, and – well; perhaps we need an Olympic Games of great gastronomy.

Within this truly fabulous variety, however, a few simple themes emerge. In great cooking the world over, the staples prevail: cereals, pulses, tubers. Vegetables and fruits, spices, herbs, fungi and fermented foods are present in huge variety and abundance. Meat, the world over (in truly great cooking) is always used sparingly; and always in the greatest possible variety – with all the bits, the livers and feet and tripe and even the lungs, minced into sausages, spiced and herbed. The basic Chinese meal is variations on a theme of rice. The Chinese banquet is endlessly varied, but heavily dependent on leaves and soups. Arabs make the most exquisite feasts from wheat, mint, honey, and wisps of goat. The southern French make wonderful casseroles with haricot beans, while the Egyptians favour broad beans. And so on. All this is truly great gastronomy. Much of what has passed as 'haute cuisine' through much of the twentieth century (although the vogue is mercifully dying) is mere vulgarity: epitomized, I think, by the fillet steak with some hideous creamy sauce: bad farming, bad nutrition, bad gastronomy; just a gratuitous show of wealth and ignorance.

In short, all we really need to do to put world farming back on course, cure and prevent the nutritional disasters of modern humanity, ensure that our descendants are well fed for ever, and eat like gourmets, is to take food and cooking seriously. The correspondence between good farming, good nutrition and great gastronomy is absolute, and wonderful. Lose sight of the principles of any one, and the others are compromised. Yet this is not just a stroke of good fortune. Traditional farming reflects nature; human beings have evolved to be adapted to what nature naturally provides; and the peasants who devised great cuisine (for the greatest chefs have merely built on what people at large worked out over decades and centuries) naturally adapted to the farming that in turn reflected nature. Seen thus, the perfect fit between good husbandry, sound nutrition and great cooking is inevitable. Truly, the future belongs to the gourmet. Here is one of the outstanding serendipities of human existence.

So why, if life can really be so easy, have we gone so disastrously astray? Some of the reasons have been described in this chapter – notably, the mid-twentieth-century over-emphasis on protein, and hence on meat. But the prime reasons are economic and political, and these underlying causes are described in Part III. Even in this, though, meat and livestock are at the heart of things. Meat and livestock are so important that they need a chapter to themselves.

4

Meat: How Did We Get It So Wrong?

As poor people grow richer, they eat more meat. That seems simply to be a sociological fact. I stress 'poor', because people who are already rich don't necessarily eat more meat if they become even richer. They may even eat less. But poor societies, and individuals within those societies, do. I have been to many an agricultural meeting where it was taken to be self-evident that the shift to meat is inevitable and 'natural': that it represents 'progress', and should be encouraged. One of the most chilling was in Washington DC in the late 1980s, where agricultural scientists outlined their ambition to establish an intensive piggery in Malaysia. Intensive pigs need cereal, but Malaysia (apart from rice) is not arable country; and, of course, it is predominantly Muslim. But then, the pork was intended for Australia.

The farmers and meat barons (and governments) who encourage meat-eating also take it to be self-evident that they have right on their side. The fact that poor people do eat more meat as they grow richer proves (for how could it be otherwise?) that they have an innate appetite for it, which in the past has been suppressed by their poverty. Once the blockage (the lack of cash) is removed the floodgates open. To satisfy people's desires is simply to act democratically. Furthermore, nutritionists were telling us through the middle decades of the twentieth century that people need huge amounts of protein – effectively, as much as possible – and that this need could be satisfied only by the flesh, milk or eggs of animals. The meat barons have grown rich, sure, but they have grown rich by giving people what they want and need.

Taken all in all, the general shift towards meat (and the saturated fat that comes along with it) is the greatest transition of modern times both in nutrition and in agriculture. If human beings really do need all that meat – and if, even if they don't need it, they truly desire it – then we will just have to live with that fact. I argued in Chapter 1 that when human numbers stabilize, around 2050, the world can stop increasing its total output of food. But if people go on demanding more and more meat then – in a democratic world – it seems only right to provide it. As discussed in Chapter 2, that would mean that overall output would have to go on rising after all.

But although the powers that be take it to be self-evident that more meat is necessary, and that people have an insatiable appetite for it, and that on both counts the demand for it must be satisfied, in reality that entire thesis is spurious. The sociological evidence has been misconstrued (again); the underlying nutritional theory is junk; and the psychology that allegedly underpins the whole case is crude in the extreme. There simply is no good evidence that people feel the need to eat more and more meat if left to themselves, and (above a small amount) they certainly have no need of it. In truth, meat has been sold, and sold, and sold again, as energetically as any commodity has ever been sold; backed by huge commercial forces which have been supported by governments who have taken their lead from out-of-date science. To some extent the commerce has been cynical (some people just want to make huge amounts of money), but to a large extent the various players have felt themselves to be virtuous: 'Doing well by doing good', as the American satirist Tom Lehrer put the matter in a slightly different context. When virtue is on the side of wealth it's game, set, and match. But it's time the game ended.

To begin with, do we really need a lot of meat? If people have a very poor diet, then modest inputs of meat can be hugely beneficial. Meat provides high-quality protein, and so enhances plant protein. It supplies esoteric fats, minerals (such as zinc, and calcium in milk) and vitamins (such as B_{12}) that can be difficult to supply in other forms. It is also a high concentrated source of energy. Fat is the greatest energy source there is; and the surplus protein in meat can be broken down to supply almost as much energy, weight for weight, as sugar does. So for people on 1,500 calories a day or less – and especially for lactating mothers in out-of-the-way villages, living on a diet of maize gruel and desperately trying to take in at least 2,500 calories a day so they can feed their baby – even a small amount of meat can be a godsend.

But 'small' is the watchword. As outlined in Chapter 3, people's protein needs are modest. Non-lactating adults at least can get all their protein perfectly well from plants (cereals, pulses, some tubers), and the prime role of meat is to guarantee its quality. Esoteric fats, minerals and vitamins are 'micronutrients' – and so, by definition, we don't need much of them. Excess protein in the diet is simply broken down in the body: the bulk of the protein molecule (the carbon-based bit) is used as energy, and the rest (the nitrogen-based bit) is simply excreted. For people who are already well fed, the saturated fat that accompanies meat is a positive menace, a prime cause (so it seems) of coronary heart disease and a definite contributor to some cancers (including some breast cancer).

What about our innate yearning for meat, which the meat barons claim they are satisfying in the name of democracy? That, too, has been hugely exaggerated. Cool appraisal of the evidence suggests that our appetite for meat is tightly circumscribed. We like it as a garnish most of the time, and enjoy an occasional blow-out. But most of us find a diet that is heavily biased towards meat oppressive. We are natural omnivores, but lions we emphatically are not.

The personal experience of any of us makes the point. At a smart buffet, of the kind that everyone gets invited to at least once per decade, where everything is on offer, do we home in like wolves on the ham and beef? Few do, in fact. Most like a few slivers of such provender, with baguettes and pickles, and then get stuck into the profiteroles. The hamburger is popular with rich and poor alike, not because it is overarchingly meaty, but because the meat is minced and spiced and diluted with gherkins and bread (which McDonald's insists on calling a 'bun'). In the 1960s and '70s, nutritionally more innocent than today, well-heeled businessmen had steak and salad for lunch – and this was considered austere. The rest of us wanted chips with it (and still do).

The advocates of meat also appeal to evolutionary psychology (aka sociobiology); the science that seeks out the evolutionary roots of human behaviour. (Evolutionary psychology has its detractors, but at its best it deals rigorously in testable hypotheses, and can properly be called a science.) One of the fundamental notions of evolutionary psychology is that modern humans have inherited the predilections of our ancestors who lived on the plains of Africa in the Pleistocene, between 2 million and, say, around 40,000 years ago. The idea goes that our Pleistocene ancestors would have found it very difficult to get enough to eat at all, and would have been chronically deprived of energy. Natural selection would accordingly have favoured the evolution of an unrestrained appetite for energy-rich foods such as honey and fat. These are the highest-energy foods to be found in the wild but both are rare in natural environments. They are also hard to obtain, for honey is protected by murderous bees and fat belongs to big animals with horns that are not keen to give it up. We evolved a frantic appetite for such foods (the thesis has it) because without such freneticism we would not have bothered to seek out such foods at all; and yet, taken all in all, it was to the body's advantage to take the risks. The same principle has been applied to salt (sodium). It is necessary in small amounts, but in very short supply in most areas, and worth seeking out. (Many wild animals will indeed walk many miles in search of salt licks.)

However, there is no evidence that any human beings in a state of nature have ever bothered to increase their meat intake beyond a rather modest amount – unless, like the traditional Inuit, they had no alternative.

Traditional people are hunter-gatherers – and generally gather more than they hunt. Yet as Richard Lee showed in the Kalahari, traditional hunters are not exactly overworked. They could catch more if they chose. But they don't want to. Evolutionary psychologists also seek insight into human nature by studying our close animal relatives, notably the chimpanzees. My primatological friend Dr Jennifer Scott (now at the Wesleyan University in Connecticut) tells me that although chimpanzees may show an alarming predilection for meat they do not generally eat it neat. They eat it with leaves: one bite meat, two bites leaf. We do much the same. The traditional standby in Britain is meat and two veg.

But if we are not truly the out-and-out meat-eaters that the meat industry pretends, why do poor people eat more meat as they grow richer? What is this, but the floodgates of appetite opening? When people are truly poor, they can indeed be undernourished – deprived not simply or especially of protein or vitamins or minerals in particular, but of energy in general; and, as we have seen, meat (with fat) is a high-energy food. For poor people on marginal diets, meat can indeed be just what the doctor ordered. How, though, do we explain the continuing rise in intake well beyond nutritional need – unless it is indeed the case, as some evolutionary psychologists have suggested, that we have inherited an all-powerful appetite for flesh?

In truth, the idea that we have inherited carnivorous leanings for reasons of physiology is old-fashioned: still promulgated in some circles, but out of date. More recent evolutionary psychologists stress that, above all, human psychology is driven by reproductive (sexual) and social pressures. We need to find mates; and whatever society we live in we need both to be sociable in a general way (or we are liable to be ostracized) and to maintain our status. If we don't have status we are liable to be bullied; and, in general, those with highest status get the first pick of the potential mates.

For many social creatures, social and sexual kudos are demonstrated by food. Herbivores, whether birds or mammals, may advertise their status by protecting territories that offer rich pickings. Carnivores do the same – and also, commonly, offer gifts of meat to their potential mates to show their prowess; as courting eagles do, for instance. For human beings, meat is definitely a status symbol. The hunter who brings home a side of venison thereby demonstrates his strength, courage and resourcefulness. He deserves to go far. The breadwinner who progresses from shop-floor to management demonstrates his status with a new house and a bigger car – and by inviting his friends to a barbecue. In a general way, too, (as chimpanzees and Homer Simpson attest) the sharing of meat is a sociable thing to do. Also many traditional societies – whether poor or not – like to demonstrate their status by their modernity. In the recent past this has meant adopting the customs of the West – and among rich Arabs and Japanese, for instance, this has included consuming more meat (and whiskey, particularly for the Japanese).

Finally, of course, meat is the ultimate convenience food. Nothing is easier to cook than a steak (although it isn't so easy to cook it well). A current ad on British TV urges us to 'slam in the lamb' (that is: to throw a chop under the grill, or a shoulder into the oven). People in this busy world have largely lost the art of cooking, and have no time for it, and anything fast is welcome (although the rise of the TV chef suggests a hankering for lost glories).

In short, meat intake increases with wealth not because people have an overweening and open-ended desire for it, and certainly not because they need more than a small amount of it, but for social reasons. People buy more because it has been perceived as the smart thing to do – and people, above all, do what is perceived to be smart. Clothes are supposed to keep us comfortable, but almost all of us will suffer – and some will endure agonies – to be in fashion. Many of us would (literally) rather die than be shown up. For all kinds of reasons, meat has been a status symbol through much of history (when it implied wealth) and prehistory (when it demonstrated strength and courage). But if we took away that social connotation, would the perceived appetite still be there? Certainly, most of us would still want the odd chop, and the occasional burger. Of course we like meat. But is there any reason to suppose that human beings would go on eating more and more of it if the social pressure was removed?

Yet there is another, powerful reason why meat consumption has risen so dramatically these past few decades. It has been sold as energetically as any commodity (including cars, cosmetics or the strips of favoured football teams) has ever been sold in the history of the world. For the ever-increasing sale of meat solves at a stroke the most fundamental economic problem of all agriculture.

To survive in any business, and to make a profit, there are two essentials. One is to ensure the greatest possible gap between the cost of production and the sale-price. This gap (the margin) contains the profit. The second absolute requirement is turnover. It is far, far better to make a 10 per cent profit on 100 items than to make a 50 per cent profit on one item. As the adage has it: 'pile 'em high and sell 'em cheap' – although of course if you can pile 'em high and sell 'em dear that's even better. The fundamental problem for farmers is that people cannot eat unlimited amounts. Rich people do eat more than poor people, but if anyone who is not riding in the Tour de France eats more than about twice as much as is needed simply to stay alive, then they start to get seriously fat or suffer in other ways. But how can farmers 'pile 'em high' if people don't want more than a modest amount? How can they increase turnover? Their market, as traditional economists say, is 'inelastic'.

Meat is the answer. People could easily be fed on a relatively modest output if they were content to eat traditional, cereal-based diets (as most people, left to themselves, clearly are). But this is bad news for the farmer. The way out of this economic cul-de-sac is to give half the cereal to livestock. This immediately reduces the total output of energy and protein per unit area of land by about tenfold. So then the total area under production (or the yield of cereal per unit area) must be greatly increased. The higher the proportion of meat in the diet, the more the output of staples can and must be raised. Simple arithmetic suggests that if people could be persuaded to eat an all-meat diet, then the output of staples could be raised ten times – and indeed would have to be. In truth, though, there is no upper limit; for when people have enough meat in general, output can be increased still further if the suppliers (producers, processors, retailers) then throw the bulk of the carcass away. This is already happening. Modern supermarkets, which in countries like Britain sell most of the nation's meat, are piled high with cutlets, steaks and fillets – the prime parts of the animal. It is still possible to buy liver, and sometimes kidney, but any request for tripe would be met with a blank stare. Tripe in the hands of accomplished cooks (in France, Spain, China, even bourgeois Switzerland) is delectable. But for modern commerce it is also too cheap and filling. The commercial powers that be would rather sell it to the Chinese, who still know what to do with it, or put it into sausages or dog food. By such institutional profligacy the wheels of commerce are kept turning, ever faster.

The downside of such commerce is all too clear. At an individual level, the world has witnessed an unprecedented and shocking rise in 'diseases of affluence'. Sugar and salt have played their part in this but so too, above all, have meat and dairy products, with their cargoes of cholesterol and saturated fat. Traditional diets are systematically corrupted: the glories of Asia, North Africa and Southern Europe – banquets produced from rice, wheat, spices, herbs, with just a few slivers of goat and mollusc – are replaced by gobbets of instant flesh. Health is compromised, and so are individual sensitivity and cultural diversity. Apart from genocide, and final extinction of our species, the loss of cultural diversity is surely the greatest tragedy that can befall humankind.

At the global level, the steady rise in meat output is a disaster. The intensive livestock units that have sprung up to meet the perceived 'demand' are cruel and polluting. But they in turn consume almost half of the world's wheat and most of its maize (two of the three outstandingly important crops), and almost all the soya (which is by far the biggest pulse crop). These staples in turn are provided by more and more intensive arable farming – more fertilizers, more pesticides, and now (God help us) more and more genetically engineered crops, expressly designed to respond to those chemical inputs. To accommodate this vast production, fields grow bigger, and woods and hedges and the creatures that live in them disappear. As outlined elsewhere in this book, when livestock are raised according to the tenets of good husbandry (the ruminants to eat the grass on the hills and wet meadows, the pigs and poultry to clear up the leftovers) they hugely increase the overall economy of farming. Agriculture that includes the appropriate number of animals, judiciously deployed, is *more* efficient, not less, than an all-plant agriculture. But when livestock is produced in vast (and ever-increasing) numbers, needing correspondingly vast inputs of cereal, they compete with the human species. As already noted, if present trends in meat-eating continue, then by 2050 the world's livestock will be consuming as much as 4 billion people do: an increase equivalent to the total world population of around 1970, when many were doubting whether such human numbers could be fed at all.

And what, in the end, is it all for? So that a few (or at least a minority) of people who are already rich, can become very rich indeed. It is for this, and nothing else, that the health of the entire human race, and the future of the entire planet and all the creatures that now live on it are being compromised.

I do not suggest that the rise in meat-eating these past few decades has been a conspiracy, or a simple confidence trick. Although corruption is all too obvious, and some modern meat barons seem altogether hateful, I still do not believe that most people in high places are evil. The farmers who have striven to raise their output of meat have in the main responded, as farmers in every age must always do, to the economic pressures of their day. Many believed, too, that by raising meat output they were doing a good thing. Indeed it seemed self-evident. The nutritionists who urged greater intake were sincere. They believed it was necessary. Politicians concluded that the increase in livestock was good for people, and was in line with people's desires, and was also good for farmers and hence for the economy as a whole – and what else are politicians supposed to do? Yet the whole enterprise has been at least as damaging in the long term, as, say, the arms industry. Individually, and en masse, human beings are capable of the most horrendous mistakes; and we go on making the same mistakes, decade after decade, even when their foolishness seems obvious.

What's the way out of this mess? We can't expect much help from the corporations who now supply most of the world's food, or from the politicians. The status quo is very much in the interests of the modern food industry, and governments these days have neither the will nor the clout to override it. But (as re-emphasized in Chapter 12), consumers can do the trick. In many fashionable circles it is no longer chic to serve vast volumes of meat, as it tended to be in the 1970s. The only statement that steak makes these days is, 'I was in a hurry'. Status is won by the care with which food has been prepared – as of course is also the case in all traditional cuisines. Risotto is nutritionally simple and economical, but it takes a day and a half to make it properly. Status can be gained too by showing intelligence, and it just is not intelligent to serve food that undermines the health of one's guests.

In fashionable circles worldwide, then, the vogue for meat and more meat is surely passé. It is perceived to be vulgar. If the countries that are now sacrificing their magnificent traditional cuisines in order to become more like the modern West perceive, first, that the West has been spectacularly wrong on the issue of diet and, second, that fashionable westerners are already realizing their mistake, then perhaps the rot might be stopped.

5
Food Unsafe

Food is intrinsically hazardous. Plants have spent a billion years evolving toxins specifically to stop creatures like us eating them. The flesh of animals is liable to carry parasites of all kinds – worms, protozoa, fungi, bacteria, viruses, prions – that are ready and able to infect us. Indeed, most of our infectious diseases arose originally as 'zoonoses' – diseases that originate in animals – for where else could they come from? In both plants and animals, too, those infective agents may produce toxins, which in turn poison us. Mycotoxins (from fungi) are a particular hazard. History records spasmodic outbreaks of fungal poisoning, in seasons that favour fungal growth – notably of ergot, produced by a fungus that infects rye, which among other things leads to abortion and to delusions, and may well have fostered the cult and the fear of witchcraft. In mid-twentieth-century China the commonest malignancy was cancer of the oesophagus, which is rare in the West, and was probably caused by mycotoxins in cabbage. Botulism, caused by toxins from *Clostridium* bacteria in rotting material of all kinds, is a (fairly) rare but constant hazard worldwide. Crops in the field, too, are always prone to contamination from atmosphere or soil – for example with heavy metals, such as airborne lead; and fish can be notorious accumulators of aquatic contaminants, notably of mercury in tuna. Eating is dangerous, in short.

So we should be grateful to the ancient farmers and breeders, who produced a range of crops that are reasonably safe – often (as with potato, tomato, parsnip, cassava, and many a bean) from wild ancestors that could be lethal. This task continues: for example, scientists at the International Centre for Agricultural Research in Dry Areas (ICARDA) at Aleppo, Syria, have just produced safe varieties of grass pea. This is an enormously hardy and useful pulse, widely grown in hot countries in times of drought; but when the traditional varieties were eaten in large quantities, they produced paralysis. We should be grateful, too, for the ingenious souls who developed the arts of drying, smoking and pickling, which allow storage without spoilage; and to their successors, the inventors of canning and other packaging. Of course, too, we would rather our food was free from the residues of pesticides, applied in the field and in store. But the danger of fungicide residues (say) must be balanced against that of the fungi that they keep at bay.

Yet even in modern societies, with modern methods of farming and storage, the hazards are far, far greater than they could and should be. To be sure, it is unfair to condemn all pesticides and fungicides out of hand, for such chemistry has often saved us from worse ills. But the ideal is to produce good food that is contaminated neither by wild mycotoxins nor by fungicides, and we should not give up on that ideal. In general the hazards are greatly reduced if standards of husbandry are high (for example, the risk of contamination can be greatly reduced by mixed farming – avoiding monoculture), and if the food is delivered fresh. But modern industrialized food production favours monoculture and long-range delivery. Besides the agents that protect food, too, there's a catalogue of additives intended to puff it out, make it shinier or a brighter shade of green, stop it coagulating, and so on and so on, as it sits in the warehouse for a month or two. To what extent are the chemicals that are now applied and added to food strictly necessary, and to what extent are they merely deployed to compensate for bad husbandry – or make it possible to freight food halfway across the world? To what extent, in short, are the chemicals that we all eat every day, added not for our benefit (as the producers, processors and retailers insist) but simply so that food can be produced by the industrial methods that are favoured by corporations?

In truth, it is hard to assess the health hazards of pesticide residues or additives. Toxicity can be gauged most easily when only one toxin is present, when the concentration is high, and when its effects are discrete and short term. But modern food may contain a score or more of different additives and residues, all in minute quantities, that are ingested every day and sometimes accumulate in the body (notably in fat – including that of mother's milk) and are not usually acutely toxic but may (for example) affect a person's mood and behaviour, and perhaps may clip a few years off their life. It is virtually impossible to measure such effects critically. We have to make do with anecdote (the fact, for example, that some hyperactivity in children clearly is due to food additives), and common sense. It isn't enough to protest, as commercial and government scientists are wont to do, that there is 'no evidence' that minute traces of pesticide, or the standard cocktails of additives, do us harm. As the adage has it, 'absence of evidence is not evidence of absence'; and common sense suggests that the less of such stuff we ply ourselves with, the better.

But some hazards are easier to pin down – and it's obvious that modern food, even in the best organized countries, is far more dangerous than it should be. In Britain alone, at least 4.5 million people are *known* to suffer some form of food poisoning each year, equivalent to one in twelve of the population; and fifty or sixty of them die. Yet it is also clear that most food poisoning goes unreported.

Most alarming of all, though, is that much of the danger of modern food is *caused* by modern practice. It clearly results directly from the modern, obsessive attempt to cut the cost of production; and above all (since labour is generally the most expensive input) to replace traditional husbandry, and the people who practise that husbandry, with machinery and industrial chemistry. Yet the cutting of costs is not intended to produce cheap food, as is so often sanctimoniously claimed, but to maximize the margin between the cost of production and the sale price. The same industry that goes to such lengths to cut costs, dedicates the rest of its energy to 'adding value'. In short, the greatest hazards of modern food are not those of nature, or of bad luck. They follow, as night follows day, from policy.

There isn't space here to catalogue all the hazards and misdeeds of modern food production (and others have done a good job on this, including Eric Schlosser in *Fast Food Nation*, Penguin, 2002). Instead I will focus on four issues which I feel are so alarming, and so scandalous, that we should all be out on the streets protesting. To be sure, some of the perpetrators have already been rapped decorously over the knuckles. But although I do not want to revert to the times when heads were put on stakes, I do feel that the casual disposal of other people's lives, and the unnecessary suffering and slaughter of millions of livestock, deserves a little more official opprobrium than it tends to receive. We are all of us – and our children and grandchildren – exposed to quite ludicrous dangers simply because the world is run by people who, above all, want to get rich; and because we have elected politicians who feel it is their job to help them to do so, and who cannot apparently envisage any other way of going about things. We are not angry enough. Not by far.

The first of my four issues is that of straightforward villainy. You may feel it unfair to blame the system for this. After all, there are rotten apples even in the best-appointed barrels. Even monasteries, presumably, harboured the occasional crook. You can't blame the authorities, and the powers that be, if unpleasant people sometimes buck the rules.

But you can. Some set-ups are more conducive to villainy than others. Town planners are rightly blamed for creating cities with no trees or amusements but with plenty of dark corners for muggers to hide. The design encourages the crime. The modern food supply chain is so convoluted and so long that it allows endless opportunities for malpractice of all kinds – including many that beggar the imagination of those who are not criminally inclined. The supply chain is impossible to police because it is so complex, and because policing is so expensive (and nobody wants to pick up the bill – certainly not the governments who win votes by keeping the price of food down). Sometimes, though, it is not at all easy to draw a line between outright villainy (like the adding of contaminants) from the standard, legitimate practices of the modern food industry.

My other three examples all relate to livestock – mass production at rock-bottom cost. First, there is the continuing mass-use of antibiotics – even though the hazards were officially spelled out way back in the 1970s (and there were warnings long before that). The potential dangers are horrendous. Then there are the two epidemics that began in Britain within the past few years – the aftermaths of which are still with us, and with the rest of the world. BSE and the epidemic of foot-and-mouth disease were not bad luck or acts of God. They resulted entirely from bad husbandry, all of it resulting directly from the overweening desire to cut production costs and some of it resulting directly from official policy. The people who made these things happen are for the most part still in charge; and although some adjustments have been made, the grand production strategy that produced these horrors is still the official line – and indeed becomes more deeply embedded by the month. No one in high places seems seriously to consider that livestock should not be churned out like boiled sweets; that the world as a whole should not be run as one giant farm, and controlled by a few corporations; that short-term profit should not be the prime or the sole consideration; or that science should not simply be the handmaiden of commerce. Until they reflect on these matters, we are all of us in danger. Better still we need different people, with broader perceptions.

Industrialization and the Paths to Villainy

It was obvious from the very start of the Industrial Revolution that the mass production and distribution of food provided new opportunities for malpractice. Indeed, as John Burnett describes in his excellent *Plenty and Want* (Penguin Books, Harmondsworth, 1966) some of the worst and most blatant examples of adulteration come from early-nineteenth-century Britain, where the industrialization of the world truly began. The principal whistle-blower was an erstwhile assistant to the great Humphry Davy: Frederick Accum, who in 1820 published a *Treatise on Adulterations of Food and Culinary Poisons*. The principles that emerge are still relevant.

Bread, beer, tea, coffee and confectionery were prime objects of abuse in nineteenth-century Britain. Not everything that was added fraudulently was toxic. Beer was commonly watered down, and sugar then added to make it seem richer. Sulphate of iron was added to give it a good head. But what is an adulteration, and what is legitimate processing? Nowadays pork and other meats routinely have water added to them, to bring them up to the legal maximum allowed. Consumers pay the same price per kilo for the added water as they do for meat. (When you fry supermarket bacon, you first have to boil off the water, then start again. I remember when this was not the case.) Accum reported that bread was routinely adulterated with alum (potassium aluminium sulphate) to whiten inferior flour and bulk it out; and a report of 1848 (*A Treatise on the Falsifications of Food*) found this practice to be universal. Though alum was generally considered safe, these days its aluminium content would surely raise concerns. But sometimes the added ingredients could be damaging even though, in principle, they seemed safe. For example, oatmeal was commonly bulked out with barley meal, which is not only cheaper but also less nutritious. The high mortality among the pauper children in Drouitt's Institution in 1850 was ascribed to oatmeal tricked out with barley, which reduced their intake of energy and essential fats even further, and gave them diarrhoea for good measure (which in the modern world, particularly in poor countries, often precipitates malnutrition).

Some of the adulterations of tea seem comic. At least eight factories in London were dedicated to recycling: skivvies were paid to collect spent leaves from pots and kitchens and put it back on the market. (John Milton, in Book Seven of *Paradise Lost*, wrote presciently of 'The black tartareous cold infernal dregs adverse to life'.) Leaves of ash and elder were curled and coloured on copper plates to look like tea. An official report from the early nineteenth century (not Accum's) found that the 6 million pounds (around 300 tonnes) of tea that the East India Company imported each year were supplemented by another 4 million pounds of junk. But some of the adulterations were terrifying. There was a vogue for both green and black tea. One grocer (or at least, one that was caught) coloured green tea with verdigris, an oxide of copper. Black tea was darkened with lead. Coffee was routinely coloured and puffed out with anything that came to hand: chicory was routinely included; ground horse liver was among the more inventive bulking agents. These are innocuous – but lead was used for colouring too. Accum reported pickles coloured with copper, the rind of Gloucester cheese tinted with red lead, and pepper eked out with floor sweepings. Confectionery was an alchemist's delight. Sweets were brightly coloured with bisulphate of mercury. Fifteen people died after eating lozenges from a Bradford market, to which (by mistake) the manufacturer had added arsenic. Twenty guests at a banquet in Nottingham were laid low by a green blancmange, alluringly aglow with arsenite of copper. Lead, copper, mercury and arsenic accumulate in the body. Perhaps, says Burnett, they helped to cause the chronic gastritis that was common among the nineteenth-century urban poor. Effective legislation was introduced in the 1870s but until then, says Burnett, 'food adulteration virtually became organized crime'. But then we might reasonably observe, without irony, that the gulf between organized crime and legitimate practice is not so broad and deep as is commonly supposed (see Chapter 12).

How did such malfeasance come about? The Victorians, after all, prided themselves on their public and private morality. John Burnett offers three main thoughts, all highly plausible and still pertinent. First, he says, in the early nineteenth century the new industrial towns separated producers and processors – farmers, butchers, bakers, brewers – as never before from consumers. Traditionally, publicans often brewed their own beer, and bakers milled their own flour, but by the early nineteenth century brewing and milling were largely centralized and many pubs were 'tied' to the brewers, effectively becoming the brewers' employees.

It was, says Burnett, very difficult for traditional growers and processors to cheat people who lived in the same village, or the same street. But when producers and consumers are widely separated, there is no immediate feedback. In Burnett's words: 'Food adulteration is essentially a phenomenon of urban life.' The distinction is as in war, between killing people in hand-to-hand combat, and dropping bombs on their cities from a great height. In addition, the poorer consumers at least were a captive audience, with no time or opportunity to shop elsewhere.

Then again, from 1266 until 1815 the quality and price of bread and ale in England had been fixed by a system of Assizes (local courts), following royal and later parliamentary edict. Local inspectors watched over foods apart from bread and ale, and whoever drifted from the straight and narrow was punished (not least in the pillory). But the early nineteenth century was the first golden age of free trade. Independent America got into its stride at that time. The British government lifted the ban on monopoly even of the East India Company, the first global corporation, and the prime creator of the most significant portion of the British Empire. The committee that abolished the Assize of Bread declared: 'Your Committee are distinctly of the opinion that more benefit is likely to result from the effects of a free competition . . . than can be expected to result from any regulations under which [the bakers] could possibly be placed.'

Commentators have offered totally opposite explanations for the failure of the early-nineteenth-century market to provide the quality and justice that were expected of it. Some have blamed the newly emerging big companies, like the new brewers, with their near or actual monopolies: for they were so powerful they could call the shots, and get away with whatever they wanted. Others (including the big companies themselves) point out that in fact, they had the higher standards. It was the smaller traders who cheated more. But, says Burnett, small traders in particular were forced to cut corners. In the days of the Bread Assizes, bakers simply did the job that was statutorily required of them. When free trade ruled, more and more bakers came into the market, and they fought each other like dogs for custom. By 1850, there were 50,000 bakers and three-quarters of them were 'undersellers': they sold their bread effectively for less than the cost of production. They could achieve this, as one disaffected employee put the matter, 'only by first defrauding the public, and next getting eighteen hours' work out of the men for twelve hours' wages'. The adulteration of the Drouitt's children's oatmeal with barley came to light when some astute inspector saw that the merchant was selling it for less than the known cost of production.

How did the traders get away with it? Most consumers, says Burnett, simply did not know what was going on. The urban poor in general had no choice: no access to information, and neither the time nor the opportunity to shop elsewhere. When the facts did become known, people who had been brought up on the tang of lead and verdigris continued to prefer it to the real thing. Such is the way food preferences are acquired. Finally, says Burnett,

> When faced with irrefutable evidence of their offence, adulterating traders put forward a number of justifications designed to prove that they were in fact performing a public service – that adulterations were only practised in response to public taste, that they constituted 'improvements', and lowered the price of foods which would otherwise have been too expensive for the poor to buy. Needless to say, it was not mentioned that adulteration was always for the profit of the seller and the expense of the buyer, and that, quite apart from the possible dangers to health, adulterated goods were dearest in the end . . . Shopkeepers continued to conduct their business on the maxim of the Common Law, '*Caveat emptor*'.

Under the present, spreading rules of globalization farmers worldwide will again be obliged to fight like dogs and undercut each other, just like Victorian England's bakers. Many commodities are now sold in supermarkets for less than the production costs, albeit bolstered by supplements and tax breaks of one kind or another. In most of the Western world, milk commonly sells for less than bottled water. Modern schoolchildren are enviable indeed compared to early-nineteenth-century orphans, yet there are many dubious reports on school meals, bought from mass manufacturers at rock-bottom prices by school boards strapped for cash. The same economic arguments are rehearsed as in early Victorian times: free trade versus control; the power of the big companies versus the desperation of small traders. Food processors take it to be self-evident, too, that the often remarkable catalogues of emulsifiers, anti-oxidizing agents, texturizers, colourants, etc. etc. etc. that now by law adorn their voluminous packaging, are entirely for the benefit of the people who buy them. If we weigh up our own times as dispassionately or as wryly as we like to look back at the England of Dickens or indeed of Accum, the differences do not seem quite so stark as we might care to suppose. We are richer, obviously, in the sense that most of us have more disposable cash. But is that all that 'progress' is supposed to mean?

Accum's report was a sensation. Soon after it was published the *Literary Gazette* commented:

> Does anything pure or unpoisoned come to our tables, except butchers' meat? We must answer, hardly anything . . . Bread turns out to be a crutch to help us onwards to the grave, instead of the staff of life; in porter there is no support, in cordials no consolation, in almost everything poison, and in scarcely any medicine cure.

Meat, though, and the means of producing it, no longer offers the safe haven that the *Literary Gazette* supposed. Three contrasting but equally disturbing features of modern livestock husbandry suggest it may now pose the greatest threats of all.

Antibiotics

Antibiotics, put into the feed of intensively raised animals to promote growth, are immensely threatening. They can and do promote bacterial resistance to antibiotics, which can spread from farmyard bacteria to bacteria that are medically significant. Worse still is the phenomenon of 'cross resistance': resistance to the kinds of antibiotics used in animal feed may also confer resistance to those used in medicine.

Pathogenic bacteria are hugely diverse, resourceful, adaptable, and little understood. They are also potentially extremely dangerous. Plague, cholera, typhoid, leprosy and tuberculosis (TB), each a source of horrendous epidemics and smouldering disabilities, are bacterial diseases. A veteran general practitioner in the 1960s told me that doctors at that time were divided into two classes: those who were old enough to remember TB, and those who were not. In his young days, it had been *the* disease: all the others were minor by comparison even though they sometimes killed people and disabled them (including whooping cough, which is bacterial, and viral infections such as mumps and measles, which currently are top of the danger list). Yet by the 1960s, thanks to vaccination and the pasteurization of milk, and the tuberculin test (TT) that detects infection in cattle, TB had become rare (at least in affluent Western countries). Now it is coming back.

Less dramatic are the many agents that cause food poisoning. Many have been identified and named: *Salmonella, Shigella, Campylobacter,*

Listeria, various strains of *Escherichia coli* (E. coli) and many more. Most, however, in this age of advanced biology, have not been identified at all. A century and a half after Pasteur, microbiological knowledge is stupendous; but clearly, relative to what there is still to be found out, it is in its infancy. On the other hand, it has been clear at least since the 1970s that much food poisoning is caused not primarily by bacteria but by viruses, including rotaviruses. When viruses are the culprits, putative antibiotic resistance among bacteria is presumably not relevant.

Antibiotics were among the great discoveries of the twentieth century, and indeed of all human history. The Scottish bacteriologist Alexander Fleming discovered the first antibiotic in the 1920s – penicillin, from the fungus *Penicillium* – and this was made into a practical therapeutic drug primarily by Ernst Chain and Howard Florey, during and immediately after the Second World War. Now there is a huge range of antibiotics. The molecules themselves are modified chemically in a hundred different ways, for different purposes; and the fungi and bacteria that produce them are often genetically engineered, and thus provide one of the most convincing demonstrations that genetic engineering really can be valuable.

Even without antibiotics, humanity would surely have made significant inroads into bacterial infections this past half-century. Better nutrition, housing and hygiene have all had an enormous impact, and so has vaccination. There are other antibacterial drugs, too, besides antibiotics. But antibiotics complete the armoury. Hygiene works so long as the basic civil engineering stands up, and the economy; but it fails, for example, in times of flood, when cholera and typhoid are particular hazards. Floods, worldwide, seem to be growing more frequent, thanks not least to deforestation and climate change. Vaccination is not universal and for all kinds of reasons is not always appropriate or applicable. So potentially dangerous bacteria will always be with us; and when they do gain entry, antibiotics are the front-line defence. I know several children even in affluent, middle-class Britain whose lives have been saved within the past few decades by antibiotics, and nothing else would have done the job.

But antibiotics are beset by three huge problems of a biological kind (in addition, that is, to the constant difficulties of cost and distribution). First, some kinds of bacteria are simply not susceptible at all to any or most of the common antibiotics. Second, many that are commonly susceptible have nonetheless developed at least some resistant strains. TB bacteria may be resistant to a wide range of antibiotics. In recent decades, resistant staphylococci have come to the fore. Typically, patients are taken into hospital perhaps for minor, elective surgery and then suddenly die from uncontrollable MRSA: 'methicillin-resistant *Staphylococcus aureus*'. Warnings that this could happen date from the 1950s, when clinically available antibiotics were still new. The basic point again is one of biology: that microbes are wonderfully adaptable. Within any wild population of susceptible bacteria there are always liable to be a few that are not susceptible: individuals that possess particular genetic mutations that enable them to produce particular enzymes that destroy the antibiotic before it destroys them. Mutation, in bacteria, is frequent; so if there are no resistant individuals to begin with, it is always possible that some will soon arise. The few that survive the first wave of antibiotic attack rapidly multiply to fill the ecological space vacated by their more vulnerable relatives.

Third, bacteria naturally practise a peculiarly promiscuous form of sex – one that can be seen as informal genetic engineering. That is, they pass on packets of genes to each other, known as 'plasmids'; or we might see plasmids as mavericks, infective agents in their own right, that spread throughout microbial populations. However we look at them, plasmids pass between bacteria of the same species – and between those of different species. E. coli is a common participant in such transactions. Most strains of E. coli are harmless but a few produce toxins, and these are a significant cause of food poisoning and even of death. Resistance can pass from harmless E. coli to pathogenic types – and beyond species boundaries into totally unrelated, pathogenic bacteria, including *Salmonella*.

Whether we are talking about antibiotic resistance in bacteria, or pesticide resistance in insects, the general logistics are the same. It's important that bacteria should not get a sniff of antibiotic (except of course for the ones that occur naturally in the wild) until the clinician hits them with it. The clinician should go on hitting them until all the bacteria are dead (meaning that patients should take the complete course of prescribed pills, even after they feel better). Then, when the bacteria are first hit they are 'naive'. They have had no previous chance to build up significant numbers of resistant types. If the screw is turned until they are all dead, then they have no chance to give rise to a new generation of resistant types. If such resistant types are allowed to develop, then of course they will live to fight another day – and they may pass their resistance to other bacteria that were previously naive. The same applies to insects: zap them with a pesticide they have not seen before, and leave them all dead. On the other hand, if for some devious reason you wanted actively to encourage the spread of resistance, then you would do the complete opposite. You would ensure that large numbers of bacteria in as many contexts as possible are constantly exposed to low levels of a range of antibiotics.

This is precisely what happens in modern farming. Intensively raised livestock are fed low doses of antibiotics through most of their lives. Because the dose is low, the bacteria that are only a little resistant survive, and over time, with the same selective pressure and a few more appropriate mutations, truly resistant populations are almost bound to arise. Via plasmids, the resistant types can spread their resistance to other bacteria, of the same and different species; and the resistance conferred may well cover other antibiotics as well. Obviously, only a person who was seriously deranged would set out deliberately to foster antibiotic resistance in bacteria. But modern farmers of intensive livestock do this routinely, not as a matter of prime intent, but in the name of commerce.

For antibiotics are given by the thousand tonne to modern farm animals to promote growth. They are given to all the major classes of livestock that lend themselves to intensive rearing: pigs, poultry and cattle. They seem to promote growth for two different reasons. First, it seems that some of them at least might directly influence the animals' metabolism, though for reasons that are largely unknown. Second, beyond any doubt, they serve to suppress infection.

If you wanted to create the maximum possible disease among animals then, in principle, you would do as the factory farmers do. You would cram as many as possible into the smallest possible space so that the population in a given place was as high as possible, so that the contact was maximal, and so that the animals were stressed, which (as is known in many contexts in human medicine) tends to suppress the immune response. For good measure, you would ensure that all the animals were genetically identical, so that any pathogen that was able to gain a foothold in any one individual would be sure to find the others agreeable too. There are many defences of factory farms (the animals like it really, some say: animals are sociable, and the housing keeps them out of the rain, and in any case, who cares?) but their prime purpose is that in the present economic climate, they are the most economical. Profit is the key and competition is the spur. Production must be maximized while costs must be cut and cut again. Intensive units are horrendously expensive to set up but still, as things are, they are profitable. Productivity per unit is stupendous because of the sheer numbers of beasts, and because the rate of growth is so high since high-energy food can effectively be shoved down the animals' throats, and they are kept warm and cannot move so they do not waste energy on body heat and exercise. Costs are kept low (after the initial capital outlay) because labour is cut to the bone, and then cut some more. One worker may oversee thousands of pigs, and hundreds of thousands of chickens, ducks or turkeys. But all this would be absolutely impossible if infection was not actively suppressed. Hence the mass distribution of antibiotics.

It gets worse. Thirty years ago, commercial agricultural companies in Britain were offering feeds for animals that in effect contained all of the main classes of antibiotics that were in front-line use in human medicine, including penicillins and various tetracyclines. A keen observer was Dr Bernard Dixon, a bacteriologist turned commentator; and he observed in his book *What Is Science For?* (Collins, London, 1973) that one company was even offering a preparation of chloramphenicol to add to the drinking water of broiler chickens. At that time, chloramphenicol was just about the only antibiotic effective against typhoid. Typhoid has been rare in Britain these past few decades, but it's still out there. It is extraordinary to take such liberties.

The first formal warning came as long ago as 1970, when a committee appointed by the British government and led by Sir Michael Swann recommended that antibiotics that are used in human medicine to cure sick people should not be used on farm animals that are not sick. Tetracyclines and penicillin could still be prescribed by a vet to treat animals that were sick, but not simply to promote their growth. Swann, though, was eminently reasonable, and did allow the continued use of feed antibiotics that are not used in human medicine, such as zinc bacitracin. The assumption was that any resistance that built up in animal microbes to these somewhat esoteric agents, would not spread to medically important microbes, and would not confer cross-resistance to the medically important antibiotics, like penicillin and the tetracyclines.

I would not presume to criticize Swann's committee; yet a purist, working on general principles of science and ethics, might well have been even more rigorous. On biological grounds, the committee might reasonably have issued a total ban. Its recommendations are clearly a compromise; sensible and 'realistic' in the face of agricultural, commercial concerns, but a compromise nonetheless.

Tetracyclines, penicillin, streptogamins, and other medically important antibiotics are now banned within the European Union as a whole for non-therapeutic purposes in farm animals. Farmers can still use such drugs if their animals are truly sick, as confirmed by a vet; and can feed other, lesser antibiotics, that have no significant use in human medicine. But Swann's principle essentially reigns and, you might think, the farmers should think themselves well served.

But there is no such ban in the United States. In the country where freedom of information is supposed to be an absolute requirement, a key component of democracy, the Department of Agriculture does not publish complete figures on the amount of antibiotic used in animal feed. In 2001 the Union of Concerned Scientists (UCS), based in Cambridge, Massachusetts, did their own calculations and produced the astonishing figure of 24.6 million pounds of 'anti-microbials' (not quite but essentially synonymous with 'antibiotics') per year – around 10 million kilos, or 10,000 metric tons (tonnes). Of this, around 13.5 million pounds (about 6,000 tonnes) are of the kinds that are expressly prohibited in Europe, and are key players in human medicine: tetracyclines, penicillin, erythromycin and others. Furthermore, total use in agriculture seems to have increased by about 50 per cent since 1985: from around 16.1 million pounds (roughly 7,000 tonnes) in the mid '80s to the present 24.6 million pounds (10,000 tonnes). American chickens these days receive three times as much antibiotic per head as in the mid 1980s. Yet, the UCS emphasizes, these figures refer only to non-therapeutic use. In short, American farmers give six times more antibiotic to their livestock purely as growth promoters than are used in human medicine; or, to put the matter more precisely, so that they can keep animals in conditions in which, without the drugs, they would almost certainly die.

If sanity prevailed in this world this egregious deployment of antibiotics would beggar belief. The stakes are extraordinarily high. Anyone with any knowledge of history, or who has ever visited a poor country and stepped outside the hotel, must know how dangerous bacteria can still be. We see very few serious bacterial outbreaks in the West these days (although sometimes some oddity like Legionnaire's disease or bacterial meningitis or anthrax or hospital staphylococcus crops up to remind us) but all the great ravages of the past are still out there. Smallpox has been eliminated in the wild (though of course it is still a potential agent of biological warfare), but smallpox is a virus. No bacterial pathogen has been more than dented. Bacteria are reproductive r-strategists *par excellence*. They can be reduced to very low numbers and yet spring back in weeks, or indeed in hours. In principle a crack in a drain is all that's needed to trigger another epidemic. Antibiotics have been a kind of miracle. Fleming himself emphasized how lucky he was to stumble across them. Yet, in full knowledge of the dangers, we take the most extraordinary liberties. What for? Simply to allow farmers of a particular commercial kind to keep livestock in ways in which no animal should ever be kept. Why do they do that? Because, in the present economic climate, this is more profitable. Why does profit matter so much? Because, in the politics that now prevails, profit is all that is deemed to matter.

The use of antibiotics is bad practice, but it is legal bad practice. Almost equally disturbing are the many opportunities provided in the modern food supply chain for out-and-out crookedness which, in reality, is all but impossible to monitor or to police.

Dangerous Livestock: Policy and Chicanery

In 1906 Upton Sinclair published the novel that became his most famous: *The Jungle*, set in the Durhams' meat-packing plant in Chicago; fictional, but representative.

> It seemed as if every time you met a person from a new department, you heard of new swindles and crimes . . . It seemed that they must have agencies all over the country, to hunt out old and crippled and diseased cattle to be canned. There were cattle which had been fed on 'whisky-malt', the refuse of the breweries, and had become what the men called 'steerly' – which means covered with boils. It was a nasty job killing these, for when you plunged your knife into them they would burst and splash foul-smelling stuff into your face; and when a man's sleeves were smeared with blood, and his hands steeped in it, how was he ever to wipe his face, or to clear his eyes so that he could see?

Sinclair wrote this about eighty years after Accum's report, and nearly a hundred years before the present. Obviously, no such malpractice would be possible today.

Wouldn't it? In 2001 there came to light in Britain a truly horrific scam involving thousands of tonnes of meat and more than 600 food businesses. We will never know (it is impossible to trace) how many people were made ill by this particular villainy, or indeed died; as noted above, four and a half million Britons each year are known to suffer significant food poisoning from bacteria such as *Salmonella* and *Campylobacter* and fifty to sixty of them die. With food, as with narcotics: for every scam that is exposed many others surely flourish, for if the chances of being caught were high, there would be no scams at all. The most obvious lesson (the one that the authorities in general prefer) is that some people are simply crooked, and that crime is a constant fact of life from which the authorities must protect us. But the larger and more important lesson is that as the food supply chain (or any other human endeavour) becomes more complex, the scope for dishonesty becomes greater. An enterprise as complex as Britain's food supply chain is impossible to police. This is certainly the case when – as in Britain – the overall requirement is to keep costs down, which is the special ambition of recent governments that have sought above all to minimize public spending and to demonstrate that private enterprise can do all that is required. Yet Britain is among the best organized and regulated countries in the world. Whatever happens here is surely echoed in all countries where the food supply chain is long and convoluted.

Several of the more alarming incidents were detailed on BBC Radio's *File on 4* in November 2001. They mainly involved chickens and turkeys – to which, at that time, many Britons had fled in fear of beef and BSE. One case involved the town of Rotherham, in the north of England, where the environmental health officer (Lewis Coates) first received an anonymous letter which claimed that several premises on a small industrial estate were trading in meat that seemed ridiculously cheap. Mr Coates and his colleagues traced some of the meat to a nearby poultry plant where they observed the goings-on through a lavatory window. They watched teams of men with loaded trays and trucks and vans coming and going at night and at weekends. In all, said Mr Coates, around 1,500 tonnes of 'unfit poultry meat ended up going out on to people's dinner plates'.

When the officers finally made their strike they found huge containers of 'badly smelling bruised poultry, containing faecal matter, flies and feathers'. They also found a large quantity of salt, used to remove the slime from the carcasses and generally to freshen them up. Via 300,000 documents they managed to trace at least some of the poultry back and forth through 'an elaborate chain of middlemen' to a firm called Wells By-Products in Newark; a company owned by a giant rendering concern called Prosper de Mulder, which supplies nationally renowned pet-food companies including Spillers and Pedigree. But a local detective, the commendably astute Craig Moon of South Yorkshire Police, found that two of the Wells managers were hacking the breasts from the birds, slicing off 'the bits that were going green or some such' and bundling them off in plastic tubs to a company called Clifftop Petfoods. Clifftop sometimes cleaned the meat a bit further, and sometimes added some packaging, and then sold it on to a company called South Yorkshire By-Products.

The owner of South Yorkshire By-Products is currently serving a five-year gaol sentence, and since people who have 'paid their debt to society' as he will soon be deemed to have done should not be hounded to the grave I will simply refer to him as 'X'. X, however, was the key player. South Yorkshire By-Products had an *alter ego*: South Yorkshire Meats. Under the guise of South Yorkshire Meats, X operated as a meat broker: a trader who sits in his office (in X's case at his own home in a Rotherham suburb) and sells meat by telephone to other brokers dotted around the country. Eventually the meat gets to a dealer who actually sells it to a customer. Astonishingly, as a broker, X 'didn't need a licence and wasn't subject to any controls'. Trading as South Yorkshire Meats, X sent the dubious chicken breasts to a local dealer who, says DC Moon, 'performed the final act of transformation with salt and water'. Then, by telephone, this dealer sold the meat on to 'a string of other brokers across the country'. Moon traced this trail to Scotland and Wales, and commented, 'I didn't realize that meat could pass through so many different hands, particularly without people even seeing it.'

In general, meat in most organized countries follows one of three paths: it is destined for human consumption, or for pet food, or is condemned. Meat for human consumption should be documented from birth to the butcher's slab. It may pass through a long string of dealers but each transition is assiduously recorded to leave an unbroken paper trail. In France this trail is summarized on the label so that French consumers can ascribe particular joints to particular farms. The paperwork is burdensome but the attractions of well-labelled meat are obvious. But in Britain at least, meat for pet food is not so conscientiously documented. Dealers are not required to record whom they sell pet meat to, or in what quantities, so as one Environmental Health Officer told the *File on 4* programme, 'There is this opportunity for a small amount of pet food to be sold to bona fide outlets, and a large amount of the product to disappear.' As pet food, the meat in the South Yorkshire scam was worth about 10p per kilo. As chicken breasts for human consumption it would finally sell at about £2.50 per kilo. As condemned meat, it would have no value at all, and indeed the producer would probably have to pay for its safe disposal. To shuffle meat from one path to another can obviously be extremely lucrative. The particular racket that DC Moon helped to expose is thought to have realized nearly £3 million.

Exposing the Yorkshire meat scam cost well over £500,000, and since it affected a fair slice of all Britain, we might suppose that the grateful nation would be pleased to pick up the bill. Not at all. Rotherham Borough Council, who initiated the exposure, spent about a third of their annual environmental health budget on it and the auditors rapped them over the knuckles for falling short in other areas.

X, as noted, was sent to gaol for five years, and others in the chain for between three and seven years. But even that, in this apparently open-and-shut case, was far from straightforward. Cases of fraud in the food supply chain may impinge on laws of animal welfare, the Food Safety Act, and the Animal By-products Order. So what charges should be brought? When this has been decided, it still isn't clear who should bear the expense of bringing them. In Britain, responsibility is commonly split between the Crown Prosecution Service, the RSPCA (the Royal Society for the Protection of Animals), the Trading Standards Departments of local authorities, and an environmental health department, while any one case is liable to involve a county council, DEFRA (the Department of Environment, Food and Rural Affairs), the state veterinary service, and the Food Standards Agency. If the intention is to employ bureaucrats and lawyers, then the system works brilliantly. Byzantium would be envious.

In practice, meat scams can be so blatant that we may wonder how anyone falls for them. Britain has spawned the phenomenon of the 'car-boot sale', where (at least as originally intended) non-professional traders sell off the contents of their attics from the backs of their cars. But distinctly cut-price meat also turns up at car-boot sales. It seems incredible that anyone would buy their meat at such a venue for such a suspect price. But apparently they do. Within the more formal market, even the legitimate trade may be fooled. Thus on 24 April 2001 the BBC's six o'clock news recorded that the giant supermarket chain Tesco had recalled 'thousands' of frozen chicken steaks 'amid fears that they may have been made with meat destined for the pet-food industry'. So, also, did Farmfood supermarkets.

Crooks may turn up anywhere, of course – but the present system seems expressly designed to foster criminality, just as it seems designed to encourage disease. First, as John Burnett commented, the kind of scams that Frederick Accum exposed in early-nineteenth-century England are 'essentially a phenomenon of urban life'. Comparable examples were reported from the ancient cities of Athens and Rome (by Pliny) but in primitive or agrarian societies, 'systematic adulteration would not have been possible'. Modern methods of supplying food do not provide the moral checks, notably the face-to-face contact of producer, processor and consumer, that were general before people lived in big cities. Indeed, the players in the modern food supply chain become ever more remote from each other. Bad people turn up in all societies; but in societies as detached as ours has become, only the most saintly, who feel and respond as George Eliot described to some inner voice of morality and in particular of duty, can avoid temptation. Furthermore, for all kinds of reasons the supply chain has become more and more complex.

There is no legislation on complexity per se. The route from farm to fork is as intricate as the various traders choose to make it. Indeed, it is so complex it is impossible to police. Ask the police themselves how many of their men and women they can spare to trace diseased chickens. Ask the food inspectors. Ask the Department of Health. Ask Rotherham Borough Council if they would risk such investment again. Then spread the net to Europe and America or anywhere else and ask them. In short, the present system has been designed or rather has evolved for reasons that are purely commercial, is removed from the moral checks that result from ordinary social interaction, and cannot be adequately monitored. Ask Mr X: or, more to the point, the hundred other Mr Xs who haven't been caught, if you know how to find them.

Second, the system as it now stands does not inspire confidence even when nobody is behaving crookedly. Even at its best, Britain's livestock industry operates on a wing and a prayer; despite being – comparatively – so well organized. Thus the catalogue of disorders among birds from bona fide British farms may well include skin lesions, septicaemia, liver disease, heart disease (pericarditis) and high counts of bacteria, plus mechanical damage caused not least by weakened bones. Of course there are 'poultry inspection assistants' to weed out the bad ones but as an (anonymous) inspector told *File on 4*, 'You can have a line going at a hundred and twenty to a hundred and seventy birds a minute and only two inspectors on a line, and if you're trying to say to me that someone can check a hundred and seventy birds a minute and find every one with a disease, I tell you that's rubbish.'

Just to emphasize again how little changes, we may note that Upton Sinclair told a similar story in *The Jungle*, albeit with respect to hogs rather than chickens. Each carcass

> had to pass a government inspector, who sat in the doorway and felt the glands in the neck for tuberculosis. This government inspector did not have the manner of a man who was worked to death . . . he was quite willing to enter into conversation with you, and to explain to you the deadly nature of the ptomaines which are found in tubercular pork; and while he was talking with you you could hardly be so ungrateful as to notice that a dozen carcasses were passing him untouched. This inspector wore an imposing silver badge, and he gave an air of authority to the scene . . .

This inspector, unlike his counterparts in modern Britain, was clearly not overworked. But the end result was the same.

Much worse than any of this, however, is that Britain has suffered two epidemics among livestock in recent years that are among the most alarming in all of history. They both began in Britain, and have largely been centred here, but they both spread to mainland Europe and they affect the whole world. The first, bovine spongiform encephalopathy (BSE) was unprecedented; no such disease had been heard of before the mid 1980s, when it was first reported in British dairy cattle. The second, foot-and-mouth disease, has been the worst-ever outbreak of its kind.

BSE

BSE came to light in February 1985. A cow died after suffering head tremors, loss of weight, and lack of coordination. Others quickly followed. Sheep show such signs when they have scrapie, which has been known for many centuries. BSE was a novelty. Brilliant work by Stanley Prusiner in the USA revealed that the cause of BSE was a new kind of pathogen. It wasn't a virus, because viruses contain genes in the form of DNA (or sometimes RNA), as well as protein. But the agent of BSE contained just protein; no genes (DNA) at all. This protein was effectively the same as proteins that occur in the normal nervous system, yet in a somewhat mutated form (although the term 'mutation' is not quite accurate, since it is conventionally applied to genes). The mutated protein, on the surface of nerve fibres, prevents the rapid and accurate transmission of nervous impulses. Such mutated proteins are also able to prompt other, normal proteins of the same kind to 'mutate' in their turn. Furthermore, they may spread from animal to animal. These new, rogue proteins Professor Prusiner called 'prions'. They mess up the system and they spread to new systems, and cause them to change in ways that result in the animal's destruction, but in detail they behave more like computer viruses than the viruses that infect living organisms. Prions might reasonably be classed as a new form of life: able to do at least some of the things that living organisms do (that is: reproduce themselves) but without themselves possessing genes.

Prions, it quickly turned out, were also the cause of scrapie, and of a strange group of encephalopathies that occur in human beings. These include Creutzfeld-Jacob disease (CJD), which again leads to rapid neural degeneration and death; and 'kuru', which occurs in New Guinea. Although prions are devastating, they are not (mercifully) easily transmitted. By far the most important route of infection is ingestion: consuming tissue that is already infected. Scrapie spreads as sheep give birth side by side in fields, and may ingest each other's placentae. Sufferers from kuru were traditionally assumed to be cannibals, which may well be the case.

So how did the apparently novel disease of BSE get into Britain's dairy cattle? Where, first of all, did the prions come from? There are two possibilities. Perhaps cattle have always been infected by prions, but in cryptic form. Perhaps they just happened to flare up, for whatever reason, in the 1980s. One suggestion is that they were finally pushed into pathogenicity by exposure to organophosphorus pesticides. There is some circumstantial evidence for this, and all possibilities must be taken seriously. Or perhaps the BSE prions mutated from scrapie – meaning that there must have been transmission across species. What is clear, though, is that the infected cattle were given the prions in their feed, and in particular in meat and bonemeal. Alarm bells will surely sound at this point. Cattle are herbivores, supreme consumers and converters of low-grade, bulk fodder, notably grass and its derivatives, such as hay. That is why they are so useful to human beings; they eat plants that grow abundantly, but which we cannot eat. So why are they eating the flesh of other cattle, or of sheep, containing these agents of disease?

The answer of course lies in the twin imperatives of modern farming: to maximize output and minimize costs. The modern cow that produces 1,000 gallons (5,000 litres) a year must average 15 litres per day over a ten-month lactation, while the 2,000-gallon animal averages 30 litres (although the output is not constant: it rises to a peak and then tails off). Milk is potent stuff, mostly water but typically containing around 3.5 per cent of fat by weight, plus protein and lactose. It is hard to sustain such output, and virtually impossible if cows eat nothing but grass. So they are also fed 'concentrate', generally as they are being milked: especially high-energy, high-protein cakes and pellets. We may object on welfare grounds that an animal should be expected to produce 2,000 gallons a year or even more (and many do object, including some veterinary professors), but in principle it's perfectly reasonable, and sound, husbandry to feed at least some concentrate to virtually all specialist dairy cows. In the end, it's the concentrate that provides most of the milk. Of course, people pick up the prion by eating beef, not by drinking milk. But as discussed in Chapter 2, most beef these days in countries like Britain is provided by the offspring of dairy cattle or indeed, by the cow-meat itself.

Conventionally, though, concentrate has been in the form of seeds: oilseeds like linseed, and oily and proteinaceous seeds like soya. The cows' herbivorousness is respected. But animal flesh, of course, may also provide fat and protein. So from the mid twentieth century concentrates have been made with protein from other beasts – including sheep and cattle. Enforced cannibalism at best seems unaesthetic, and in cattle it is positively bizarre.

Of course, the flesh was not fed raw. The manufacturers of concentrates followed strict rules of preparation, boiling the material at temperatures and for periods that killed all known pathogens. But there were two snags. Firstly, the specifications for boiling were reduced somewhat in the 1980s, not least (inevitably) for reasons of cost. Secondly, nobody knew about prions in those days, and least of all about prions in cattle. Prions are not disarmed by the kinds of temperatures that kill farmyard bacteria. But what people don't know about, they can't legislate for. So Britain's dairy cattle, in the 1980s, were fed the flesh of other cattle (and of sheep) that evidently contained active prions.

The big question then was whether these prions could spread to human beings. There was reason to think not. After all, people have lived in close company with scrapie'd sheep for centuries, often consuming the

BIRMINGHAM COLLEGE OF FOOD

eyeballs and the brains, which have widely been prized as delicacies, without apparently contracting kuru or more than their share of CJD. In Iceland in the early 1970s I found odd bits of sheep appearing at almost every meal. On the other hand, the prion might have leapt from sheep to cattle, and if so this would show that it could indeed cross species barriers. Perhaps this was a new, more versatile 'mutant'.

The UK government was first officially told about BSE in 1987, and in 1989, under Prime Minister Margaret Thatcher, it convened a committee under the chairmanship of Oxford zoologist Sir Richard Southwood, to ask whether BSE in cattle could spread to human beings. The committee's deliberations were bound to be speculative since BSE was a new disease, but it finally concluded: 'From present evidence it is likely that cattle will prove to be a "dead-end host" for the disease agent . . . [and] most unlikely that BSE will have any implication for human health.' However, the committee recognized full well that this could only be a guess and added: 'Nevertheless, if our assessments of these likelihoods are incorrect, the implications would be extremely serious.' (*Report of the Working Party on Bovine Spongiform Encephalopathy*. Joint Report of the Department of Health and the Ministry of Agriculture. Chaired by Sir Richard Southwood. London, 1989.)

Should the committee have been more emphatic? Should they simply have said, 'Frankly we don't know whether this pathogen can spread to people, because we have never seen anything quite like it in the past. It is a bit like scrapie, and scrapie doesn't seem to cross species barriers, so perhaps BSE won't either. On the other hand, it is at least possible (and indeed most likely) that BSE did originate in sheep, either as a transformed kind of scrapie, or as ovine BSE which has simply gone unrecognized. If this is the case, then it has already crossed a species barrier. But, cattle are fairly closely related to sheep, and human beings are not; so perhaps it will not leap such a wide species gap. But then again, many zoonoses do leap enormous species barriers – cowpox, tuberculosis, brucellosis, and anthrax are just a few of the infections (albeit very different from BSE) that do pass from cattle to people. So – we just don't know. Everything we know about prions (which isn't much) suggests they are not likely to pass from cattle to people. But still, it is an enormous risk.' Or, more simply: 'We don't think BSE will affect people. But it's impossible to tell. If we're wrong, there'll be hell to pay.'

Perhaps they should have said this; and perhaps if they had, Britain's politicians would have been more cautious. But scientists don't talk like that. Such straight-from-the-shoulder vulgarity is not their style. As it was, Thatcher's ministers seemed to take the Southwood committee's more detached comments as carte blanche. Their priorities are clear from their actions. It might be risky for people to eat beef but to Britain's politicians it was much more important to 'reassure the public' and to 'support the beef industry'. Most notoriously, in May 1990 at an agricultural show the then Minister of Agriculture John Selwyn Gummer in a televised fit of bravado encouraged his four-year-old daughter to eat a beefburger, to prove its safety.

The first human cases were reported three years later: not of BSE in its original form, but of a variant form of CJD. The first four victims, perhaps inevitably, were dairy farmers. The first death from vCJD was in 1995: a young man, aged nineteen. In March 1996 the UK Health Secretary Stephen Dorrell told the world that there was a 'probable link' between BSE and vCJD. That same month, the European Union imposed a worldwide ban on all exports of British beef. In April, with what seems like truly wonderful cheek, the UK government challenged the export ban on legal grounds. It also introduced a scheme to slaughter and destroy all cattle more than thirty months old. Cattle younger than that (it was reckoned) should not have been fed the infected meal, which was intended for lactating cows. Later (April 1997) it became clear that BSE could be passed from mother to offspring, so the thirty-month barrier was less absolute than it might have seemed; and that calves inherit a genetic susceptibility to the disease.

The Government continued to flap, as British soldiers used to say in the Second World War when things were clearly out of hand. The BSE prion primarily infects nerve tissue, so in December 1997 the (new, Labour) Government banned the sale of beef on the bone. Bones, after all (particularly the spinal cord, which features in ribs of beef) tend to contain nerves. Some enterprising butchers in far-flung places continued to offer prime ribs and T-bone steaks in their pristine state as dog food, though not at dog-food prices. The occasional hotelier was prosecuted. The beef-on-the-bone ban remained in force until November 1999. In August 1999 the European Union lifted its ban on British beef though France continued to enforce the embargo. (The French are Britain's traditional rivals, and have often been our enemies. But sometimes you can't help admiring them.)

The horrible saga continues to be played out into the twenty-first century. In February 2000 a baby girl, born to a mother with vCJD, was also found to have the disease. The prion is evidently transmitted 'vertically' in humans, as well as in cattle. In October 1997 a man of seventy-four died of vCJD, showing the disease is no respecter of age. In November 2000 France banned beef on the bone and also banned livestock feed containing meat. Italy then banned the import of adult cows and of beef on the bone from France. That same month, the first case of BSE turned up in Spain, and then Germany. In December, the French announced that there was more BSE in their country than they had previously thought. Then the World Health Organization announced plans to convene a worldwide meeting. Then Thailand, and then Australia and New Zealand, banned beef imports from a whole raft of European countries.

The world is still working through the aftermath of BSE. The mass slaughter of infected herds and the widespread ban on feed containing meat and bone might have brought the infection to a close, but this is not actually likely. For one thing, farmers are thrifty, and whenever something is banned, whether it's a feedstuff, or drug, or pesticide, then some farmers, somewhere, will first work through their stocks. Even if BSE itself is halted, vCJD has still to run its course. It isn't known (how could it be?) how long the prion can hang around in the human nervous system before it causes disease, and as yet there is no way of detecting the prion until people are ill, so it is impossible to know how many people became infected in the 1980s or early 1990s, when infected meat was still on sale, and have yet to become ill. The disease is invariably fatal, so the stark question is, will hundreds die during the next few years or decades, or thousands, or tens of thousands? The most pessimistic epidemiological projections have suggested that 130,000 could die, although more recent trends suggest that such a toll is unlikely. In November 2002, however, came suggestions that the prions may cause more than one kind of neural disorder in humans: not just vCJD, but also 'sporadic CJD', the extent of which is unknown since it has only now been recognized. The present death toll from vCJD itself is now (end of 2002) approaching 200. In the face of all this misery the financial cost is secondary. But, so far, BSE is estimated to have cost Britain around £5 billion.

In December 1997 Britain's government set up a committee of inquiry, to be led by one of the country's most senior lawyers, Lord Phillips of Worth Matravers, Master of the Rolls. The committee convened on 12 January 1998 and their report, sixteen volumes of it, was published with commendable speed on 26 October 2000. It criticized the civil service for its poor communications and the Government for its 'culture of secrecy', which caused it to withhold information from the electorate. The report had some good phrases, such as: 'Officials and ministers followed an approach whose object was sedation.' It also admonished politicians for erring on the side of optimism in their desire to avert panic: 'It is now clear that [its] campaign of reassurance was a mistake.' The report named some individuals but found no scapegoats.

Some politicians behaved well in the face of Phillips's comments. Tim Yeo, minister of agriculture under Thatcher's Tory government when BSE first began, publicly apologized to those who had suffered (including the families of the eight who by then had died of vCJD). John Major, who was Prime Minister at the height of the BSE crisis, told the House of Commons that 'All of us must accept our responsibilities for the shortcomings'. Overall, the press described the report as 'damning'. 'I don't think we have pulled our punches,' said Lord Phillips, 'and I don't believe the report is a whitewash.' The report cost £27 million, which perhaps is small beer compared to the overall cost.

Lord Phillips is widely recognized as one of Britain's and hence of the world's most competent lawyers, and indeed as one its wisest judges. He was and is exactly the person most able to get to the bottom of things; and his report is generally agreed to be as thorough, clear and outspoken as could be asked.

Yet, one might humbly suggest, it was a waste of time. Lord Phillips was simply not asked to address the important question. The punches were not pulled, but neither were they aimed at the right target. His committee's remit was merely 'to establish and review the history of the emergence and identification of BSE and the new variant CJD in the UK, and of the action taken in response to it up to 20 March 1996'. Thus he was invited to inquire into the behaviour of the Government and the civil service after BSE was firmly established in the British herd, and during and after its transference into human beings. What really matters is why and how BSE got into cattle in the first place.

The proximal, biological cause is known: dairy cows were given the flesh of cattle and sheep, which contained prions. The deeper and more important question is why such a bizarre and self-evidently hazardous practice was not only permitted, but was positively encouraged; indeed made necessary by the economic climate. The question that Lord Phillips should have been asked to address – and might still be asked to address – is: 'To what extent did the agricultural strategy of the British government contribute to BSE?' The answer to this I suggest (in advance of any such report) is 'Entirely'. BSE was caused entirely by cut-price husbandry; and to cut the cost of production is one of only two coherent themes that has run through Britain's agricultural policy over the past thirty years. (The other theme, linked to the first, is the need for globalization.)

We might even argue that Lord Phillips's report, in net, was a smokescreen. The fact that a person of his stature was invited to chair the inquiry signalled to the world that the Government was sincere in its desire to get to the truth. The net effect, though, was to focus the world's attention on a few politicians who may have misjudged the situation somewhat (but then, they were doing their best in difficult circumstances) and on a few civil servants who may have dragged their feet. We are all supposed to feel older and wiser as a result. A line can be drawn under an unfortunate setback, and now we can all move on. But all this is largely irrelevant. It tells us little that was not obvious (with all due respect) and next to nothing that will be of use in the future (since politicians will always be over-optimistic and civil servants will always make mistakes because, after all, they are human beings).

The lesson that should have been derived is the importance of good husbandry, and the real cost of overriding it. The first rule of husbandry is that chains of infection should wherever possible be interrupted, or should not be allowed to arise in the first place. In this case, a chain of infection was being deliberately created which simply does not exist in nature. Cattle were fed the flesh of other cattle, and of sheep. Cattle are not carnivores, and they are certainly not cannibals. Not all infections can be spread by cannibalism, but a great many can be. No good, traditional farmer, practising sound husbandry, would do such a thing. But in modern, industrial farming, the kind that is now called 'conventional', this became standard practice. The simple fact is that modern industrial farming is not rooted in sound husbandry. Its central thrust is to maximize profit. This is done in general by maximizing output (which in this case is achieved through concentrated feed) and by minimizing costs (sheep and cattle flesh that would otherwise go to waste is cheap protein). Since this is what maximizes profit, this is what has to be done. When competition rules, those who don't take the most profitable course miss out to those who do ('If I don't do it, somebody else will.' How often do we hear that sentiment?) Good husbandry is simply flouted. Science helps in this process, in this case by helping to provide the technology that provides the concentrated feed. Here is another unpleasing but salutary lesson:

that science is increasingly deployed in modern farming not to abet good husbandry, but to override it; and to do so in the express interests of profit.

There is a second lesson, at least as chastening, though of a quite different kind. The BSE episode reveals the innate inadequacies of governments, and of the experts who advise them, and of the ways in which the two communicate with each other; the formality, not to say pomposity, in which meaning is lost. It also shows that people at large can be much more sensible than their ostensible leaders and advisers. Thus, from the first (public) hints of BSE, many people simply stopped eating beef, and some school canteens dropped it from the menu. This was a wise move. People were right to be suspicious of the beef, and of the politicians who so blandly or zealously sought to assure us that it was perfectly safe. The butchers who defied the Government ban and continued to sell beef on the bone were right too. The meat sold on the bone typically came from specialist beef herds raised on grass, which had never been fed contaminated concentrate; and the butchers knew this perfectly well. Besides, there was no deceit in these sales. The butchers were not trying to sell a pig in a poke which, in effect, is the way that most processed food is sold in supermarkets, perfectly legally. What the customers saw was what they got. Everyone knew about the ban, and if thinking people chose to buy beef on the bone, then this was up to them. This was what the butchers and their customers felt; and fair enough. We may also ask what right a government has to pass such nit-picking laws when it itself has caused the problem in the first place. Their absolute confidence in their own right to rule and their competence to do so is sublime.

Foot-and-mouth Disease

Hard on the heels of BSE came foot-and-mouth disease. Indeed foot-and-mouth overlapped BSE (although there was still time to sandwich a quick and nasty epidemic of swine fever between the two). By the time Britain's epidemic of foot-and-mouth disease had run its course between the start of 2001 and the first months of 2002, it had affected 10,000 farms, cost £2.7 billion ($3.8 billion), and led to the slaughter of 6 million animals, mostly cattle and sheep. It also spread to Holland and France, although both those countries contained it rapidly. Britain probably needed to reduce its population of livestock, but this random stroke that knocked out some of the finest herds, which had been built up over decades, as well as those that were not so special, was clearly not the way to go about it.

The foot-and-mouth virus affects all cloven-hoofed animals, which among typical British livestock include cattle, sheep, goats, pigs and deer, and nowadays may also include llamas and vicunas. British zoo directors feared for their beasts too: all the above plus antelope, camels, and even elephants. Horses are not affected, but can help to spread the virus. It is extremely contagious. It is carried by wind, by contact, by residues in fields and buildings and on vehicles, and incubates for only a few days before typically erupting in ulcers on the tongue and between the toes. It makes the animal ill, although it is not generally lethal. Some countries, such as India, have traditionally been content to let it run its course, like a childhood illness. But it sometimes makes the animals very sick indeed, and often leaves them permanently weakened. Thus in its general epidemiology foot-and-mouth has much in common with measles (although the two viruses are not related): both infections spread quickly and easily, are not usually too awful if the victims are healthy to begin with, but sometimes leave serious, permanent damage in their wake, and are not to be taken chances with. On grounds both of welfare and economics, countries like Britain which need their livestock to grow at a steady pace and to hit particular markets at particular times and weights, cannot afford to allow their animals to lose so much condition, and fall so far behind. I argue in this book that animals nowadays are expected to grow too rapidly, and are too hassled. But nobody sensible would suggest that foot-and-mouth should simply be allowed to run its course in this country. Modern Indian farmers vaccinate when they can afford to.

In general, there are two main ways of controlling foot-and-mouth; both, of course, with their advantages and drawbacks. The ideal – and the approach adopted by most modern countries wherever feasible – is to keep the virus out of the country altogether. The import of meat and animals is strictly controlled to ensure that none are brought in with disease; and at the first sign of any outbreak, the affected farm is isolated and the herd slaughtered. Sometimes, as in Britain in 2001 when the epidemic seemed to be getting out of hand, herds on neighbouring farms and all other known contacts may be slaughtered too. The alternative policy is to vaccinate animals; preferably the entire national herd.

So which strategy is better? Ideally, total exclusion must be preferable.

Our experience of human medicine tells us that although it is good to be vaccinated against unpleasant diseases, it is better still never to come into contact with them at all. Through most of the twentieth century Britain had a total exclusion policy against rabies (there were serious penalties for anyone smuggling dogs and cats), and no cases originated in Britain in all of that time. In many other countries, including the United States and India, the fear of rabies (spreading from wildlife to pets and people) has always been there. But border controls have to be very good indeed to control foot-and-mouth by total exclusion. If the virus does get in, then the unvaccinated animals are totally vulnerable. The nation's farmers, vets and markets have to be constantly alert to isolate, diagnose and slaughter at the very first sign of an outbreak. Such vigilance is difficult and expensive. Mass slaughter raises its own problems, too. It can be as traumatic as war and the point has at least been mooted that mass burial can affect water supplies.

But vaccination raises problems of its own, as vaccines invariably do. The most effective vaccines are 'live'. They contain live virus, which in general promotes a strong immune response but can also cause outbreaks in unvaccinated animals. The killed vaccines are safer in this respect, but offer less than perfect protection. The fear in either case is that vaccinated animals may carry the disease in a cryptic form, and infect others over time. In general, a country that practises vaccination admits that the virus has in effect become permanently resident. Vaccination is also costly; and if Britain did vaccinate, it would have to give up the export of red meat, which at present earns it around £200 million per year.

In the latest epidemic, foot-and-mouth was first confirmed in a pig in Essex in the east of England on 20 February 2001, but this case was traced back to a small farm in Northumberland, in the extreme north of England. Within days it was reported in Devon in the south-west of the country, nearly 400 miles away. The Northumberland farm where the epidemic apparently began was in many ways unsatisfactory, and the standard story, at least at first, was to treat the farmers – two brothers – as scapegoats. But there was clearly more to it than that. There have been many suggestions, that have not been laid to rest, that foot-and-mouth had been lurking in Britain for a long time – perhaps effectively for ever, and just happened to break out in Northumberland. Its signs and symptoms can be very mild, ambiguous, and difficult to spot among sheep in particular; and many sheep in Britain (as in the world at large) live in remote and difficult places and are not minutely inspected day by day. It was clear, too, that the country's border controls were not all they should be. The laws to exclude rabies from Britain were rigorously enforced. Visitors to present-day New Zealand are warned by huge and strident notices that anyone bringing anything live into that country, down to and including lettuce sandwiches, is liable to be fined to the point of bankruptcy. But (as some enterprising journalists showed) it is quite easy to bring unsavoury meat into Britain, and a surprising number of people do so for their own recondite reasons. Customs simply don't have time to inspect every bag, and in any case are interested mainly in drugs. But unless borders are secure then sooner or later the virus is bound to get in.

Even this would not be the end of the world if the virus was tightly contained on entry. This implies that transport of live animals around the country should be minimal, and that animals should have as little contact as possible with others outside their own farm. Some vets have suggested that livestock production should as far as possible be arranged by regions, and that live animals should not cross regional boundaries except where this is essential for standard husbandry. Many (not just vets) have argued for reasons of welfare as well as of disease that the journey from farm to slaughter should be as brief as possible. But over the past sixty years the number of Britain's abattoirs has fallen steadily and about three-quarters of those that remained were closed in the 1990s: total numbers fell from around 1,300 to just over 300 at the end of 2000. Most of those that were closed were small and local. The general reason given was to raise standards and sometimes, doubtless, this was justified. Commonly, however, 'raising standards' is official-speak for introducing some new practice or criterion that is not relevant to small units, but which they have to conform to anyway, and cannot afford. The underlying reason for all such measures is to cut costs.

The closures meant that animals must now travel huge distances between farm and slaughter. Or they may travel to some commensurately centralized and enormous market to be sold on for fattening (or 'finishing') on another farm, where they mix with hundreds or thousands of other hapless beasts. The strategy of raising animals in one part of the country and fattening them in another is ancient. Until well into the 1830s drovers walked their cattle down from the Scottish hills and through the border country down to Lincolnshire or to Wales. But that journey could take weeks. In the unlikely event that any of the animals had foot-and-mouth when they began their journey, the disease would have run its course before they arrived (and in any case, the first clear-cut cases of foot-and-mouth date from 1835, when such heroic droving was already on the way out). Transport these days is rapid: a matter of hours. So transmission from one end of the country to the other, and virtually to all points of the compass, is possible even within the short incubation period of foot-and-mouth. The last epidemic of foot-and-mouth, in 1969, was largely confined to the cattle country around Cheshire, in the north-west of England. Local abattoirs still flourished in those days. The animals did not have to travel. The present strategy seems designed to spread foot-and-mouth around the whole country as quickly and efficiently as possible. At least, if it were designed specifically for that purpose, it could hardly be more effective than in fact is the case.

In addition, many farmers, vets and scientists feel that the Government's all-out slaughter policy was misguided. Once slaughter had failed to contain the disease within, say, six weeks, then, say the critics, the animals should have been vaccinated. If the £2.7 billion lost through slaughter and compensation had been soundly invested, the interest alone would have made up for the £200 million a year sacrificed through the loss of exports. That may be true, and it may not. But the general case seems again unequivocal: that the 2001–2002 foot-and-mouth epidemic is another toll exacted for farming on the cheap. You can't have an exclusion policy with leaky borders; and you can't allow putatively diseased animals to be whisked to the far corners of the country just to save a bit of cash. In general, animals should travel as little as possible in the course of their brief lives, and should be slaughtered locally. This would put up the cost, but if meat is to be produced safely and humanely, then it needs to be more expensive. It is not a human right to eat meat in vast amounts. My impression is that people at large are well aware of all this. But governments are afraid of the idea. Dearer food – or at least, *ostensibly* dearer food – means loss of votes. Or so they believe. And what is more important than loss of votes?

In Hong Kong in 2000 I visited a market where chickens already plucked and dressed and ready for sale lay alongside the cages where their sisters were still alive and clucking. In Yunnan in tropical China a few weeks later I watched a butcher hold a live chicken over the gutter and hack off its head with a penknife, while his customers chatted at his elbow. European health inspectors, with their diplomas and clipboards and nylon hats, would surely close down the whole of South-East Asia if they got half a sniff. But this Yunnan chicken was doubtless raised on some local family farm, a bicycle ride from the market; and would be plucked within minutes and cooked within hours of its death. There would be little opportunity for bacteria such as salmonella to accumulate in the flock, as they fossicked around the fields, and very little time for any bacteria that were present to multiply between slaughter and consumption. Animal welfare in most of the world leaves much to be desired but I'm not sure that the Yunnan chickens, which are brought up in human company and seem affronted rather than frightened by their perfunctory demise, are worse off than the beleaguered beasts of Europe. Overall, there is almost no opportunity for sharp practice; and in any case, the farmers and butchers would not cheat their customers, who are their neighbours, partly because most people are by nature honest and partly for the reason that Adam Smith acknowledged: you can't cheat people for very long who know who you are and where you live. For all the committees and the high tech and the self-righteous finger-wagging of officials and experts in modern Britain, I would as soon take my chances in tropical, rural China.

Overall, the food-safety laws of Britain are extensive and intricate and more and more detailed, so that it's becoming very difficult even to keep a few chickens or pigs for local use, or to run a village shop, or to sell cakes at the church bazaar. Men and women in suits or with nylon hats and clipboards descend like flies to point out ways in which small farmers and traders could in theory poison their customers. At the same time, the government that makes these laws presides over policies that seem designed expressly to maximize the spread of disease. Animals are raised in the greatest numbers in the most crowded quarters that are physically possible. Infection is suppressed with low levels of antibiotics, as if intended to encourage resistance. The beasts are shipped routinely en masse over vast distances at high speed, making maximal contact with other animals along the way. The overall food supply chain is as long in time and distance, and as convoluted, as commerce finds convenient, and in practice may beggar belief. Ostensibly 'fresh' produce in the modern supermarket has often been harvested weeks before. Industrial chemistry is deployed not quite without restraint but certainly liberally to keep the worst dangers at bay – and then may raise problems of its own.

Along the way, there is maximal opportunity for chicanery, and ample evidence both that this takes place and that little of it is detected. To be sure, there are crooks in all walks of life. But some environments are more conducive to crookedness than others; and again, the present food supply chain seems designed expressly to encourage malpractice. The producers, consumers and various people in the middle no longer meet face to face, as they did in Adam Smith's day, so the kinds of restraints that modify normal social intercourse no longer apply. Yet the chain is also, in practice, impossible to police. If the present food supply chain was policed adequately, the cost of food would surely be increased several-fold. In short, food is as cheap as it is because in effect it is produced and distributed on a wing and a prayer. Still, though, it is nothing like as cheap as it should be, because the cutting of costs is matched by the adding of value; both by gratuitous processing but mainly by the maximum production of livestock.

All this, the kinds of governments that like to think they are modern, preside over. People die as a result, and millions of animals are put to the captive bolt or the bullet and the funeral pyre long before their time. The role of science, meanwhile, is not to enhance good husbandry so that we may all be better fed, but to aid the commercial process: to find ways of overriding the basic rules of husbandry, and to minimize the use of human labour (and the skill and intelligence that goes with it) and so to keep costs down even further. Scientists in high places should think again. Meanwhile, little old ladies are forbidden to sell their home-made cakes, lest they should chance to break the salmonella-laden eggs on the side of the mixing bowl. It would be amusing if the reality were not so tragic.

To put the matter another way: on the one hand we have hygiene laws apparently designed to maintain food in a state of asepsis, which would probably not be desirable even if it were achievable; yet these laws are superimposed on a system of food production and distribution that seems specifically intended to generate and spread infection, or at least could hardly do the job better if it had been. Yet this process is presided over by very good scientists, who if the rules were different could truly be helping to improve the lot of humankind and of all other creatures besides. Thus is absurdity piled on absurdity. How this extraordinary state of affairs came about is the subject of the Part III.

III

Science, Money and World Power

It seems wrong that . . . the science related to producing food has to be used in a competitive fashion: the essence of science is its universality, and freedom from hunger should be the birthright of all mankind.

Sir Kenneth Blaxter, 'The Options for British Farming', in *Agricultural Efficiency* (The Royal Society, 1977)

6

Craft, Science, and the Growing of Crops

Science is one of the greatest creations of humankind. It does not provide us with absolute truth, and certainly offers no royal road to omniscience, but it does provide us with wondrous and unending insights into the workings of the material universe. As the great Viennese-British philosopher Sir Karl Popper pointed out in the 1930s, the modus operandi of science is to pose a hypothesis, or as Richard Feynman later put the matter, to make a guess. Then the hypothesis (or guess) is tested formally, by experiment. If the hypothesis survives scientists' best attempts to test it then it is said to be 'robust'; and when hypotheses are robust enough, they qualify as 'theory'. So although the theories of science should never be mistaken for absolute, all-embracing truth they do provide us with the clearest and in many ways most useful account of the material universe that we are ever likely to enjoy – and indeed of some non-material matters, including psychology and perhaps even sociology, and bits of economics. Science is an aesthetic pursuit. Karl Marx said that the purpose of philosophy is not to understand the universe but to change it; but I would like to invert that adage and suggest that the prime purpose of science is not to change the universe but to enhance our appreciation of it. The seventeenth-century founders of modern science, including Isaac Newton and the naturalist John Ray, would have had no trouble with such an idea.

But the insights that science provide can and do help us to change the universe nonetheless. For its insights give rise to a constant stream of new technologies – and it is the specific purpose of technology to change the world, and in general to make it more comfortable for us to live in. Technology defined broadly is as old as humankind – indeed older, because many other animals also make use of tools, and all tools are technology. But the particular devices that emerge from science can reasonably be called 'high technology'.

Human beings transformed entire landscapes and created great empires on the back of technologies that had no science in them at all. The aqueducts of Rome, supreme creations that they are, take precious little account of modern hydrodynamic theory, apart from the idea that water runs downhill. The windmills that transformed the economy of medieval Europe were built without reference to recognizable aerodynamics. The Japanese made the most wonderfully supple swords without modern metallurgy, while ancient Chinese ceramics beggar belief – with each passing century they brought new ingredients, methods of firing, glazes and pigments into play. Over ten thousand years (at least) agriculture changed the whole world before bona fide science came on board. It is wonderful what can be achieved by craft and experience alone. The craft of potters and seamstresses and musicians and cooks never ceases to amaze. But when we add science, traditional technologies can leap ahead and brand-new technologies emerge that traditional craftspeople could not have envisaged. The high technologies of the past two hundred years that have turned the world on its head include the internal combustion engine, the telegraph, the telephone, electronics of all kinds, the computer, the laser – and a range of biological high technologies that collectively are called 'biotechnology'.

In short, science itself should be seen (I suggest) as its first creators saw it: not as a mechanical pursuit to make our lives more comfortable, but as 'natural philosophy'. But out of science, nonetheless, through the revision of traditional technologies and the creation of completely new ones, have come the most powerful agents of change the world has seen, or is liable to see.

The insights of science, and the special technologies to which it gives rise, can help us in principle to achieve absolutely anything we might care to envisage, particularly (though not quite exclusively) of a material kind. There is nothing we might want to do that could not in principle be done better or more sure-footedly, with greater insight of the kind that science can help to provide. Agriculture should be a prime beneficiary. It is so intricate: there are so many different aspects to be integrated, including for example the whole of ecology and the physiology of plants and animals, including that of human beings. It is also of course a practical pursuit, innately technological, with endless niches for high tech, from the stainless steel of modern ploughs, to computer-controlled micro-irrigation, to genetically engineered vaccines.

But if the power is misdirected, then science and its high technologies are commensurately damaging. For the rest of this chapter I want to outline the things that science has done to and for agriculture over the past few centuries, focusing on soil fertility and the control of pests and diseases. In truth the contributions have been momentous. But the shortfalls have also been salutary.

Life as Chemistry: Fertility and the Growth of Plants

Biologists say that plants are 'autotrophs' – Greek for 'self-feeders'. They acquire the raw materials from which to build themselves in the form of minerals: simple inorganic compounds such as nitrates and phosphates. They acquire their energy from the sun, through the process of photosynthesis. More specifically, they use solar power to break down molecules of water and so release hydrogen (water is H_2O: hydrogen combined with oxygen) and combine the hydrogen with carbon dioxide (CO_2) from the atmosphere, to make sugars. They then break down some of the sugars to provide energy to build the entire range of carbohydrates, proteins, fats, and their derivatives – combining and recombining some of the sugars with the various minerals that they draw in through their roots. Then, when they have finished this heroic process, they get eaten by animals. Nature is neat, but not necessarily just.

Animals, in sharp contrast, are heterotrophs. They require their food ready-made. For plants, nutrients (minerals and so on) and energy (sunshine) are separate inputs. But for animals, food is the source both of nutrients *and* of energy. As heterotrophs ourselves, we take this for granted.

Before the second half of the nineteenth century, this distinction was not clear. Biologists tended to assume that plants feed themselves just as animals do. Why else, after all, would they benefit from manures? Did they not acquire rich, complex foods from manure just as animals do from their somewhat more appetizing provender? Aristotle assumed that plants sucked up organic material from the soil, effectively like a housefly sucking up semi-digested soup; and Aristotle's natural history was still widely believed in the early nineteenth century. Such persistent belief may seem somewhat odd since the great Dutch (now Belgian) Catholic natural philosopher Jan Baptista van Helmont (1580–1644) had shown in the seventeenth century that as plants increase in dry weight, the dry weight of the soil goes down only slightly – which surely would not be the case if plants literally consumed the soil as a cow consumes hay or a fly sucks up yesterday's dinner. There are various salutary tales here. One is that in olden times, news in science did not always travel quickly, or very far. Another is as the English philosopher A. N. Whitehead said: that although scientists must trade in facts, they do not always find facts easy to accommodate. Van Helmont's results seemed anomalous and the human psyche, even the psyche of scientists, tends to put anomaly on one side.

But the late eighteenth and early nineteenth centuries were the first great age of recognizably modern chemistry. In his *Elements of Agricultural Chemistry* (1813), Humphry Davy pointed out that plants derive their carbon, oxygen and hydrogen not from the soil, but from the atmosphere, via photosynthesis. (In truth, as is now clear, the hydrogen comes from water, which plants derive mainly from the soil.) Yet, as Lloyd Evans comments in *Feeding the Ten Billion* (Cambridge University Press, Cambridge, 1998), Davy remained 'ambivalent', for he also referred to soil organic matter as 'the true food of the roots'; and in this there are still clear echoes of Aristotle.

Enter now, stage far right, Justus von Liebig (1803–73). It would be unfair to German scientists as a whole to suggest that Liebig was their archetype; but he was certainly their stereotype. He was extremely clever, energetic, confident, irascible, and – as reflected in the flocks of acolytes who came to work at his laboratory at Giessen – charismatic. Science is supposed to be about *ideas*, which are supposed to be arrived at objectively, by the most assiduous observation and the coolest possible assessment, all laid out clearly for the world's inspection: as rational and, in its openness, as democratic as it is possible to be. Any suspicion that science is arcane and not accessible to all (at least in principle) seems to betray its fundamental tenet. This is in sharp contrast to religion, whose prime truths are attained through revelation, and are revealed only to individuals with special training or with special gifts. Religion relies on authorities – priests, gurus, lamas, shamans, prophets – while science needs only its honesty and intellect: 'Here are the observations, this is the reasoning, there is the maths – see for yourself.'

This is not how human beings operate, however. We *like* authorities.

We like people to tell us, with deep certainty, what is true; and, once told, we cling to our newly privileged insights doggedly. In reality, science is full of people who are personally modest, and content to let their ideas speak for themselves; but it is equally packed with prophets manqué, who make their own journeys into truth and reveal what is the case to the rest of us as confidently as any patriarch. Science, in short, though widely perceived as the antidote to the authoritarian theology of many traditional societies, commonly operates in precisely the fashion that it affects to supersede. I am labouring this theme not through whimsy but because it is still of huge sociological importance, reflected in the way that scientists now present GMOs (for example), and governments seem slavishly to accept whatever they say. Liebig was the archetype of the scientist qua patriarch; wonderfully talented by the normal standards of humanity but deeply flawed and, in particular, in his own eyes never wrong. But as a scientist his task was to study nature; and nature will not be second-guessed.

In the late 1830s Liebig was invited by the British Association for the Advancement of Science (still going strong) to lecture on new ideas in chemistry and he chose to speak on organic chemistry and its application to agriculture. Out of all this, in 1840, came his *Organic Chemistry in its Application to Agriculture and Physiology*. This may seem somewhat esoteric but in truth, Liebig's boldness came as a shock, and is of profound philosophical importance; a pivotal point in human understanding of the universe and of human ability to manipulate it. For scientists, theologians and philosophers alike had generally taken it for granted that life is special. Alchemists and the chemists who superseded them could play with minerals in their crucibles and retorts, but the processes of life remained innately mysterious, imbued with some extra factor denied to mere minerals. Indeed the idea of 'vitalism' – that life required some extra vital spark – was particularly strong in Germany, and persisted well into the nineteenth century and even into the twentieth. But in 1828 the great German Friedrich Wöhler (1800–1882) was astonished to find that when he heated ammonium cyanate, which was supposed to be 'inorganic', he obtained urea, which is found in mammalian urine and is therefore classed as 'organic'. Until then, no one supposed that organic materials *could* be synthesized from inorganic. Thus it began to seem that living systems might not be beyond the scope of the laboratory investigator; that they might indeed yield to mundane analysis. In all this we see the two meanings of the term 'organic': with mystical, vitalist connotations on the one hand, and with its down-to-earth reference to the compounds of carbon on the other. Liebig, twelve years on from Wöhler, was already discussing plants not as the lilies of the field, as inveterately mysterious and godly creations, but as chemical systems. What had been sacred was becoming not profane exactly, but accessible. It was a huge transition in human thinking. Mary Shelley anticipated modernity perfectly in *Frankenstein, or the Modern Prometheus*, written in 1816, although her point has been coarsened and largely lost in the retelling. She was observing absolutely correctly that while mystery may be demystified, the demystification can never be complete. There is no final unravelling.

Unlike Davy, Liebig did not equivocate. He buried for ever the idea that plants directly consume humus. What plants did derive from the soil, said Liebig (accurately) was all the other minerals they need apart from carbon, oxygen and hydrogen: the minerals that are found in the ash that remains when plant material is burnt. These account for the diminution in the dry weight of soil as a plant grows that had been noted by van Helmont. These essential elements were already known to include iron and sulphur (and others), and Liebig added phosphorus to the list, in the form of phosphate. Phosphorus had been hard to analyse before Liebig's day. To Liebig more than anyone belongs the idea that all these specific, essential minerals need to be returned to the soil after the crop is removed; and the modern realization that the border between biology and chemistry is not absolute; that at least some aspects of biology can usefully be analysed and understood in chemical terms.

Liebig, however, was spectacularly wrong about nitrogen. Carbon, oxygen and hydrogen are the basic components of all carbohydrates, fats, proteins and nucleic acids. But proteins and nucleic acids are also compounded of nitrogen; and since living tissues contain proteins in particular in significant amounts, plants require a great deal of nitrogen. They need less nitrogen than they need carbon, hydrogen and oxygen, but they require a great deal more nitrogen than any of the other minerals. Compost contains a great deal of carbon (since it is made up of living materials, or at least of materials that were once part of living organisms), but if it is truly to act as a fertilizer and not simply as a soil conditioner (improving soil structure) then it must also contain nitrogen. Nitrogen is the chief component of artificial fertilizers and sometimes it is the element that is most lacking in the soil (and so lack of nitrogen is commonly the 'limiting factor'). The abandoned fertilizer bags that decorate many an English hedgerow typically bear the legend 'NPK': N is nitrogen, P is phosphorus and K is potassium. Phosphorus (as is now known, though of course not in Liebig's day) is an essential component of nucleic acids, and potassium contributes among other things to the delicate mechanisms of cell membranes. (Sulphur would also be a common component of fertilizers since it features in some amino acids and hence is an essential ingredient of proteins. But until very recent times, at least in industrial countries like Britain, plants have received all the sulphur they need in rain, derived from the smoke of fossil fuels.)

But Liebig did not get to grips with nitrogen. He started off on the right foot, stating in the first edition of *Organic Chemistry* that to supply crops with nitrogen is 'the most important object of agriculture' – which, if you plant in the right season and take care of diseases, is essentially correct. But he got it into his head that plants derive most of their nitrogen from the atmosphere, just as they derive their carbon and oxygen: not directly from the nitrogen gas which makes up 80 per cent of the atmosphere, but from gaseous ammonia. Ammonia is formed as lightning strikes, and constantly wafts from the soil and the sea as a product of decay; and the main reason that there isn't a great deal more ammonia in the atmosphere is that it is extremely soluble, and is quickly absorbed by surface water, and because it combines with oxygen to form oxides of nitrogen. But Liebig's calculations, somewhat wild as it turned out, convinced him that there was enough free ammonia at least to nourish forests, although crops which generally produce more biomass in a year than trees do, would need extra ammonia.

In principle Liebig's ideas on nitrogen were far from stupid. The biggest mistakes in science do not usually spring from stupidity. Modern forests in industrial countries typically derive all the nitrogen they need from nitrates and other nitrogenous compounds released from car exhausts. But, as it happens, in this instance Liebig was wrong. In the same year that Liebig wrote *Organic Chemistry in its Application to Agriculture and Physiology* (1840), Jean-Baptiste-Joseph-Dieudonné Boussingault of Paris showed that plants obtain nitrogen from the soil in the form of nitrates. But he, too, was treated to the standard and much-feared Liebig ridicule. Liebig overlooked nitrates completely; and, for good measure, he downplayed the importance of guano, the powerful and highly effective manure derived from the ordure of seabirds.

Instead, Liebig focused on phosphates, and released a patent fertilizer in 1846. But Liebig's fertilizer did not work; or at least, it did not justify the expense. His students became disillusioned and drifted away. Liebig himself clung to his *idées fixes*, as they had become, and refused to accept new evidence as provided for example by Boussingault. His cardinal mistake, as Lloyd Evans comments, was to underestimate the complexity of the problem. He assumed, as the ancient philosophers of Greece tended to do, that because his theories were so impeccably logical and internally consistent, nature was bound to conform to them. Not at all. Nature does what nature does and all scientists, even the most prodigious German autocrats, must be humble in the face of it.

For all that, Liebig was supremely important in the history of agriculture. He more than anyone established the key importance of experimental science, and the model of the modern chemical laboratory. He also focused on the key issue: the fertilization and the growth of crops. He buried once and for all the ancient Aristotelian notion that plants eat humus. He emphasized the essential role of soil minerals. By the 1860s it was known that plants require large amounts of nitrogen, phosphorus, potassium and sulphur; and also significant quantities of magnesium and iron. Then scientists in several countries effectively worked through the periodic table of elements (which itself was formally outlined only in the 1860s), providing plants with different nutrient solutions each of which lacked just one element, to see which ones they really needed. This was easier said than done because some elements, though essential, are required only in spectacularly small amounts (which is why they are called 'trace' minerals); and they tend to be present in nutrient solutions even when heroic attempts are made to eliminate them. Others are vital in tiny amounts but toxic when the dose is increased even slightly. Manganese was not pinned down until 1897 because it is a common contaminant of iron. Early tests of copper were confounded because the ostensibly 'pure' water used to prepare the nutrient solutions had been distilled in copper vessels; yet copper in amounts only slightly greater than necessary becomes a significant toxin, commonly used as a fungicide. The fact that plants do need a little copper was not finally revealed until 1931. Their absolute need for chlorine and – surprisingly – for sodium was not pinned down until the 1950s; and nickel was not added to the list until 1983. So now the complete list of additional trace minerals known to be required by plants includes iron, boron, manganese, zinc, copper, molybdenum, chlorine, sodium and nickel. Liebig initiated the formal inquiries that have produced all these insights: although they have taken a century and a half to work out and the game, surely, is not over yet. Again we see that nature is complicated and research can take an awfully long time.

Although trace minerals are required in such tiny amounts, they are essential. Plants may suffer specific diseases when they lack any one of them, and they die if totally deprived. Thus copper-deficient citrus fruits suffer 'die-back' and manganese-deficient oats suffer 'grey speck'. The trace minerals can in fact be limiting factors. A farmer may supply all the nitrogen in the world, but if the plant lacks the required trace minerals it will not grow; just as it will not grow with all the water and sunshine in the world if it lacks nitrogen. Worldwide, zinc is the trace mineral most commonly lacking: the farming fields of the USA receive 40,000 tonnes of zinc per year (compared with 10 million tonnes of artificial nitrogen). But my favourite example is from the Terai zone of Nepal, where for many years chickpeas were tending to drop their flowers, and so failing to set seed. Grey mould (botrytis) was thought to be the culprit. But observers from ICRISAT (the International Crops Research Institute for the Semi-Arid Tropics) noticed that flower drop was worst on acid soils, suggesting some nutrient deficiency. Boron turned out to be the problem. Additions of boron at just 1 kilogram per hectare increased chickpea yields by 42 per cent up to an astonishing 92 per cent; almost double, or an extra half tonne of chickpeas per hectare. Chickpeas are an extremely important staple in the dry tropics, and boron is cheap. Here is a classic example of good science providing an easy solution.*

Less spectacular than Liebig but in the end more solid, were the English landowner and philanthropist John Bennett Lawes (1814–1900) and his assistant the chemist J. H. Gilbert. They were theoreticians too, of course, for all experiments must be rooted in ideas, but they were also, literally, more down to earth. Always they asked, as Liebig apparently disdained to do, what are the facts of the case. Lawes established the Rothamsted Experimental Station on his family estate in Hertfordshire and there, year after year, they carried out experiments to see exactly which combinations of fertilizers had what kind of effect on different crops. A century and a half later, useful data are still coming in from those original experiments.

Lawes began his studies when, as a young man not long down from Oxford, he was challenged by a neighbour to explain why crops respond

* *Chickpea Improvement at ICRISAT* (Andhra Pradesh, India, August 2000), p. 16.

only erratically to applications of crushed bones: a favourite fertilizer of the day, at least in England. Bones are an excellent source of phosphorus (the hard material within them is calcium phosphate) but the phosphorus is not readily released to the plants and, of course, in any one crop, lack of phosphorus is not necessarily the limiting factor. But Lawes made the phosphorus more soluble and accessible to the plants by treating bones (and, later, mineral rocks) with sulphuric acid to produce 'superphosphates'. These worked particularly well on turnips, which are greedy for phosphorus; and turnips in turn were key players in the Norfolk rotation, which played such an important part in nineteenth-century farming. So, in 1842, Lawes acquired a patent for his new superphosphate fertilizer.

With the money thus earned Lawes was able to employ Gilbert, who had trained with Liebig, and he expanded the studies in laboratory and field (Liebig disdained the field) to look at all the major nutrients in a variety of different crops, pastures, and rotations. Soon Lawes and Gilbert showed that unless crops are manured, they need significant applications of nitrogen, for example in the form of sodium nitrate or of ammonium sulphate (which is largely converted to nitrate by bacteria in the soil). The one set of crops that did not seem to benefit from extra nitrogen were the legumes (peas, beans, clovers and so on) – the crops that had long been known to increase soil fertility, and enhance the growth of subsequent crops. So they announced that Liebig was wrong to insist that plants obtain their nitrogen from atmospheric ammonia. They need it presented to their roots, in the soil. Gilbert, one feels, would have enjoyed stealing a march on his former, uncompromising mentor. Lloyd Evans comments that 'Liebig responded with equal vigour, but with less constraint by the facts.'

Lawes and Gilbert finally settled the matter in 1857 with help from an American chemist, Evan Pugh. They demonstrated that plants did not remove any nitrogen gas from the air as they grew. Clearly, then, they obtained all the nitrogen they needed from the soil. Liebig was shown to be wrong on this key issue, once and for all.

The first clue to the puzzle posed by legumes, which grow perfectly well even when the soil is deficient in nitrogen, was provided in France by Louis Pasteur, who was beginning to show not simply that bacteria exist (which had been known in a general way since the seventeenth century) but that they are all over the place, and do a great many different things. Indeed they, in effect, run the basic ecology of the world. In 1862 Pasteur suggested that bacteria in the soil could perhaps create soluble nitrogen compounds, notably nitrate, by capturing or 'fixing' gaseous nitrogen from the atmosphere. This has proved to be true: indeed there is a host of soil bacteria that convert atmospheric nitrogen gas into ammonia and then into nitrites and nitrates and back again. Legumes make special use of these bacteria.

Now, more than a hundred years later, it's known that many bacteria fix nitrogen, including some cyanobacteria – the creatures formerly known as 'blue-green algae'. Indeed, many trees are more or less self-fertilizing because the cyanobacteria that grow like dark slime on their trunks (not to be confused with lichens) drip surplus fixed nitrogen down into the soil. Many plants, including some important crops such as maize, seem to create a chemical zone around their roots that favours nitrogen-fixing bacteria, which in turn enhance local fertility. The paddy fields in which rice is grown are commonly coated with tiny floating ferns known as *Azolla*; and in spaces within the leaves of the *Azolla* live nitrogen-fixing cyanobacteria of the genus *Anabaena*. Thus the *Azolla–Anabaena* combination is a very significant source of fertility for one of the world's most important staples.

Even more important though, because of their worldwide significance, are the symbiotic relationships that leguminous plants form with bacteria of the genus *Rhizobium*. The pink nodules on the roots of healthy legumes are the custom-built boarding houses in which their rhizobial collaborators reside. The investment in these structures is well worthwhile.

Because their roots harbour rhizobia, legumes can grow where other plants cannot; and as farmers have known for many centuries, other plants can grow very well in soil recently vacated by legumes, as they leave surplus fixed nitrogen behind (in the form of nitrates). The same effect is achieved of course if the leguminous crop is ploughed in as a form of green manure. Even before nitrogen fixation was formally demonstrated in 1886, one-fifth of Europe's farmland was down to legumes, which probably contributed more to overall fertility than manure did. Alternatively, in many mixed cropping systems worldwide, legumes are grown in between the main crop. You see this in thousands of traditional settings, with hundreds of combinations: pigeon peas with millet or sorghum; alfalfa with maize; broad beans among the olive trees in Mediterranean countries. Mixed swards of grass and clover form one of the world's great traditional pastures. It's another pleasant serendipity (discussed in Chapter 3) that pulses and cereals together provide such fine protein. One hundred and twenty years after its discovery, nitrogen fixation remains a prime target of agricultural research, and so it must. No process is more important to the survival of the human species, and the ecology of the world as a whole.

There is one final and vital chapter in the fertilizer story. In 1908 the German chemist Fritz Haber combined nitrogen directly with hydrogen to form ammonia, with the gases at high temperature and pressure and in the presence of a catalyst; and five years later Carl Bosch scaled this up into an industrial process. The initial intention was to aid Germany's war effort by making TNT but here, of course, was an obvious source of fixed nitrogen for fertilizing crops and in essence, the Haber–Bosch process is still the basis of modern nitrogen fertilizer production. The nitrogen comes from the atmosphere which contains around 10^{15} tonnes of it, which is 1,000,000,000,000,000 or a thousand million million tonnes, so it is not going to run out (and in any case it is constantly recycled). The hydrogen comes mainly from fossil fuels including, in the US, natural gas. Most of the nitrate in nitrate fertilizers is derived from ammonia, produced by the Haber–Bosch process. In the US, ammonia itself is commonly applied to the soil, either in solution or in anhydrous form, and this practice may spread since it is cheap and efficient. Plants can absorb nitrogen in the form of ammonia (as Liebig supposed was the norm) but most is presumably oxidized to nitrate in the soil before it is absorbed.

The annual turnover of nitrogen in the world's crops (the amount that gets in, and is removed as the crops are harvested) is an estimated 175 million tonnes; and nearly half of this – around 80 million tonnes – is nitrogen that has been fixed by the Haber–Bosch process. Of the rest, about 30 million tonnes of nitrogen is supplied by biological fixation (mainly legumes), 30 million tonnes comes from mineral sources, and the final 15 million tonnes, less than one-tenth of the whole, comes from farmyard manure. The total nitrogen supplied by artificial fertilizers is equivalent to about 70 or 80 per cent of the amount supplied by natural nitrogen fixation in all the land of the world, both agricultural and wild. Wild soil is normally fairly low in nitrogen. This tremendous increase in soil fertility is an extraordinary transformation of the biosphere.

At this stage of history, with the present population, it is hard to see how the world could do without the massive contribution from artificial fertilizers, which is perhaps the main reason for doubting whether organic farming, for all its unquestionable virtues, can take over entirely from 'conventional' (now meaning 'industrialized') farming. World average yields of rice and wheat have risen dramatically since the 1960s: the former from less than 3 tonnes per hectare to around 6; the latter from less than 1.5 tonnes per hectare to around 4. These rises almost exactly parallel the increase in application of artificial nitrogen – although as described later it would not have been possible to add such an amount of nitrogen unless the breeders had produced new varieties of rice and wheat that were able to make good use of it.

It is hard in all this to doubt the contribution of agricultural science and indeed of industrial science to human survival. The artificial fixation of nitrogen is a prime reason why humanity has been able to double its numbers in less than eighty years, and it will allow us to double again in the next fifty. Haber and Bosch both received Nobel Prizes. They were archetypally in the spirit of Alfred Nobel, who stated specifically in his will that he wanted to reward 'inventions' of the kind that had made a difference to humanity.

The Haber–Bosch process is also one of the most significant steps ever taken on the path to the industrialization of farming. Field fertility is the sine qua non of agriculture. Before Haber and Bosch, fertility had mostly been a matter for the farmers themselves, with a little help from the odd low-grade adventurer, digging guano or quarrying nitrogenous rock in some distant continent. The farmers decided whether to grow clover, and how to balance stock against crops. Now the single greatest input (apart from water and sunshine and air) came courtesy of the fertilizer factory. Food processing and distribution were already well on the road to industrialization by the start of the twentieth century, but the production itself was not. After Haber and Bosch, the entire food supply chain had been brought within the purlieus of industry; and in particular, the chemical industry was firmly on board. The late nineteenth and early twentieth centuries were the great age of industrial chemistry, just as ours is the age of electronics and biotech. In agriculture, though not in most of the rest of the world's economy, the glory days of industrial chemistry continue, confident and insouciant as ever. Some might see Haber and Bosch as the saviours of humankind while others might regard them as arch destroyers of the planet, but in either case it is hard to nominate anyone in modern history who has had a greater influence on the way we live – or indeed on our ability to live at all.

To summarize: the nutrient that crops require in greatest amounts (apart from carbon, oxygen and hydrogen which come from air and water) is soluble nitrogen, and of this there are four great sources. Most important by far at present, in sheer quantity, are the artificials, ultimately supplied by the Haber–Bosch process for fixing nitrogen. Mineral forms of nitrogen in the soil and nitrogen fixation by bacteria and particularly by *Rhizobium* in the root nodules of legumes are next in importance in farming (although the latter is still the greatest source of added nitrogen fertility in the world as a whole); and farmyard manure plays a relatively small but nonetheless noble role. Qualitatively speaking, each source has its advantages and disadvantages. All in all, we need a sensible balance between the four. It has often been suggested that artificials are too cheap for the world's good and are applied excessively, so that they run off and pollute waterways (or oxidize and pollute the atmosphere). So, beyond doubt, they have often done. But in Britain at least, studies at Rothamsted have shown that grassland ploughed for arable in the Second World War accounted for most of the nitrogen pollution that was manifest after the war. Grass is itself rich in nitrogen, mainly in the form of protein, and when it is ploughed in it dies, and that nitrogen is released into the soil. If the land is then left bare for any length of time, the nitrogen begins to run away into the groundwater. Farmyard manure carelessly applied can be at least as damaging. Artificial fertilizers come in convenient forms and with care can be applied in exactly the right amounts exactly when and where they are needed, and so in principle can be least polluting of all, since they are taken up by the crops almost as soon as they are applied. Yet it is clearly pernicious to use artificials to the exclusion of all else simply because they are cheap and because they seem to represent modernity, and idle to deny the advantages of natural bacterial nitrogen fixation and of farmyard manure. All in all, then, the world needs to aim for balance. Sometimes this is achieved, sometimes not.

Pests, Plagues, Poxes and General Murrains

Crops and livestock die for all kinds of bizarre reasons. Shepherds say that sheep spend their hours in the field thinking up new ways to meet their maker. I once stayed in Yorkshire at Easter and it snowed unseasonally. The sheep had lambs and were hungry and the snow in a nearby churchyard weighed down the yew trees so their branches came within reach of the sheep in the neighbouring field. Yew is poisonous. The sheep ate the greenery and died, half a dozen of them, splayed out and bloated in the morning sun. The vet, clever man, was not fazed for an instant. He stabbed open their sides with a penknife, plunged in his hand, and brought out handfuls of the dark bristling leaves. Who could have anticipated such a thing? Small animals were traditionally hauled off by foxes and larger ones by wolves and big cats. Animals have commonly died in storms and sometimes have even fallen over cliffs (as in a climactic scene in Thomas Hardy's *Far from the Madding Crowd*).

But the greatest and steadiest toll year by year on both farm animals and crops is taken by pests and pathogens (organisms of disease); conveniently lumped together as 'parasites'. The world's literature from Genesis onwards (probably before, but this is the first I know about) shows that pests and diseases have always been a problem. Traditional husbandry emphatically did not keep them at bay. The half-dozen pretty cows of traditional farms did have liver fluke, and mastitis, and TB, and brucellosis, and warble fly, and endured a hundred other major and minor disasters besides, some of which they passed on to the people whom they helped to feed. Those in the landscapes of the great seventeenth-century Dutch painter Albert Cuyp have the angular, skin-and-bone look that simply denotes heavy milking but also the arched back that suggests fluke. People did die and go mad from ergot poisoning.

Sometimes parasites exact a steady tithe, with some crops or animals taking the infestation or infection in their stride and others succumbing. Often parasites reduce the crop far more than it can be increased by improving fertility. Sorghum growers in the Sahel traditionally expect to lose up to 50 per cent of the field to mildew. Every now and again disease gets out of hand and then there is epidemic. Epidemics have always been with us; wild animals suffer them too. The pharaohs in Exodus 9 speak of plagues, for example of boils in cattle, probably anthrax. One of the most terrible devastations in recent history was the Potato Famine of Ireland, western Scotland, and parts of Europe in the 1840s caused by potato blight.

Another famous, recent wipe-out of a major crop was that of maize in America in 1970, caused by southern leaf blight. There were huge outbreaks of rinderpest in South African livestock in the early twentieth century, which went on to lay waste the hoofed animals of the wild. Some remember the foot-and-mouth outbreak in Britain in 1969. No one can forget the epidemic of 2001–2002 which was the biggest on record and became world news.

Some disease organisms smoulder year after year in particular animals, like tuberculosis (TB) and brucellosis in cattle. Others are highly contagious, run their course, kill a few animals while others recover and become immune, and then sweep on. Foot-and-mouth is of this kind (like human measles). But if the susceptible population is large enough, then diseases like foot-and-mouth or measles will persist year after year, hopping from population to population, infecting each new generation of susceptible animals or children as they come of age. Thus foot-and-mouth has long been endemic in the cattle of India. Some parasites remain confined to one particular species or group of species; others are more catholic. The unpasteurized milk of cattle was a significant, traditional source of TB in people. Brucellosis causes abortion in cattle and can infect the people who look after them, including vets, hanging around in the body and not least in the joints, where it causes chronic arthritis. The remarkably robust and versatile bacteria of anthrax have long been recruited into biological warfare. Country people in pig areas used to be infected with the pig roundworm, *Trichinella*, as a matter of course. Human beings may also suffer from diseases and infestations of plants. Mycotoxins, caused by fungi (such as ergot) on crops, are a huge source of ailment and death worldwide.

Biologically, the parasites that cause these diseases come from all branches of the evolutionary tree. They include vertebrates such as mice, pigeons, crows, and the African quelea which swarms like locusts in flocks that seem to whirl across the plain like tumbleweed but at breakneck speed. There are scores of insects – beetles (which include weevils), scale insects, aphids, warble fly. Mites are a huge problem. Worms of many kinds are implicated, including the round nematodes, which infect animals and plants alike, and the multifarious flatworms, including tapeworms and flukes. Molluscs, including many a snail and slug, can cause their own forms of havoc, some because they too may carry additional parasites. Fungi of many kinds cause a host of diseases: mildews, rusts, smuts. Some flowering plants are severe pests, like *Striga*, a relative of the foxglove, which parasitizes other plants and lays waste many a hectare of maize and tropical pulses. Many others, of course, are serious weeds.

But the greatest and collectively most versatile of all parasites both in crops and in livestock, are bacteria and viruses. Anthrax, TB, brucellosis, and the 'scours' (diarrhoea) that claims the life of many a calf are all caused by bacteria. All the major food crops have their catalogues of viruses (commonly carried by aphids, which is why the threat of them can be so serious) and a host of animal diseases, including foot-and-mouth, swine fever and fowl pest, are viral too.

Overall, the toll of pests and diseases in the world's crops and livestock is beyond measure. Economically, they claim a significant slice of the world's agricultural output, with the destruction continuing well beyond harvest. If one guessed that they account for a third of farming's costs, this surely would be reasonable. Many traditional societies routinely expected to lose half their grain in storage, while mycotoxins and bacteria of many kinds, often acquired post-harvest or post-slaughter, ensure that food poisoning is a constant threat. The cost is compounded, of course, by the expense of control.

Biologically the impact of disease is immense, not only on the crops and livestock but also often on wildlife. The flow of infection runs both ways: rinderpest spread from cattle to hoofed animals; badgers are slaughtered in Britain because they apparently transmit TB to cattle (although it would be more humane if more expensive simply to lock up the cattle at night when the badgers are active). The pests and diseases of livestock and crops form a significant subtext of all human history and have often caused entire societies to change direction (as with the potato famine of Ireland). Unsurprisingly, and indeed as a matter of necessity, entire professions, university departments, dedicated research stations and industries are committed to the control of farming's pests and diseases. Here is another esoteric and mundane (many might feel) pursuit that is among the most crucial of all human endeavours.

So how can we control all these pests and diseases? In practice there are three broad strategies. The first is essentially ecological: to ensure as far as possible that the parasites are kept away from their intended hosts. (The post-harvest equivalent is to keep the food supply chain as simple and short as possible, in time and distance, leaving the organisms of spoilage as little time and opportunity as possible to get stuck in.) The second is to maximize the resistance of the hosts by good feeding and general health, by building specific resistance by judicious exposure to induce immunity, by vaccination, or by breeding specifically resistant varieties. The third is specifically to kill the parasites.

Simple logic tells us that the best approach in principle, by far, must be the first. To take a couple of human examples: if bubonic plague were to come again to England as it did so spectacularly in the fourteenth and again in the seventeenth century, we could presumably control it with modern antibiotics and to a large extent by killing the (black) rats that carried the bacteria that cause it. But the status quo is far preferable: not to have the disease organisms in the country at all. Plague still exists over much of the world, but the danger of catching it on the streets of Britain is effectively zero. This is the ideal: not to wallop infection with heroic measures, but to avoid contact in the first place.

Scientific Pest Control

Farmers here and there have attacked insect pests with ad hoc agents since ancient times. Homer refers to sulphur as an insect repellent, while the Chinese of the fourth century AD attacked pests of rice with silkworm droppings. In the eighteenth century, when formal chemistry was beginning to take shape, and alchemy was fading into the mists, and the bases were being laid for agricultural science in general, Matthew Tillet in France treated bunt in wheat, a fungal disease, with salts of copper; and in 1807 Benedict Prevost of Montauban controlled bunt in wheat by steeping diseased seed in extremely dilute copper sulphate, which aborted the spores. This is the first formal example of seed dressing, which is now common practice. (Patrice Leconte's *Ridicule* (1996) beautifully reflects the rise of French and particularly agricultural science and engineering at the end of the eighteenth century: contrasting wonderfully with the aimless decadence of the French court.)

But again, chemical control of pests and diseases on a scientific basis – formal investigation of possibilities on the one hand, testing specific hypotheses on the other – had to wait for the mid nineteenth century. The science of chemistry was well in train: chemists had by then inferred that there is only a finite number of elements (the 100 or so of the periodic table) and were steadily working through them to see how they all behave. Louis Pasteur in particular was initiating the modern science of microbiology. He showed that organisms too small to see with naked eye – 'microbes' – including bacteria and yeasts (single-celled fungi) have an immense influence on the world's affairs, both in causing disease and in more benign spheres, including the fermentation of wine. The formal science of entomology was well established by the mid nineteenth century (the natural history of insects and spiders was a great pursuit from the seventeenth century onwards, and Darwin was among the many beetle fanatics), and the science of physiology, too – how the bodies of animals and plants actually work – had also come on apace since the seventeenth century. So by the mid nineteenth century agricultural science proper was ready to take off and it began with the leading formal science of the time, which was chemistry. It is more than a little unfortunate that it has to a large extent stayed with chemistry: that the high tech of agriculture is so firmly rooted in industrial chemistry. To a large extent the modern bias towards chemistry in agriculture, and the modern agriculturalists' inability to think more broadly, is a nineteenth-century legacy. (I do not wish to denigrate, but merely to observe.)

But again it's worth noting that agricultural science is astonishingly new: hardly 150 years, out of a total history of at least 10,000 years, and (as I suggest in Chapter 2) probably another 30,000 years of accumulated know-how before that. If we also reflect that ideas in science do not slot into place routinely and instantly, hot from the press, but take decades or centuries to unfold, we get some feel, not for the presumptuousness of science itself (which I continue to believe is wondrous), but at least for the naivety of some scientists and in particular the industrialists and politicians who seek to deploy their findings. There really is an underlying belief that the latest high technology, whatever it is, not only can but should be encouraged to transform tradition – all the crafts and traditional rules of thumb that have taken millennia (at least) to unfold – as quickly as possible. Thus, from the nineteenth century onwards, the chemical industry was encouraged to create a new kind of agriculture, and any system that did not partake of its inputs was officially deemed to be recalcitrant to the point of perversity, not to say wickedness; and now in this new age of biotech we are being told that without genetic engineering in particular, we will all starve in our beds, and that those who protest even *sotto voce* are the enemies of humankind. Yet the new ways that are being brought to bear so energetically are frightening in their newness. Liebig's mistake is repeated in principle time and again: ideas that work in theory, and seem to pan out in the laboratory, are assumed to tell us all we need to know about nature and are applied as a matter of supreme urgency, as if the world would otherwise collapse. It is, when you think about it, several kinds of madness. The naughty thumb of science has indeed prodded the earth, as e. e. cummings observed; but science has been horribly prodded and misshapen in its turn by the naughty thumbs of politics and commerce.

Anyway: the chemical control of pests and diseases first became an orderly and formal pursuit in the mid nineteenth century. But again we see how easy it is to be led astray: that wrong ideas often seem the most plausible, and are very difficult to dislodge. So although it was obvious that insects destroy crops (you can see them doing it) the role of fungi and particularly of bacteria was far from clear, as the tragic Potato Famine of the 1840s demonstrated.

As is so commonly the case, Ireland's famine was rooted in political injustice. The English dominated all of Ireland, typically as absentee landlords, content to rule from afar. They also disliked and feared Catholics in general, with whom they had been in conflict for at least three centuries. Irish Catholics typically survived on small farms – half of which, in 1845, were of five acres or less. Potatoes were the main crop not least because in that wet climate they provided about three times as much energy per acre as cereals. An acre of potatoes could support an entire family. But for about 3 million people – nearly half the entire population – potatoes were not only the main crop, but often were virtually the *only* source of food. Potatoes are among the very few crops that can support human life virtually unaided – although to get enough calories, a labourer had to eat his way through an astonishing 14 pounds per day (around 5 kilos). Clearly, too, such a diet was highly precarious. Furthermore, most of Ireland's potatoes were of just two varieties – the 'lumper' and the 'cup'.

In the summer of 1845 the potatoes failed. Within a few days of harvesting they turned black and slimy. Now the cause is known: a fungus-like organism called *Phytophthora infestans* – colloquially known as 'potato blight'. But although investigators saw the blight 'fungus' clearly enough they did not think it was the prime cause: merely a latecomer, feeding on the decaying flesh. The disease itself, they vaguely opined, was caused by 'mortiferous vapours'.

Over the following ten years, 750,000 Irish people died from the famine while another 2 million emigrated, not least to Canada and the United States. Many more were condemned to the workhouses of Ireland and Britain, where many of them died. Overall, the famine reduced the population by a quarter. The ruling English government debated the matter, of course, and some were genuinely concerned, but then, as now, the powers that be put their faith in the free market. The oats that grew abundantly in Ireland were for export. Trade could not be perverted by giving them to the local people. The Corn Laws, which for some decades had excluded cheap imports of grain from abroad, were repealed in 1846. But this wasn't enough. The price of grain came down but the starving Irish had virtually no money at all to pay for it.

In 1845, while the Potato Famine was still taking its toll in Ireland and elsewhere, the Reverend M. J. Berkeley in England suggested that the blight was caused by the *Phytophthora* 'fungus' that invaded the stricken leaves. In this, as we now know, he was absolutely right. If *Phytophthora* is controlled, then blight does not happen. Yet Berkeley was ignored. *Phytophthora* was taken to be a fungus; and fungi were supposed to be inveterate saprobes – not parasites on living flesh, but creatures that feed on dead (or dying) organic material. This is what mushrooms do, after all: the most famous of fungi. So it was supposed that *Phytophthora* did not *cause* the blight. The disease itself must have been caused by something else. The *Phytophthora* was merely feeding on the already moribund tissue. Perfectly logical; in line with prevailing theory; but wrong. Only after Pasteur had shown that mysterious microbes (including yeasts, which really are fungi) can cause disease did the idea become plausible that the organism associated with blight was indeed the cause of it. Scientists are not content with observation. They like plausibility as well.

Deliberate, chemical control of a disease of a kind that can truly be called 'scientific' finally came on board in the 1880s. The grapes of France were suffering horribly from downy mildew, which had arrived from America in 1878. Alexis Millardet (1838–1902) was commissioned to investigate and in October 1882, in the vineyards of St Julien in the Médoc, he noticed that the vine leaves by the side of the path were free of the mildew although it was rife elsewhere. The leaves, he perceived, had been treated with some chemical, which left a whitish-blue deposit; and this turned out to be verdigris, which the growers applied to discourage passers-by from stealing the ripe fruit. Verdigris is basic copper acetate, which the growers prepared by covering sheets of copper with fermenting grape husks. Millardet also found that fungal spores would not germinate in water from his garden well, which was served by an old copper pump. So he set out to create a specific fungicide based on copper, and by May 1885 had produced the famous Bordeaux mixture: copper sulphate mixed with lime. It proved its worth in the summer of that same season. Bordeaux mixture wasn't quite as novel as Millardet liked to think. Copper had already been employed against bunt, as we have seen; and Paris Green was already on board – copper aceto arsenite – and was widely deployed in the 1860s in the US not to control fungi but the Colorado beetle, notorious scourge of potatoes. It is not obvious, either, why the Médoc wine growers did not notice themselves that the copper salts they used to deter passing scrumpers also killed the fungus that was destroying their entire crop.

But again, credit where it is due. Millardet is a key player in the history of scientifically based chemical control. As the nineteenth century unfolded the chemical armoury grew. Copper and arsenic, the key components of Paris Green, were prominent. Organic compounds came on line too, including nicotine, a significant insecticide (which plants of the genus *Nicotiana*, or tobacco, produce to protect themselves from insects). Some of these traditional agents were highly toxic to human beings, including of course that favourite standby of classic detective fiction, arsenic (simple to slip into the afternoon tea). Gardeners in the increasingly exotic glasshouses of the late nineteenth century were short-lived, like the miners of coal and tin. It is ironic that organic farmers, rightly seen as the guardians of purity, still may make use of these traditional agents, while eschewing for example DDT, which has become so notorious and yet, in principle at least, is far safer.

Weedkillers

Before the 1880s weeds were controlled by hand: by 'roguing' (walking through the arable fields and pulling out the wild oats or whatever) and by hoeing. Then farmers in France began to use copper sulphate (developed for use on mildew) on some weeds. By the 1920s chemists were supplying vicious and corrosive inorganic agents to kill everything, including chlorates and perchlorates; and by the 1930s had provided organic agents, based on benzene.

The road that led to the herbicides based on plant hormones began as so much of biology has with Charles Darwin in 1880 (when he was in his seventies). The great biological theorist of modern times, John Maynard Smith, commented that if Darwin had not, in 1859, transformed all human thought with his idea of evolution by means of natural selection, he would still be remembered as the finest field naturalist of all time. But on top of all that he was also a fine experimentalist, and among his concerns in his later years were the tropisms of plants: the mechanisms by which they grow towards particular stimuli. To find out (at least in principle) how seedlings turn towards the light, he and his son Frank fitted the shoots of grasses and cereals with hoods, or painted them, and found that they turned their tips no longer. From such studies he concluded that 'some influence is transmitted from the upper to the lower part, causing the latter to bend'.

Then in 1926 a young chemist in Utrecht, Frits W. Went, isolated this 'influence' – which in modern parlance is described as a plant hormone; and some time later (1934) chemists at Utrecht showed that the hormone was indolyl-3-acetic acid or IAA. It soon became clear that IAA has a wide range of functions. It seemed likely, too, that if IAA was applied to plants in high doses it would disrupt growth, and thus could serve as a herbicide. Furthermore, hormone-based herbicides could be selective. They tended to have more effect on broad-leafed 'dicotyledonous' plants than on narrow-leafed 'monocots'; and since dicots include most of the world's weeds while cereals are monocots, they might in theory eliminate the former without harming the latter. So it proved, in principle: IAA was first used experimentally in England to suppress charlock in oats. After the Second World War chemists in the US and Britain synthesized a shortlist of hormone-like chemicals that acted as selective disrupters of growth, of which the most important were 2,4-D (2,4-dichlorophenoxyacetic acid) and MCPA (with an even longer and even more forgettable chemical name).

The principle by which these hormone weedkillers were created is common in all pharmacology: some of the most effective drugs operate by imitating and thus enhancing or subverting the action of hormones or other such chemical messengers. So to base weedkillers on the plant's own hormones is good science; much more sophisticated than simply knocking them sideways with a general toxin like chlorate. Hormone weedkillers are also extremely effective. So again, they have had an enormous influence, some of which is unequivocally good and some of which has unfortunate side effects. For instance, hormone weedkillers make it possible to simplify rotations, which is not bad in itself; but this in turn further encourages monoculture and reduces the need for labour, which again pushes agriculture down the road of industrialization and into the hands of corporates.

On the other hand, effective hormone herbicides also reduce the need to plough, often and deeply, simply to control weeds: and ploughing, though popularly perceived to be benign (the plough has often been seen as the symbol of agriculture) seriously disrupts the flora and fauna of the soil. With effective herbicides, 'minimum tillage' becomes possible: wipe the surface clean, chemically, and then sow directly into the unploughed soil. Or indeed if the herbicides are selective enough, the young crop and the weeds can be allowed to grow up together and the latter can be zapped only when they threaten to overwhelm. The initial growth of the weeds encourages insects and hence birds, and in general must be good for wildlife. Of course, if the wild creatures are zapped in their prime along with the weeds, then the gain is equivocal. But with good and sensitive timing – if the farmer perceives chemical weed control as part of a broader wildlife policy – then we can begin to have the best of all worlds: maximum productivity with minimum disruption of wildlife.

In short, modern hormone-like herbicides offer a prime example of industrial chemistry potentially on the side of enlightenment; or at least they can be benign if conscientiously deployed. Again, it's a pity that the abuses of industrial chemistry have caused so many to reject it altogether, so that indeed it has become a symbol of all that is evil in modern farming. Those who seek alternatives to industrial ways tend as a matter of principle to prefer mechanical techniques to chemical. Yet wildlife may well prefer the latter. Again we see that the polarization of opinion, between the zealots of the 'scientific'/industrial approach and its detractors, is so regrettable. But the fault lies mainly with the zealots, who in general have geared the new and often fabulous techniques to the narrow ends of commerce, and not to the broader goals of aesthetics and good biology.

The Age of Chemical Pesticides

But the greatest and sharpest cause of conflict prompted by industrial chemistry has been in the area of pesticides. The influx of artificial nitrogen fertilizer can be seen as the first age of agricultural industrial chemistry; and the development of organochlorine pesticides (and later of organophosphates) can be seen as the second age.

By the 1930s farmers were already using pyrethrum (a natural product prepared from a species of African daisy) and rotenone to kill pests, and there was a range of mineral agents as we have seen. But all these agents had obvious disadvantages. Some, like nicotine and the arsenicals, were obviously toxic to people and other animals. Others, like pyrethrum, were not readily available. Some (like pyrethrum again) were not stable in sunlight or air, and lost their effectiveness almost as soon as they were applied. So in 1935 the Swiss company J. R. Geigy commissioned the Swiss chemist Paul Muller to come up with something that would be easy to synthesize but more effective than the existing pesticides. He succeeded remarkably quickly. By September 1939 (the very month that Britain and Germany went to war) he had synthesized DDT. He tested it on blowflies, and it worked remarkably well: it remained on the walls of the cages after spraying and if the flies so much as landed on it they fell stone dead to the floor. It was soon shown that DDT has very low toxicity in humans (in the short term, apparently none) and before long it was used to protect stored grain and timber, the pests of wool, and vectors of human disease including the lice that carry typhus and the mosquitoes that carry malaria. Switzerland was of course neutral in the Second World War, and by 1942 it had apprised both sides of the virtues of DDT as an insecticide. In Britain, production of DDT became high priority and at the end of the war the British used it to control typhus in Italy and along the Rhine, and malaria in South-East Asia. Newsreels from the mid 1940s show DDT applied like snow to the people of Italy, inside and outside their clothes, and the delight with which they received it is unfeigned and obvious. Typhus is caused by a strangely simplified kind of bacterium known as a *Rickettsia*, and is carried by various arthropods including lice. It is a horrible disease, which begins by destroying the small blood vessels so that the blood leaks out, and is often fatal by the second week. DDT really did kill the lice that were carrying it. Beyond doubt, in that campaign alone it saved many thousands of lives. DDT soon proved its worth in agriculture, too, for which it was originally intended: the Swiss used it to control Colorado beetles during the war, and soon afterwards it was widely applied to crops of all kinds, including vegetables and fruit. Absolutely rightly, Paul Muller was awarded the Nobel Prize in 1948: not for Agriculture (which Nobel did not single out for reward) but for Medicine.

Enthusiasm ran high, as indeed it should have. Between 1945 when the war in Europe ended and 1953, twenty-five more synthetic organic pesticides came on board, including chlordane, toxaphane, aldrin, dieldrin, endrin, heptachlor and parathion. Nineteen fifty-three was a symbolic year in Britain. Queen Elizabeth II was crowned and schoolchildren were encouraged to think of themselves as 'the new Elizabethans' (although I was among them and I don't recall that we ever did). The end of world famine and of horrible insect-borne epidemics, the rise of new and invigorating science on all fronts, and the appearance of a new generation that hardly remembered the war at all, were all seen as harbingers of a new and better age. There was a sense of rebirth.

But of course, life is not so simple. Nature is complicated, and cannot be turned on and off as complaisantly as the pesticide enthusiasts supposed. Of course it is easy to say this with the wisdom of hindsight, but in fact some commentators expressed reservations at the time. As early as 1945 Britain's Kenneth Mellanby urged field research on DDT 'on all manner of apparently unimportant insects and other forms of life, to ensure that there was no serious effect on the "balance of nature" with subsequent disastrous effects'. But in practice these quick, apparently easy and certainly cheap agents of chemical control began an industrial boom and they were sprayed into every conceivable corner in every halfway relevant context as if the new millennium had indeed arrived. In agriculture itself, as Lloyd Evans comments, crops were sprayed heavily and often 'by the calendar rather than by the need'. The cost was small relative to the value of the crops, so why not? To be fair, the farmers who sprayed in this way were not necessarily stupid, or lazy. They saw their new routines merely as prophylaxis. In medicine, after all, it is widely accepted that 'prevention is better than cure', and it seemed merely sensible to cover crops in insecticide before the pests arrived so that they never became established at all. Nowadays, with a few decades' experience behind them, farmers generally believe that it is better to monitor the crops carefully, and not to apply extraneous chemicals unless a particular pest looks as if it is liable to get out of hand. The crops can tolerate minor infestations, or the natural predators will keep them in check. But that lesson has only lately been learned.

Paul Muller, when he accepted his Nobel Prize, listed the features that made DDT so attractive. It was, he pointed out, very toxic to insects and very rapid in its action and had a very wide spectrum of activity – that is, it killed many different species. It was also persistent: DDT did not simply fade away in the sun and air, and once an area was sprayed it stayed sprayed. Hence the 'knock down' of blowflies in cages that had been sprayed earlier. Yet it had little or no apparent effects on mammals or on plants. All that, and still it was cheap. But of course, nature is not entirely predictable. There were some unfortunate side effects that were not anticipated: on beneficial insects, of course, as Kenneth Mellanby had warned, and if not on mammals then on birds. Nature fought back rapidly, too, as more and more target insects developed resistance. And, ironically, some of the qualities that Muller identified as virtues soon proved to have disastrous consequences; even DDT's inexpensiveness.

The 1960s can be seen as the golden age of organochlorine pesticides: production of DDT alone in the US peaked in 1963 at 80,000 tonnes. Yet the drawbacks were already becoming apparent by the early '60s, and were excellently summarized in 1962 by Rachel Carson in *Silent Spring* (Houghton Miflin, New York). She was herself a professional biologist (primarily marine) and a fine writer, and she brought the unfolding disaster to the world's attention as few would have had the ability to do. *Silent Spring* was one of the most influential science books of the twentieth century.

Thus it was becoming clear by the early 1960s that beneficial insects were indeed being lost, just as Kenneth Mellanby warned could happen. These benefactors included pollinating bees and butterflies but also the predators which normally suppressed the pests; and when the pests themselves became resistant to the pesticides the loss of the predators that otherwise would keep them in check was a disaster.

DDT (and the other organochlorines) are soluble in body fat, too, which means that once ingested (for example on sprayed food) they are not easily eliminated. In fact they remain in the body, and so become more and more concentrated the further you move up the food chain. Thus, one of Rachel Carson's most telling examples was from Clear Lake in California, where DDT was used in abundance to control the gnats that annoyed the anglers. In the water itself, the concentration of DDT seemed fairly low: 0.02 parts per million. But in the plankton the concentration rose to 5 parts per million, and in the fish that ate plankton and plants it was between 40 and 300 parts per million, while the carnivorous birds and fish that ate other fish accumulated an extraordinary 2,500 parts per million. At the top of the food chain, the western grebes simply died off. Similarly, accumulated organochlorines in the bodies of birds of prey such as peregrines and eagles caused them to lay eggs with thin shells that broke during incubation. If things go on as they are, said Carson, the springtime will indeed be silent, as birds of all kinds are killed, robbed of their food supply, or are unable to reproduce.

Targeted insects of all kinds developed resistance remarkably quickly. Among the first were the major pests of cotton, boll weevil and boll worm; but soon there were more than 200 other resistant strains, including house flies, mosquitoes and lice (among the major bearers of typhus). Resistance (as subsequent studies have shown) does not, for the most part, involve new genes. It's just that among all wild populations of all kinds of creatures there is some genetic variation. Quite by chance, in many populations of insects, there are often some individuals that carry mutant genes that enable them to overcome particular toxins, including DDT and similar compounds. Typically (though not invariably) these 'resistance genes' produce enzymes that break down the toxin in question. When the toxin is not present, these mutant genes confer no particular advantage and the individuals that carry them are likely to remain rare in the population. Indeed they may be disadvantaged, since it takes energy to produce enzymes that are not usually needed. But when the relevant toxin is present, the individuals with the metabolic ability to disarm it are at a premium. All their fellows, who would otherwise be competing for food, die. The resistant types suddenly find themselves with a clear field. Reproductively speaking, most insects – and certainly all the kinds that become serious pests – are r-strategists: that is, they have an enormous number of offspring, although normally most of them will die. But when a resistant insect finds itself in an environment where most of its rivals have been killed off, its offspring – also resistant, since they inherit its resistance gene – survive en masse, and breed en masse in their turn. After all, they no longer have any competition. Thus a heavy attack with DDT (or dieldrin or whatever) can indeed lay waste a population of insect pests effectively within hours. But a few (the resistant ones) may pull through. If they do, then within a season or two the same insects can be back in force, but this time round the pesticide does not work.

Two of the qualities in particular that Paul Muller singled out for praise seem tailor-made to encourage resistance. The first is persistence. When the wall of a house is sprayed to eliminate mosquitoes (which was a common strategy in the early days), the DDT stays there. Thousands and thousands of mosquitoes are thus exposed to it in the ensuing weeks. This constant low-level exposure creates the very conditions that will kill off non-resistant individuals but allow the resistant types to flourish in their place (since the competition has now been wiped out). The second fatal flaw is the inexpensiveness. Because DDT was so cheap, it could be and was sprayed all over the place for all kinds of reasons, in tens of thousands of tonnes. Accumulating toxicity and resistance came about in spades.

As a biologist, Rachel Carson wrote: 'Single-crop farming does not take advantage of the principles by which nature works . . . Nature has introduced great variety into the landscape, but man has displayed a passion for simplifying it. Thus he undoes the built-in checks and balances by which nature holds the species within bounds.' This kind of idea is now universally acknowledged (at least among biologists), but it wasn't at the time that she stated it. Again I remember propaganda films, produced for example by agencies of the United Nations and shown on British television in earnest black and white, that spoke of the 'insect threat', and effectively warned that if such-and-such a pest were not jumped on instantly then it would quickly multiply to take over the whole world.

The question as to why no insect population or any other creature had ever taken over the world in the past (give or take the odd plague of locusts) was glossed over. But the obvious answer was, as Rachel Carson said, 'checks and balances'; or as Kenneth Mellanby put the matter, 'the balance of nature'.

But Carson was also a realist, born in the early twentieth century (1907) in a newly industrial town surrounded by farming and forestry. As she made clear in *Silent Spring*, 'It is not my contention that chemical insecticides must never be used.' She simply warned against their abuse. For her pains, she was in her turn abused by the new industrialists whose fortunes she was threatening, with insults of the kind you can imagine: that she wasn't a 'proper' scientist (though she had a good degree and had held university posts); that she was, horror of horrors, a woman, and therefore 'hysterical'. She was also ill, as it happens, and died in 1964 from cancer. But of course she was right; both to warn against the excesses, and to suggest that a total ban of all chemical pesticides was probably not the most sensible strategy.

But many citizen groups brought legal actions after *Silent Spring*, and William Ruckelhaus, administrator of the Environmental Protection Agency, soon banned all use of DDT in the United States except for essential public-health purposes. Now DDT and other organochlorine pesticides are banned for most purposes in most developed countries. The rise of DDT was meteoric (new science rarely becomes applied technology so quickly) and its fall has been equally dramatic. What, though, should we conclude? First, of course, that nature is complicated, and we can't expect to throw agents as powerful as the organochlorines into the midst of it and hope by such crude and simple measures to bring it into line: to solve our own perceived problems while leaving the rest intact. That is ludicrously bad biology. On the other hand, the potential advantages of DDT are still true, just as Paul Muller listed them; and if DDT and its fellows were used delicately and precisely, exactly when and where they are needed, then they could still be beneficial even in some agricultural contexts. Resistance to pesticides can soon decline when the pesticide itself is withdrawn, since the resistant types are no longer at an advantage and the non-resistant types (which in other respects may be stronger than the resistant types) get going again.

But humanity, at this stage of our history, seems unable to handle difficult material sensibly. Industry competes to develop new technologies as rapidly as can be contrived, and is encouraged by governments and international agencies to spread it around the world as quickly as possible: and when the whole precipitate exercise goes wrong, as it surely must, then total ban is imposed for all kinds of reasons that are partly political (the voters demand action) and partly logistical (for partial, restricted use in farming would be hard to police). Farmers are far from innocent in all this, and commonly make use of banned materials, including pesticides and animal feeds, to solve short-term problems. Most are honest, of course, but it takes only a few mavericks to queer the general pitch. Maybe in the future we will get better at handling technologies. But at the moment, with present-day politics and commerce (and average standards of honesty), we don't seem able to get it right.

Biological Controls

There is one other way to zap specific pests without recourse to chemicals: attack them with predators and parasites. This is 'biological control', an expression first coined in 1919. The helpful predators include vertebrates of all kinds – mammals, birds, reptiles, amphibians and fish (used for example to feed on aquatic mosquito larvae); insects such as ladybird beetles and parasitic wasps; arachnids including spiders and mites; and snails and worms of various kinds. Microbial agents of control include fungi, bacteria, and sometimes viruses (like the myxoma virus of myxomatosis, first developed in Australia to control rabbits).

Again, in principle, conscious biological control of specific pests is an ancient concept. The Chinese seem to have been first in the field, encouraging ants in citrus groves to control caterpillars and boring beetles; with their typically down-to-earth yet poetic ingenuity they even built runways of bamboo to help the ants from tree to tree. Again, science has come on board only lately but again, it has made enormous inroads very quickly. Carl Linnaeus was on the case in the eighteenth century (quoted by Lisbet Koerner, *Linnaeus: Nature and Nation*, Harvard University Press, 1999, p. 83): 'Until now no one has thought about exterminating insects with insects. Most every insect has its lion which persecutes and exterminates it; these predatory insects ought to be tamed, and taken care of, so they can purge plants.' In this way, farmers could 'easily drive away snails from the fields, butterflies from the meadows, ants from the garden'. Charles Darwin's hugely fat and hugely able grandfather, Erasmus, suggested in 1800 in his *Phytologia* that aphids (commonly called greenfly, although they may also be black, white, pink, or tan) might be controlled in greenhouses with predatory fly larvae.

Agricultural pests become pestilential for two kinds of reason. Firstly, farmers obligingly provide them with provender and habitats which, in the wild, may be rare. Thus aphids are not particularly common in the wild; but juicy, uninterrupted crops of cereals or potatoes provide them with effectively endless paradise.

But secondly, and historically of enormous importance, pests of all kinds are also carried in all kinds of ways from country to country by human beings – in ships, in sacks of grain, on clothing, or sometimes like the Australian possums in New Zealand or the European foxes in Australia, brought in effectively for fun. Worldwide, creatures introduced to new countries, known as 'exotics', are widely recognized as the second greatest cause of wildlife extinctions, second only to habitat loss. Many exotics threaten agriculture or have done so in the past – like the prickly pear cactus in Australia, introduced from America, or bracken in New Zealand, brought in from Europe. Many of agriculture's worst pests, too, are exotics. A prime reason for this is that when a new species is introduced, it may leave its own predators and parasites at home: the creatures that constrained it in its own country. In general, after all, predators and parasites need time to evolve a relationship with their hosts, and if a new host suddenly turns up in a new environment there may be nothing around to attack it. In fact, organisms that seem extremely modest not to say feeble when on their own native territory, constrained as they are by pests and predators that have evolved to feed upon them, may become the most vigorous of invaders when they escape their enemies. This fact alone shows the immense importance of natural pests and predators; and also illustrates how immensely dangerous it is to play with the balance of nature, either by taking organisms away from their natural enemies or by eliminating the enemies themselves, for example by profligate use of pesticides.

In practice, the first deliberate application of biological control in modern farming was in California in 1889. The pest was the cottony-cushion scale insect that was devastating citrus trees. It had first appeared in California in 1868, probably imported from Australia. Fumigation and the pesticides of the day didn't work. So the strategy that has since been adopted time and again was essayed: return to the pest's own country of origin, find its natural enemies, and bring them back. Albert Koebele left for Australia on behalf of the USDA in September 1888 (although

Congress was reluctant to spend the money on what they apparently saw as a frivolous expedition) and by October he had singled out the ladybird *Vedalia* as the cottony-cushion scale insect's arch enemy. He shipped the beetles off to California where they were released in two commercial orchards. By 1889 the ladybirds were multiplying rapidly and by 1890 they had effectively done the job. The £5,000 that Koebele had spent on travel and postage was returned a thousandfold. Cottony-cushion scale stayed under control until 1946 when DDT killed off the ladybirds while the scale insect itself became resistant. When DDT was withdrawn, *Vedalia* came back, and the pest again met its nemesis.

Biological control can be just as effective on weeds, too; a weed being loosely defined as any plant growing where people do not want it to. It has been extremely effective in controlling the prickly pear cactus of the genus *Opuntia* in Australia. Prickly pears found their way to Australia in the early 1800s – introduced first as attractive house plants, as indeed they are, but easily grown in Australian gardens, whence they took off for the wide open spaces. By 1925 they had infested 25 million hectares, obviously at the expense of the native flora and fauna but also (which is the thing that governments tend to care about) at the expense of potential farming. The call went out for predators, and about fifty different candidates were introduced into quarantine from the United States, Mexico and Argentina. Argentina came up with the goods in 1925: the cactus moth, *Cactoblastis cactorum*. It was released in 1926 and had done its job by 1930.

Since then there have been scores of successes in scores of crops using predators and parasites of all kinds. Among the most successful and widely applied are the tiny, parasitic ichneumon wasps which lay their eggs in the bodies of aphids, where their larvae hatch and eat the insects from the inside. Parasitic fungi and bacteria are also widely applied. When biological control works it can work very well indeed: thorough, rapid, highly specific, and leaving no residue. Ideally, when the pestilential host has been killed off, the controlling agent dies off too.

But it is too easy to be carried away by the outstanding successes. Biological control has many theoretical difficulties, and has led to some monumental disasters, sometimes at least as bad as the problems they were supposed to contain. There are obvious logistical difficulties. For example, ichneumon wasps (say) cannot multiply and so destroy aphids until there are aphids to feed upon. In other words, the pests must already be multiplying (with some danger of damage) before the biological control measures come into play. Logistically speaking this seems at first sight to be less satisfactory than prophylactic spraying with chemicals, which prevents the pests becoming established in the first place (until they become resistant). Such problems can be overcome: for instance, with good timing, the initial pest attack should be less than devastating (the infected plant soon gets over it), so if the predators are introduced early enough there is no crop loss. Biological control in general requires good, alert husbandry: there is little or no scope here for blind routines or farming-by-numbers.

A much more serious issue is that the agents introduced to control pests have sometimes gone on to lay waste other creatures that they were not intended to attack. In practice, the logistics of biological control to some extent militate against this. For example, biological control is particularly applicable in greenhouses (it is the norm in organic fruit and vegetable growing), and greenhouses provide a different (generally warmer and often damper) environment than the wild fields outside, which of course is the point of them. So an agent of biological control that does a very good job inside the greenhouse may languish when it gets outside, which means it should never be a menace to the wild creatures all around.

Yet there have been some spectacular disasters in the cause of biological control. The cane toad is one, centred yet again on Australia which, since the Europeans 'discovered' it in the eighteenth century, has become the world's biological dumping ground. Cane toads are huge (about the size of a dinner plate), toxic and ridiculously tough. They were introduced to the sugar cane fields of Queensland in the 1920s from Hawaii to control rats; which itself shows how very big and tough they are. They have remained, bred and spread to lay waste the native marsupials and rodents.

Looking to the Future

Much traditional good husbandry has evolved to reduce the contact between vulnerable crops and livestock, and their parasites. Farmers qua craftsmen and women knew that some strategies and techniques produced healthy crops, while others led to disease. Nowadays, deeper knowledge, if we care to apply it, can put the traditional crafts in a formal context and suggest even more approaches. Traditionally, for example, small-scale producers of eggs moved their chicken flocks around in portable runs and cages (as in the 'Moran' system) so that parasites could not build up in the soil in any one place. Cattle farmers are happy to graze mature cows after calves but not the other way around. Cows can generally take in their stride the parasites that the latter leave behind (including coliform bacteria and liver fluke); but calves, being much smaller and with immature immune and digestive systems, may succumb rapidly to the same parasites left behind in much greater numbers by the cows. Livestock traditionally, for the most part, did not travel far. At least, there were some spectacular migrations, for example of cattle from state to state in North America and from the Scottish hills to the pastures of Lincolnshire; but the movement was slow (the cattle walked). Most traffic was between farms of the same region, and the animals were slaughtered and consumed locally. Rotation of crops – different species in different seasons – helps to ensure that parasites, such as the spores of fungi, cannot so easily build up in the soil and lie in wait for the next crop. Traditionally, fields were small, and farms were mixed: no one crop was grown in huge quantities in any one place, and dairy farmers typically had just half a dozen cows (rising to thirty or so in England by the mid twentieth century).

The tricks of traditional husbandry – rotations of crops and grazing, small fields, small herds, general variety – all helped traditionally to ensure that no one parasite could ever get stuck in seriously to any one crop. The antithesis of this is monoculture: a single crop grown over a whole area. The perils of such strategy were spectacularly revealed by the Irish Potato Famine. As in the nineteenth century, so in the present day. Modern Africans starve because the system of land tenure and the general economic climate make it impossible for them to farm properly – but the blame is placed, as in nineteenth-century Ireland, on the people themselves: their 'fecklessness' or their 'ignorance'. The free market is supposed to solve all ills but (surprise surprise) it does not. In the name of modernity, traditional mixed farming is replaced by monoculture – which in general is simpler and cheaper – and the simplest rules of biology suggest that monoculture, every now and again, is bound to fail. We should learn from history, but on the whole it seems more convenient not to.

Ideally the world should combine excellent husbandry with excellent science. Increasingly, in modern industrialized farming, the two have become alternatives. Only traditional farmers need respect the tenets and the logic of good husbandry, or so it seems to be supposed. The up-to-date farmers – the ones now known as 'conventional' – have been encouraged simply to reach for the bottle and the spray-gun, and to take their lead from the corporation expert. This is what gets science a bad name. This is why so many people who would like the world to be a better place, and indeed would like human beings and our fellow creatures to survive in the long term, see science as the enemy. It enables malpractice. Increasingly it is paid for by the fruits of malpractice. This is why politicians who have no science in their education, but believe the instruction they receive at company lunches, are so dangerous. Non-scientists lead the protest; but scientists themselves should see how they have been compromised, and refuse to be party to such nonsense.

Of course there are also good scientists working on more broadly based approaches, of the kind known as 'holistic'. In the past few decades many biologists have developed the concept of 'integrated pest control': subtle combinations of good husbandry and mechanical, chemical and biological methods for attacking pests. I have discussed various scenarios with Dr John Pickett, entomologist at the Rothamsted Experimental Research Station (although the following are in part extrapolated from our discussions and he may wish to disown them). One is to spray the underside of leaves with 'pyrethroid' pesticides – synthetic derivatives or analogues of natural pyrethrum. With modern spraying machines, it is possible to apply the chemicals exactly where they are needed (a far cry from the catch-all spray planes featured in Hitchcock's *North-by-Northwest*). The toxins could be constructed deliberately to be particularly unstable in sunlight, so that if any escaped from the dark side of the leaves, or when the crop is finally harvested, they would immediately break down into harmless derivatives such as carbon dioxide. Pests such as aphids tend naturally to congregate on the shady side of leaves. This kind of chemical application could be very specific indeed; and would be 'biological' insofar as it exploits the specific behaviour of the pest.

An alternative is to create chemical lures based on the insect pest's own mating pheromones. The offending insect could be lured to areas of toxin; or indeed might be seduced away from the growing fields altogether and, say, into hedgerows, where natural predators such as spiders would be waiting. In short, scientists have only just begun to explore the possibilities of chemistry, and indeed of the behaviour of insects and of the chemical signals that pass between them. We should not be put off by the innate nastiness of nicotine and arsenic, which were simply crude early stirrings; or by what proved to be the fiasco of DDT. In the future, with astute chemistry artfully deployed, pests and their predators could be positively orchestrated, and with minimum disruption of the world at large. That would be great husbandry and great science truly in combination. It is a huge mistake to give up on either.

7

Better Crops, Better Livestock: The Craft and Science of Breeding and Genetic Engineering

Improvement of husbandry is only half of farming. The other half is to 'improve' the crops and the animals themselves: create breeds and varieties that are easier to handle, or less toxic; more nutritious or more succulent; more resistant to disease and other local stresses – salinity, drought, whatever; and are also more responsive to whatever improvements in husbandry the farmer chooses to make. Improvements in husbandry and improvements in livestock and crops must proceed in dialogue; usually there is little point in one without the other. Thus, if a wild cow were given concentrates of soya or whatever, she might get fat or she might get sick but almost certainly she would not produce more milk, as a modern Holstein does. If wild plants are given extra fertilizer they either ignore it or grow lanky and lush and succumb to disease – and are quickly replaced by species of a weedy kind, that are more adapted to high fertility. On the other hand, if a modern Holstein is raised on low-grade scrub, then its prodigious ability to turn nutrient into milk is squandered; and modern wheat that breaks all records when plied with extra nitrogen may look ordinary when inputs are niggardly. Farmers should not invest more in husbandry than their crops and livestock can respond to or, of course, than their customers are prepared to pay for. Commercially speaking, they always tread a fine line between over-investment and under-investment.

Since the Palaeolithic, breeding has moved through four phases. In the beginning 'breeding' was, presumably, just an extension of natural selection: people grew what grew best because – well: that was what had survived from the previous season and so was available to replant. Then, as farming itself became more established and widespread, farmers began to apply true craft: consciously shaping their crops and livestock by methods based on experience and ad hoc theory. In this, overall, they were extremely successful. The crops and animals of biblical and Greek and Roman times had already come a very long way from the wild, and the cattle and corn of traditional landscape painting from the seventeenth and eighteenth centuries are what most of us imagine cattle and corn ought to look like.

In the third phase, breeders began formally to apply the theories of 'classical' genetics. This era could have begun in the mid nineteenth century when the Moravian friar Gregor Mendel carried out the key experiments which effectively established the ground rules of heredity, but since his work was seriously under-appreciated in his own lifetime, formal genetics in practice had to wait for the twentieth century.

The fourth and final phase, the one we are in now, got under way in the 1970s when genes or at least bits of DNA, which is the stuff of which genes are made, were physically transferred from one organism (a bacterium) to another. Thus began the age of what is formally known as 'recombinant DNA technology' but is colloquially (and somewhat unfortunately) called 'genetic engineering'. Genetic engineering is now abetted by a series of subsidiary biotechnologies which between them are now promising, or threatening, to provide us with 'designer' organisms: crops or livestock (or, conceivably, flesh that is neither one nor the other) built from scratch in the way that Ferrari builds motor cars.

It is not the case, however, that each phase has simply replaced the ones that went before. Whatever was done in earlier eras has continued in the later ones, and the inheritance passed on from stage to stage was in all cases tremendous. Thus after many a millennium the world's first farmers and proto-farmers had produced the kind of primitive (in the non-pejorative sense) varieties known as 'landraces'; so the pre-Mendelian breeders of ancient times, the Assyrians, Romans and the peoples of the Old Testament, absolutely did not begin their own endeavours with wild creatures. Mendel himself, the founder of modern genetics, began with a stupendous range of traditional varieties and with breeding lore that dripped with theory (much of it wrong, but serving nonetheless as working hypotheses). In his seminal paper of 1866 he acknowledges his eighteenth- and nineteenth-century forerunners. The present time is seen as the age of molecular biology and of biotechnology but in truth, improvement of crops and livestock still depends overwhelmingly on classical, Mendelian genetics (and to some extent on pre-Mendelian craft) and, beyond doubt, this will always be the case.

In the following account of breeding since ancient times I will focus largely though not exclusively on wheat partly because it is first equal (with rice) among all that we grow and partly because it illustrates a great many principles of fundamental biology and of practice.

Phase I: From the Wild to the Landrace

The first creators of crops and livestock partook directly, as all of us are bound to all the time, of the principle that Charles Darwin identified as the great shaping force of nature and is now widely acknowledged as one of the fundamental principles of the universe: natural selection. Most obviously, and basically, the first farmers must simply have sown the seeds and raised the offspring of the crops and livestock that had survived the previous season. This simple and even unconscious process would quickly produce strains of plants and animals that were significantly different from their wild progenitors, and in each case were particularly well adapted to the conditions that the farmer provided. Such locally adapted strains, which effectively are primitive varieties and breeds, are known as 'landraces'. Landraces produced by this informal process still form the basis of a great deal of agriculture worldwide, particularly in difficult conditions (high altitude, semi-desert, whatever) where modern crops typically languish. In short, the world's remaining landraces are a tremendous asset although their worth has been properly appreciated only in recent decades.

The changes wrought by such season-by-season selection would for the most part have been gradual. But among crops in particular – and especially among some of the most important ones, like wheat, maize and potatoes – we can also discern a series of happy biological accidents that from time to time produced radical leaps forward.

Notably, the wild ancestors of two of the world's most important crops – wheat and maize – shed their seeds when they are ripe, as is usual in all flowering plants; for how else would they be scattered? But in modern domestic wheat, the seeds stay firmly attached even at harvesting, and have to be forcibly removed by threshing. In modern maize, the seeds are tightly bound in cobs, and have positively to be excavated. For farmers, grains that simply drop their seeds are a serious nuisance. The seeds then have to be picked up off the ground by 'gleaning', which is what Ruth did in the fields of Boaz (Ruth 2 refers to barley rather than wheat, though the same principle applies). But some mutant individual plants would hold on to their seeds at season's end. The seeds on such plants would be more likely to be harvested than those that fell off; and so they would be more likely to be sown again the following year. In short, cereals with loosely held seeds are favoured in the wild; but among domestic grains, natural (or artificial) selection favours tight-bound seeds.

Then there is the phenomenon of polyploidy. Genes, as has been known in principle since the nineteenth century (even before the concept of 'gene' had been finally clarified), are carried in the nucleus of each body cell on string-like bodies known as 'chromosomes'. Most organisms carry two sets of chromosomes per cell and are then said to be 'diploid'. But in plants the entire double set of chromosomes may suddenly double again; and the proud possessor of two double sets of chromosomes is then said to be 'tetraploid'. This can happen repeatedly, for example to produce 'octoploid' lineages (as is basically the condition in sugar cane); and any multiplication of the basic chromosome number is known generally as 'polyploidy'. Polyploidy presumably happens in animals too but for some reason animals seem very intolerant of it; so known examples of polyploid animals are rare and generally controversial. (The golden hamster used to be offered as an example of it, but zoologists seem to have changed their minds on this.) In plants, however, polyploidy is common.

A plant that simply doubles its chromosomes is known as a tetraploid but more specifically it is an 'autotetraploid'. Tetraploids cannot generally mate successfully with their diploid parents and so they form a new, instant species. Physically, too, the tetraploid offspring may be significantly different from their diploid parents. Thus, some modern grasses bred for pastures are tetraploids.

Even more interesting, though, are the allotetraploids. In general, creatures of different species cannot mate successfully with creatures of other species. This in effect is a matter of definition: if two creatures of any kind, plant or animal, can mate to produce viable offspring then biologists decree that they must be of the same species. Often, successful mating between different species is impossible because the two would-be parents, being of different species, have different numbers of chromosomes. This matters because the chromosomes have to cooperate; first to create viable body cells for the new offspring and then, if this hurdle is overcome, in the formation of gametes (eggs, sperm, ova, pollen). If the chromosomes from the two parents are simply not compatible (as would be the case, say, in a mating of cow and horse) then no offspring is produced at all.

If the chromosomes are compatible enough to produce body cells but not compatible enough to produce gametes, then we finish up with an offspring that is viable but sexually sterile (as in a mule, the mating of horse and ass). Incompatibility of chromosomes is not of course the only reason why horses and cows cannot be mated, or cabbages and rice; but it is one of the reasons and, even taken alone, would be enough.

In plants, though, polyploidy sometimes (surprisingly often) comes to the rescue. A hybrid offspring inherits one set of chromosomes from a parent of one species, and a different, unmatched set from the other parent of a different (though invariably closely related) species – but then the entire cargo of chromosomes doubles. Thus, in one stroke, each chromosome acquires a matching partner; and hybrid offspring in which this doubling occurs can produce gametes, and are therefore fertile. The prefix 'allo-', in 'allotetraploid', implies that the hybrid offspring contains chromosomes from two parents of different species: it is, in fact, an 'interspecific hybrid'. But since it can breed with other allotetraploids of the same type, but not with either of its original parents, these allotetraploid hybrids must be recognized as new species in their own right.

Autotetraploids and allotetraploids arise in nature, and they clearly arose in the fields of ancient farmers, as brand-new entities, long before anyone knew anything about chromosomes. Thus, all the traditional varieties of potatoes grown in Europe and North America (the 'lumper' of nineteenth-century Ireland, the King Edward, Maris Piper and so on) are variations on a theme of the species *Solanum tuberosum*; and this clearly is an autotetraploid derivative of some diploid ancestor from the Andes.

Allotetraploids that arose spontaneously in farmers' fields include the swede (which the Americans call rutabaga). Botanically the swede is known as *Brassica napus*; but it is an allotetraploid hybrid between the cabbage (*Brassica oleracea*) and the turnip (*Brassica rapa*), and it apparently arose in gardens somewhere in Europe (possibly Poland) sometime before the seventeenth century. Boysenberries, loganberries, youngberries, and so on are polyploid variants, both auto- and allo-, of blackberries and raspberries.

The polyploid history of wheat is more intricate. Wild wheat, growing on the hills of the Middle East, is a grass of the species *Triticum monococcum*, and is known as einkorn. It is diploid, with 14 chromosomes (2 sets of 7). Sometime in the distant past, however, a *Triticum monococcum* received or donated pollen from or to some wild goat grass. No one apparently knows which goat grass, but *Triticum speltoides* (also known as *Aegilops speltoides*) is a strong candidate. *Aegilops speltoides* also has 14 chromosomes (2 x 7), but these are not compatible with those of *T. monococcum*. Deep in the past, however, polyploidy took place in some hybrid offspring of these two species, to produce an allotetraploid with 28 chromosomes (2 x 2 x 7). Thus was born an instant new species known as *Triticum turgidum*, which is commonly known as 'emmer'; and one lineage of wheats derived from *T. turgidum* are of the kind known as *T. durum*, which provide the world with the flour from which to make pasta.

That wasn't the end, however. Evidently the tetraploid *T. turgidum* then mated with yet another wild goat grass, *Aegilops squarrosa* (otherwise known as *Triticum tauschii*). *T. tauschii* was another diploid (2 x 7) and so contributed 7 chromosomes to the union; while the tetraploid *T. turgidum* of course supplied 14. The resulting embryo had an uncomfortable 21 chromosomes, but these apparently doubled again to produce a hexaploid, with 42 chromosomes. Thus appeared the brand-new species known as *Triticum aestivum*, which has become the most important of all. It provides all the modern varieties of bread wheat.

Durum wheats and bread wheats have long been subject to tremendous breeding programmes, involving many billions of individual plants (literally) and millions of breeding lines; but the first tetraploid durums and the original hexaploid bread wheats arose in the fields of ancient farmers (or perhaps nearby in the wild) without the farmers having to do a thing. Still, though, they must have been astute to recognize that the newly arisen tetraploids and hexaploids were superior to the general mass, and to hang on to them. Humanity ever since has benefited from their alertness.

We should also be grateful for our ancestors' courage and persistence. The wild ancestors of many of today's most favoured crops are toxic, and in some cases seriously so, including many wild beans, wild parsnips, and many wild relatives of potatoes and tomatoes (both of which are, after all, related to the deadly nightshade: and toxins run in families). The wild ancestors of today's cattle and pigs are lethal creatures, or certainly can be. But by persistence and ingenuity, the first farmers at least by classical times had produced a wide range of crops from many botanical families that were and are delicious and safe (at least when eaten in sensible amounts); and of livestock which, though sometimes stroppy (I don't envy vets), usually are pussy-cats compared to their free-living relatives.

Phase II: Formal Breeding before Genetics

We cannot know what went through the heads of the world's first farmers as they began to transform wild plants and animals into crops and livestock. They may from the first have dreamt of more succulent roots, or fatter and more fecund beasts. Or they may not. Either way, they certainly could and almost certainly did effect significant changes even without any clear idea of what they wanted to achieve – or indeed without realizing that the recalcitrant roots and beasts they strove to control and propagate *could* be transformed by deliberate strategy into something more agreeable.

But it's also clear that at some time in the distant past, farmers did realize that they could change the creatures they sought to manage; moreover, that they could nudge them step by step towards some specified ideal; and so their inadvertent inroads processed into conscious policy. The shift had clearly taken place by classical times: the peoples of the Mediterranean and Middle East who relied so heavily on horses, for example, certainly knew what they wanted, and how to select and match the stallions and mares appropriately. Farmers who strive consciously to tailor crops and livestock can properly be called 'breeders'. Those who operate(d) without the aid of modern genetics theory can be called 'craft breeders'.

The essence of craft breeding is common sense – plus endless experience and knowledge of detail, for nature is not in the end constrained by common sense and sometimes it pays to go down routes that are counter-intuitive (sometimes feeble-looking individuals that instinct tells you to throw out provide just the complement of genes that's needed). Common sense says that if you want to breed a bigger ear of corn, or a fatter pig or a fancier goldfish, then you save those individuals in any one generation that come closest to the ideal: allow them to breed, and eat, sell, or otherwise dispose of, the rest. This is (artificial) 'selection'. Common sense says, too, that if one individual shows one set of desirable characters (biologists use 'characters' to mean 'features') and another individual has different characters that are also desirable, that a marriage between the two could produce offspring that combine the best of both. This is 'crossing'. Judicious selection and crossing are the essence of all improvement programmes.

In practice, there are two main strategies. One in essence is simply to select the best individuals in any one generation, and match them with others who have complementary qualities: 'Cross the best with the best and hope for the best', as breeders say, in self-disparagement.

The other is 'mass selection': the breeder (or farmer) has some ideal in mind, and allows all those individuals who roughly conform to the ideal to interbreed freely with each other, and throws out the rest. Henry VIII was proposing mass selection when in 1541 he decreed that stallions under 15 hands (142 cm) in height should not be allowed to graze on common land in the Midlands and south of England (thus effectively signing their death warrants). Fifteen hands is not big: Tudor horses were obviously undersized. Mass selection works especially well on populations in which the required characters are genetically complex and in which there is significant variation from individual to individual. It still has much to contribute to the improvement of crops like millet, both to increase the size of the seed heads (select the biggest 50 per cent or so in each generation) or disease resistance (throw controlled doses of the disease at the growing plants, and keep the ones that survive). This is commonsensical, but common sense has sometimes been perversely overridden. Thus Mark Overton notes in *Agricultural Revolution in England* that eighteenth-century livestock farmers in England at one time got it into their heads that the differences between the best and the worst animals were brought about not by inheritance, but through environment alone; so they sent the best animals to the butchers and kept the worst, in the hope that better feeding would bring them up to scratch.

'Crossing' raises a whole new raft of difficulties. In general, if two individuals of distinctly different types are mated, then the result is a 'hybrid'. When two different species of creature are mated and manage to produce offspring, the result is an 'interspecific hybrid'. Usually such interspecific hybrids are sexually sterile – like mules. But, as we have seen, interspecific hybrids between plants may become fertile through polyploidy, and the polyploid hybrid is a new and discrete species in its own right. Much more common, however, are 'intraspecific' hybrids, which are produced by crossing different varieties or breeds of the same species. Thus (to take one example among many thousands) Friesian cows, which are big but predominantly dairy animals, are commonly impregnated by Hereford bulls to produce hybrid, Hereford x Friesian calves, to be raised for beef. Note that the breed of the sire is stated first.

Life is never simple, however. Thus we might suppose that if two individuals of the same variety were crossed, there should be no problems with breeding. Of course they would produce viable offspring. Growers of plums, cherries, pears and apples for example (all of which are members of the rose family, Rosaceae) know that many varieties will not set fruit unless they are pollinated (effectively mated) by individuals of a different variety. Those that will not self-fertilize are said to be 'self-incompatible'. Other plants that are not strictly self-incompatible nonetheless prefer to be pollinated by individuals who are not their close relatives. These are said to be 'outbreeders'. Maize, cabbages and runner beans are natural outbreeders. More generally still, any 'consanguineous' mating (mating between close relatives) either in plants or animals is more likely to produce monstrosities, or in general to produce offspring that are feeble: and this is called 'inbreeding depression'. Contrariwise, a little judicious outbreeding leads to what Darwin called 'hybrid vigour' (which in modern genetics parlance is called 'heterosis'). But outbreeding does not always produce good results. Matings between some pairs of varieties that might seem to be a good match sometimes in practice produce rather poor offspring: for example, the hybrid puppies of St Bernards and Great Danes do not fare well. And of course, if the outbreeding is too extreme (for example between horse and cow), there are no offspring at all.

On the other hand, some plants are naturally tolerant of inbreeding. Wheat, for example, does not object to consanguineous mating (though even wheat tends to produce even stronger offspring when different but compatible varieties are mated; hybrid wheats are now common). Garden peas are inbreeding as a matter of course: a crucial fact of which Gregor Mendel made good use. All these observations and many more the traditional breeders knew, lived with, and worked around – even though they lacked the unifying theory that would make sense of it all. The successes of the craft breeders are rather miraculous: as clever and unfathomable as all craft in all spheres, whether that of great potters or makers of furniture or violins or Japanese swords. Skill, honed by nothing more than experience and theory that fits where it touches, is always extraordinary to behold.

In practice all breeders, both the pre-genetics craftsmen and women and the Mendelian geneticists, must seek to strike a balance. On the one hand (broadly speaking) they want crops and livestock that are all the same, so that they know exactly what they are getting and, indeed, so that all individuals perform as well as the best. But on the other hand if the crops or animals are too uniform, then they are all equally prone to the same stresses, and one pathogen kills all; and there is no opportunity to reap the benefits of outbreeding in general and of hybrid vigour in particular, since each individual is always being mated with another that is almost exactly like itself.

In essence, this dilemma is always with us, but there are various ways around it. One is to focus on inbreeding crops, like wheat or peas, which don't mind being inbred. Such crops tend to be uniform and true-breeding, but do not suffer too much inbreeding depression. Another is to ensure that although the variety is reasonably uniform – otherwise it would be not a distinct 'variety' – it nonetheless contains a reasonable amount of genetic variability. This is the case with traditional varieties of runner beans, cabbages and maize, for example.

The third tactic is to keep two pure-bred and uniform varieties or breeds but as breeding stock only. Then when the males of one kind are mated with the females of the other *all* the offspring are hybrids and so benefit from hybrid vigour, but they are all very similar to each other since they all derive from similar parents. This is how breeders of maize proceeded through much of the twentieth century: they maintained parent lines that often looked rather feeble but whose hybrid offspring were especially vigorous. Maize breeding was transformed in the United States in the 1920s by the introduction of such hybrids; and then again in the 1950s, through the introduction of 'double crosses'; the offspring of parents which themselves were both hybrid. More generally, modern plant breeders commonly raise two different parent stocks and then sell the first generation as 'F1' hybrids.

Finally, many plants can be produced asexually or 'vegetatively'; and then all the offspring are genetic facsimiles of the parent. Together they form a clone. Thus when breeders have produced a good-looking potato by sexual means, they simply multiply it up by means of tubers. 'Seed' potatoes are tubers: not seeds at all (which are the products of sex). When breeders of roses or apples produce an especially good plant they multiply it by cuttings. If a gardener allowed his or her potatoes to flower and produce true seeds, they would not breed true; the resultant plants would be a mess, with hugely various tubers. If apple growers simply planted pips, they too would squander the fine qualities of the Cox's or Egremont Russet or whatever they started with. But so long as they stick with cuttings, they can multiply their prize plants with no loss of quality at all. Animals (at least the kind that are kept on farms) offer no such option. Only in Greek mythology is it possible to produce a new individual from, say, a piece of leg.

All this was known to the traditional breeders, so that when they wanted to shift their crops or animals in particular directions, they had the skill and the knowledge to do so. British horse breeders are known to have introduced Arab blood since the twelfth century (the first known example was Scottish) and the eighteenth-century crosses between the swift, svelte Arabs and the sturdier British thoroughbreds gave rise to modern racehorses. Eighteenth-century English farmers also brought in extremely fat pigs from China, because in those days people liked their bacon fatty, not least because the weather and the houses were cold and they needed the calories. Yorkshire bacon was especially fat. In Italy lard is still served as a delicacy in the more discerning restaurants, and very fine it is too. In Britain, Robert Bakewell (1725–95) stood out as a breeder of sheep (notably the Leicester) and of longhorn cattle; and in America, Joseph Cooper (1757–1840) as a breeder of crops. More generally, breeders even before the nineteenth century produced a wonderful range of breeds and varieties that still for the most part survive, mercifully, thanks to the enthusiasm of rare breeds societies (often in the teeth of official opposition, for example from the European Union). I now live in an area where butchers sell local meat by the breed. The beef is mostly Aberdeen Angus cross, and the pig is typically saddleback. I have found for myself that supermarket pork, compared to what we might call 'the real thing', is chalk against cheese. The real thing, bred by traditional breeders specifically for its flavour and texture, and not simply for its rate of growth and the 'efficiency' with which it turns feed (largely barley) into flesh, is in a quite different league. When rich farmers get together to plan their next lurch towards industrialization they like to treat themselves not to the products of their own high tech but to joints and cuts from traditional breeds, as initially created by pre-scientific breeders and farmers, and raised by traditional means. The carcasses that today's rich farmers provide for the urban masses are quite distinct from what, given the chance, they consume themselves.

Phase III: Mendel and the Age of Genetics

The rise both of chemistry and of scientific breeding in the late eighteenth century (and a lot more besides) strongly suggests that what was really afoot was a new way of thinking: across-the-board realization that through careful observation, the application of logic and mathematics, and a series of imaginative leaps of the kind that are indispensable, it should be possible to work out the rules that underpin all the apparent caprices of nature.

In the matter of crop improvement (plants are easier to work with than animals), a surprising amount was coming into focus long before Mendel. In the 1760s the German naturalist J. G. Kolreuter attempted to pin down the difference between 'variety' and 'species' through a programme of hybridization between no fewer than fifty-four species of plants from thirteen genera. ('Genera' is the plural of 'genus'. Closely related species are placed in the same genus, as in *Equus*, which includes both horses and donkeys, or *Canis*, which includes dogs and wolves.) Kolreuter showed that the different characters of plants 'segregate' in the hybrid progeny: so that if a parent with white flowers and green pods (say) is crossed with a parent with yellow flowers and grey pods, the offspring might have white flowers and grey pods, or yellow flowers and green pods, or indeed any combination. He reported segregation almost exactly 100 years before Mendel showed the same phenomenon in his pea experiments. He also established the general principle that formal experiments in hybridization could be highly instructive.

Outstanding too was the Englishman Thomas Andrew Knight (1759–1838), who perceived that such experiments would prove nothing unless those parents were pure-bred and of the kind which, when crossed with their own kind, were true-breeding. And, he said, of all the plants he had worked with, '. . . none appeared so well calculated to answer my purposes than the common pea'. Peas come in all shapes and sizes and colours, so there were plenty of different types between which to make crosses. But each kind remained pure-bred and true-breeding 'because the structure of its blossom, by preventing the ingress of adventitious farina [pollen], has rendered its varieties remarkably permanent'. In fact, the flowers of peas are closed, so the peas must fertilize themselves.

Knight observed, as Mendel was to do, that the first generation (F1) crosses of two dissimilar parents were all similar to each other; but when the hybrid F1s were crossed with other hybrid F1s, then the second generation (F2) were a tremendous mixture. Among those who picked up on Knight's work were the Englishmen J. Goss and A. Seton who in 1824 described the phenomenon of 'dominance': that sometimes when two dissimilar varieties are crossed, a particular character in one of the parents may completely override the corresponding character in the other parent. Thus, for example, if a yellow-flowered variety was crossed with a white-flowered type, then in the first generation *all* the offspring might have yellow flowers. But white-floweredness might crop up again in later generations.

There were many other straws in the wind, many of them blowing across Central Europe, whose blossoming economies were rooted in agriculture. In Hungary in 1819 Count E. Festetics made formal descriptions of patterns of inheritance which he intriguingly called *Genetische Gesetze* – 'genetic laws'. (It is often said that the term 'genetic' was not coined in the context of biology until the twentieth century. But here it was, in the early nineteenth.) In 1853 R. Wagner commented that the rules of heredity would be understood only by analysing masses of data statistically, although he warned that this would be expensive and time-consuming. This early emphasis on the importance of statistics is highly significant.

Yet as early as 1820, the German breeder G. C. L. Hempel commented that heredity could be understood only by 'A new type of naturalist . . . a researcher with a profound knowledge of botany and sharply defined powers of observation who might, with untiring and stubborn patience, grasp the subtleties . . . take firm command . . . and provide a clear explanation.'* Just two years later, Mendel was born. If anyone has ever been born on cue, and tailor-made for a specific and momentous task, it was he.

Mendel is remembered as Gregor, but that was the name he took when he entered the Augustinian friary of St Thomas in Brunn (now Brno) in Moravia (now the Czech Republic) in October 1843 at the age of twenty-one. As his finest biographer, Viteslav Orel, tells us, he was born on 22 July 1822 at Heinzendorf (now Hyncice) and was given the name Johann. Heinzendorf had just seventy-two households and Moravia was part of the Austro-Hungarian Empire which until 1848 was officially feudal: Mendel's father, Anton, with his two horses, was officially obliged to work for the local baron for two days a week. But young Johann was obviously bright and although the regime was formally feudal it was also in many ways enlightened, and various teachers and

* Viteslav Orel, *Gregor Mendel, the First Geneticist*, Oxford University Press, Oxford, 1996.

people of rank took an interest in him. So he was well educated first at Opava and then, after 1841, at the Philosophy Institute at Olomouc. He entered St Thomas's, it seems, not primarily through piety (although many scientists of the mid nineteenth century were extremely devout) but through poverty. Much later, at the age of twenty-nine (and as a member of St Thomas's), he enrolled full-time at the University of Vienna.

Little is known in detail of Mendel's life, and a lot of myth has grown up around him which Orel in particular has done much to scotch. Notably, Mendel clearly was not self-taught as is sometimes apparently supposed, but extremely well educated. He did not, however, focus primarily on biology and agriculture in his formal studies: he aimed to become a teacher of physics and maths (and meteorology was another of his abiding interests). Neither was he without influential connections. In physics, he certainly knew and was respected by no less than the great Christian Doppler, of the Doppler effect (which enables modern cosmologists to calculate how rapidly the stars are retreating from us). But he was also introduced to serious botany at Vienna by the renowned Franz Unger; in turn an associate of the great J. M. Schleiden, one of the founders of modern cell theory.

Brno was not a backwater, but a thriving, newly industrial town, the capital of its region. St Thomas's often ran foul of the highly conservative Catholic diocese, but by no means was it an intellectual desert. To a significant extent it continued the general tradition of the Middle Ages, when monasteries were the prime centres of learning and, depending on the predilection of the abbot, they could virtually function as research stations. The economy of nineteenth-century Moravia was absolutely rooted in agriculture – sheep, fruit, vines – and livestock and crop improvement were effectively a national obsession. Mendel's abbot at St Thomas's was F. C. Napp, who was himself a keen plant breeder and established the garden and greenhouse where Mendel did his work. The mystery, in short, is not how Mendel conceived and managed to carry out his experiments but why, given that he had been given such support and had such good connections, his work was not given the credit it seems so obviously to have deserved. Napp remained abbot until his death in 1867, when Mendel took over. Mendel was a conscientious abbot, vigorously fighting the monastery's corner against the Catholic hierarchy, but he continued with some remarkable researches, not least into the hybridization of bees. Along the way (it really wasn't a cultural backwater) he also employed the composer Leoš Janáček as organist.

So Mendel was not a primitive among backwoods nobodies, but an intellectual among intellectuals in the midst of a thriving pan-European economy. Furthermore, he was the right kind of intellectual, precisely as prescribed by Hempel. He was primarily a physicist (and well educated as such): inclined to think of hereditary factors (his term) not as amorphous fluids but as discrete entities. In modern parlance, his view of heredity was 'digital'. He was versed in maths, as the incomparably great Darwin was not (and regretted), and he knew that to produce robust results he had to quantify. He was wonderfully patient; if he had been more devout he would have been the perfect monk. But he also had gardening and plant breeding in his bones, and during his time at Opava and Olomouc he had become a keen naturalist. In short he had all that was needed intellectually and emotionally, plus time and opportunity. The one quality he possibly lacked, albeit to his credit, was pushiness. He seems to have been too modest, too self-effacing for his own good.

I will not dwell on Mendel's experiments and the science of genetics that finally emerged from them since they are the subject of a thousand textbooks. The following is the briefest sketch intended mainly for the further scotching of myths. It is based mostly on Mendel's own paper of 1866, *Experiments in Plant Hybridization*, which he had (twice) read to the Brno Natural Science Society in the previous year. Mendel did not leave much for us to read – a fellow monk inexplicably destroyed most of his papers after his death – but he did leave this crucial text.

He began the preliminary work in 1854 with 'thirty-four more or less distinct varieties of peas . . . obtained from several seedsmen.' Myth says that he was lucky to have chosen peas, which are true-breeding. If he had worked with, say, runner beans, he would have found life much more difficult. But actually, later, he did work with runner beans, and with another sixteen species as well. He began with peas because (as Knight had already pointed out) he knew perfectly well that they were true-breeding, and that if anything could yield clear results, peas could. Still, though, he did not take the seedsmen's word for it that the peas he worked on were true-breeding. Instead, before he began the experiments, he 'subjected [the peas] to two years' trial', just to make sure. Then, out of the original thirty-four varieties, 'twenty-two were selected and cultivated during the whole period of the experiments'. In short, his brilliant choice of peas, and the simplicity of his final results, had absolutely nothing to do with luck.

Critics say too how fortunate he was to have studied characters that happen to have a simple pattern of inheritance: characters such as the colour of the seeds or the form of the unripe pods. But again, as he records in his paper:

> The various forms of peas selected for crossing showed differences in length and colour of the stem; in the size and form of the leaves; in the position, colour, size of the flowers; in the length of the flower stalk; in the colour, form, and size of the pods; in the form and size of the seeds; and in the colour of the seed coats and of the albumen [by which he means the seed leaves]. Some of the characters noted do not permit of a sharp and certain separation, since the difference is of a 'more or less' nature, which is often difficult to define. Such characters could not be utilized for the separate experiments; these could only be applied to characters which stand out clearly and definitely in the plants.

In short, he knew perfectly well that some characters in peas show a simple, all-or-nothing pattern of inheritance, while others are more complicated. So he chose deliberately to work with the former. This is exemplary experimental strategy. As the British zoologist and philosopher of science Sir Peter Medawar commented a century later, 'science is the art of the soluble': in other words, you don't take on problems that are liable to prove too difficult. The art of science is to put the insoluble problems on one side, and come back to them later. In fact Mendel studied seven clear-cut characters, including the structure of the ripe seeds (smooth or wrinkly); the colour of the seed coat; the colour of the unripe pods; and the overall height of the plant. So he looked at the pattern of seven characters in twenty-two different, true-breeding varieties. A formidable line-up. He carried out the experiments themselves between 1856 and 1864 and in 1865 was ready to announce: 'after eight years' pursuit [the work is] concluded in all essentials'.

Thus, for example, when a round-seeded variety was crossed with a wrinkled-seeded variety, *all* the offspring of the F1 generation had round seeds. But when the hybrid F1s were crossed with other hybrid F1s, some of the F2s had wrinkled seeds. Similarly, when green-podded peas were crossed with yellow-podded peas, all the F1 generation had green pods. But when F1s were crossed with F1s, some of the F2s again had yellow pods. And so on, with all seven characters. In Mendel's words:

Henceforth in this paper those characters which are transmitted entire, or almost unchanged in the hybridization, and therefore in themselves constitute the characters of the hybrid, are termed the *dominant*, and those which become latent in the process *recessive*. The expression 'recessive' has been chosen because the characters thereby designated withdraw or entirely disappear in the hybrids, but nevertheless reappear unchanged in their progeny.

He also showed that it did not matter whether the quality of round-seededness was inherited via the ovum (female) or the pollen (male). The pattern of inheritance, and the dominance or recessiveness, were not affected. (Mendel had previously had a run-in with a professor of botany called E. Fenzl, who insisted that all the hereditary material was passed entirely through the pollen, and the job of the female, the ovum, was merely to supply nutrient. This was Aristotle's view, too.)

Mendel, though, was not content merely to note that some characters are dominant over others (as Goss and Seton had done in the 1820s). He quantified (as Wagner had recommended). He showed, in fact, that when the F1s were crossed with other F1s, the F2s with the dominant character (such as round seeds) outnumbered the F2s with the recessive character (such as wrinkled seeds) by almost exactly 3:1. If a plant with two dominant characters (say, round seeds and green pods) was crossed with one with two corresponding recessive characters (wrinkled seeds and yellow pods) then all the F1 offspring had round seeds and green pods. If the hybrid F1s were then crossed, the famous Mendelian ratio of 9:3:3:1 was manifest in the F2s. That is: for every sixteen F2 offspring, nine (on average) would have round seeds and green pods; three would have wrinkled seeds and green pods; three would have round seeds and yellow pods; and just one would have wrinkled seeds and yellow pods.

So what's going on? With these figures to hand, Mendel saw immediately. Each 'character' (such as green pods or wrinkled seeds) is coded by a single 'factor', of the kind that in the twentieth century was called a 'gene' (the term I will employ from now on). Each parent contains two full sets of genes (which are carried on the two sets of chromosomes, as outlined above). Each offspring inherits one set of genes from each parent. But – and here is the law of segregation – the genes are inherited separately; and we now know that during the formation of gametes the chromosomes are broken up and then re-form, which allows genes on the same chromosome to be passed on individually.

Now it is known that each species contains a specific number of genes: so I, being human, have the same numbers of genes as you do, and my genes correspond to yours (although women have slightly more genes than men. Every statement about nature comes with conditional clauses). But any one gene in any one species may come in more than one version; for if it were not so, then all individuals of the same species would be genetically identical, which of course they are not. The different versions of each gene are called 'alleles'. The allele that makes a pea round-seeded is dominant over the allele that gives a pea wrinkled seeds.

If an offspring inherits the same kind of allele for a particular gene from both of its parents, then it is said to be 'homozygous' for that gene; and if it inherits a different allele for any one gene then it is 'heterozygous' for that gene. If a pea inherits the allele for round seeds from both parents (homozygote) then it of course has round seeds. If it inherits the round-seed allele from only one parent and a wrinkled-seed allele from the other (heterozygous) it is still round-seeded, since the round-seed allele dominates the wrinkled-seed allele. Only if it inherits a wrinkled-seed allele from both parents (homozygous for wrinkled) will it have wrinkled seeds. Intuitively you can see why the cross of a round-seeded parent with a wrinkled-seeded parent produces the 3:1 ratio of round to wrinkled in the F2s.

In practice, many characters are 'polygenic': several or many genes combine to produce them. When many genes are involved, the character in the offspring may well be a blend of the parents, rather than 'all or nothing'. So critics again say, how lucky Mendel was not to have encountered polygenic characters, for they would have confused his results no end. But he did encounter them. In his experiments on peas, he neatly side-stepped the problems: these are the characters, such as leaf-size, which he noted have a 'more or less' quality, and he deliberately chose not to work with them. But later he worked on runner beans, *Phaseolus*, in which (as he noted) the flower colour is highly variable. In his 1866 paper he writes: 'from the union of a white and a purple-red colouring a whole series of colours results, from purple to pale violet and white. The circumstance is striking that among thirty-one flowering plants only one received the recessive character of the white flower, while in *Pisum* [peas] this occurs on the average in every fourth plant.' But Mendel was not fazed for an instant: 'Even these enigmatic results . . .

might probably be explained by the law governing *Pisum* if we might assume that the colour of the flowers and seeds of *Phaseolus multiflorus* is a combination of two or more entirely independent colours, which individually act like any other constant character in the plant.' Here is a perfect description of polygenic characters. Most of his twentieth-century critics don't seem to have read his paper.

Clearly, the modern science of genetics might have begun in the late 1860s, if only the world at large had realized the significance of Mendel's work. But the world at large did not. The people who should have thrown up their arms in joy instead either ignored what he said (Darwin had a copy of Mendel's paper, but did not read it) or presented him with perverse exceptions to his rules (as did Carl Nageli, a very distinguished Swiss botanist, with whom Mendel was advised to discuss his work). Thus, in one of the most bizarre episodes in the history of science, Mendel's research was more or less ignored for the next thirty-five years until it was rediscovered in 1900 by three research scientists working independently: Carl Correns in Germany; Hugo de Vries in Holland; and Erich von Tschermak in Austria. After that, things proceeded apace.

My own critics will say that I have spent too long on Mendel. To describe the developments of the twentieth century at the same pace would take a library. But in this I am happy to take my lead from the Cambridge philosopher A. N. Whitehead, who commented on the grandest possible scale that 'All philosophy is footnotes to Plato'. By the same token, all genetics is footnotes to Mendel. In his experiments at St Thomas's from 1856 to 1864 lies the essence of it all. Grasp that, and the rest falls into place. So here are the key events and insights of twentieth-century genetics presented with brutal brevity:

Very soon after Correns's, de Vries's, and von Tschermak's rediscovery of Mendel, the basic vocabulary of genetics was established: Mendel's hereditary 'factors' became 'genes'; the word 'allele' was introduced; an organism's total apportionment of genes was its 'genome', and all the alleles shared by a sexually reproducing population of creatures was the 'gene pool'. After some initial doubts, it soon became clear that Mendel's rules of inheritance fitted very well with the observations of chromosome behaviour that dated from the late nineteenth century: that is, it began to seem very likely that the chromosomes actually carry the genes, and as the chromosomes divide, break up and re-form during the creation of gametes, so the genes are shuffled. This alone would ensure that each offspring did not simply share the characters of its parents, but was bequeathed a random selection of characters from each parent. Thus, one of the key problems that had aggravated breeders from earliest known times (and even flummoxed Darwin) was cleared up at a stroke.

Add in the phenomenon of dominance and recessiveness, and we see how the characters of the parents can be mixed randomly even though the genes are distinct entities, and without any messy mixing of hypothetical hereditary fluids. The characters of the parents might indeed be combined in the offspring like the pigments on a painter's pallet – but in a way that enables them to be disentangled again in subsequent generations. Nature's subtlety beggars belief. Throw in, then, the phenomenon that Mendel clearly glimpsed – that any one character might be 'polygenic' (in modern parlance); underpinned by more than one gene – and almost all the puzzles seem to fall away.

In America, Thomas Hunt Morgan (1866–1945) established his genetic laboratory at Columbia University in 1908 and probably did more than anyone to establish genetics as a formal science. He introduced the fruit fly, *Drosophila*, which breeds quickly and has just four chromosomes, as a prime subject for fundamental study. Soon it became clear that one of Mendel's rules needed modification. Characters were not always inherited absolutely independently of each other. Some characters were sometimes associated with others more often than chance alone would dictate. This, it transpired (very logically; genetics in essence is a very straightforward subject) was because of 'linkage'. In 1913 one of Morgan's students, Alfred Henry Sturtevant (1891–1970) suggested that genes were likely to be inherited together if they were on the same chromosome. Indeed, the closer they were together, the more likely they were to be linked. Thus Sturtevant began the process of 'gene mapping': from the frequency of linkage, it was possible to infer whether any two genes were or were not located on the same chromosome, and if they were, how close they were together. The Human Genome Project, the first stage of which has been so recently completed, partook of Sturtevant's early-twentieth-century insight.

One of Thomas Morgan's star if eccentric colleagues, Hermann Joseph Muller (1890–1967) then demonstrated the vital phenomenon of 'mutation': that genes may physically change, suddenly and radically, and thus produce physical changes in the organism that contains them. This was already suspected, but in 1926 Muller showed how mutations could actually be induced, with X-rays. This largely explains another of the puzzles faced by classical breeders and naturalists: the sudden appearance of monsters or sports. It also provides a key component in Darwin's idea of evolution by means of natural selection: mutations that are less dramatic than those that lead to monstrosity are the ultimate source of variation on which natural selection bears down. Mutations, arising spontaneously in captivity or left over as rarities from nature, also provide many of the extreme characters produced artificially in, for example, fancy goldfish and pigeons. Mutations induced artificially, for example by X-rays or chemicals, have become a key source of the variation needed in modern breeding programmes.

The phenomenon of 'pleiotropy' also became apparent: that any one gene can have many different effects in the same organism, sometimes apparently quite unrelated one to another. The phenomenon of pleiotropy complicates all breeding endeavours, including those of genetic engineering: add a gene for one purpose, and you may produce a whole range of uncalled-for side effects. Put pleiotropic genes with polygenic characters and we can see immediately what complexities might arise – and yet they arise from remarkably simple principles. It became clear, too, that dominance and recessiveness may not be all-or-nothing phenomena. Sometimes one allele may dominate its corresponding partner only partially. Often, the degree of dominance is influenced by the presence and absence of other genes in the genome. Thus the genome as a whole conducts a running dialogue with itself.

One surprise was that genes are not entirely located on the chromosomes, within the cell nucleus. Some exist in 'organelles' (structures with special functions) in the body of the cell: specifically, in the mitochondria (which control cell respiration) and in the chloroplasts of plants (where photosynthesis takes place). These genes may greatly affect the function of the main, chromosomal genes. For instance, some mitochondrial genes produce sterility in males: a fact that plant breeders in particular have made great use of.

This makes it possible to produce 'female lines' that will not pollinate themselves, and so facilitates the production of F1 hybrids.

The basic mechanism of 'differentiation' also became clear: how it is that all the different cells of an animal or plant, in liver or muscle, root or leaf, presumably contain the same genes, and yet the tissues are so different. Answer: different permutations of the genes are turned on ('expressed') in some cells, and turned off in others. Over a creature's entire lifetime, the genome as a whole acts as a programme: at each stage, suites of genes are expressed that are appropriate to that stage, and are turned off at other times. Embryos look very different from adults because they are expressing different parts of their genomes. Nowadays, with modern molecular techniques, biologists can actually watch the different genes turning on and off.

All these insights took decades to unfold, although most of the above was in place by the 1960s; but it all seems to follow from Mendel's initial insights as night follows day. You can see intuitively how by playing with a few simple ideas (and they are simple in essence, now that a succession of very bright people have made them clear) it is possible to explain all the astonishing caprices of inheritance. The insights of genetics did not replace the methods and lore of the old, pre-scientific breeders. But they made it possible to see why the tricks of the craft breeders were sometimes successful, and why they sometimes failed, and in general explained the huge raft of difficulties. In general the twentieth-century geneticists built on the techniques of the craft breeders but, because of their greater insight into the underlying rules, they were able to devise far more sophisticated programmes that produced dramatic results far more quickly, and were far more directional. For instance, one of the outstanding triumphs of early-twentieth-century breeding was with maize: the creation of parent 'lines' that were not particularly promising by themselves, but produced excellent F1 hybrids when crossed, which far out-yielded the earlier varieties. (However, aficionados are wont to point out that bourbon whiskey made from hybrid maize is nothing like so good as the kinds that were brewed and distilled from the old-fashioned varieties. Nothing is for nothing.)

In short, Mendelian genetics has taken the world a very long way, and those who believe that it will be superseded by the new biotechnologies, notably genetic engineering, are simply mistaken. Those who believe that genetic engineering already dominates world farming, as some politicians apparently do, simply have no knowledge of history or of present-day actuality. They rely too heavily for their information on business lunches. Nonetheless, the twentieth century has given rise on a broad front to a range of what can properly be called 'biotechnologies'; and although these do not and should not dominate agricultural strategy, they have made a difference.

Phase IV: The Age of High Science and Biotechnology

We can reasonably define 'high technology' as the kind of technology that emerges from science; and 'biotechnology' can then be seen as high technology of a biological kind. Thus traditional agriculture involves many technologies with biological intent but it is not 'biotech' because it does not, traditionally, partake of formal science. In practice, the twentieth century and particularly the late twentieth century has given rise to two main categories of biotech. The first has to do with body cells, including reproductive cells, and with the tissues compounded from cells. The second is the high technology of biological molecules.

The Biotech of Cells and Tissues: Cloning and the Rest

The biotechnologies of cells and tissues include cell culture, which dates from the early twentieth century; artificial insemination, which was first recorded in the eighteenth century but became a practical proposition in the mid twentieth when it became possible to freeze the sperm of cattle and humans in particular without killing it; embryo transfer, which for example has been used for rapid upgrading of cattle herds; in vitro fertilization, or IVF, which first came seriously on board in the late 1970s (in humans, but also in other species); and, finally, various forms of cloning, including cloning by means of nuclear transfer. Cloning by nuclear transfer was first achieved successfully in mammals in the 1970s, and culminated in the mid 1990s at Roslin Institute, near Edinburgh. There, Ian Wilmut and Keith Campbell first cloned two young sheep (Megan and Morag) from cultured embryo cells; and then, more famously, created Dolly from a cultured adult sheep cell. A year after Dolly (in 1997) Keith Campbell and others created Polly, who was not only cloned but was also genetically transformed. In other words, Polly has been genetically engineered, and indeed qualifies as a genetically modified organism, or GMO.

In the 1980s scientists (particularly in the United States) dreamed of producing cloned animals not to abet human medicine, which is the motive behind Megan, Morag, Dolly and Polly, but simply so as to replicate outstanding ('elite') livestock, particularly cattle. To a limited extent they succeeded, though with techniques that were far cruder than those developed at Roslin, but this early venture was not a commercial success. Some people don't like the idea of cloning at all, while others clearly feel that if any available technology can be made to work, and will make money, then why not? My own position is a compromise: in general, that the cloning of sentient creatures is not to be undertaken lightly, but that to some limited extent ends may justify means. The research that produced cloned sheep at Roslin could revolutionize huge areas of medicine, for treating and preventing conditions as diverse as diabetes and Parkinson's disease, and I feel it would almost be a sin to hold up such studies (provided they are carried out with proper decorum, which is the case). But such heavy-duty interventions should not, I feel, be deployed simply to reduce the price of milk and beef (or, rather, to reduce the cost of producing it, which is not the same thing at all).

But the modern trend is to turn all available technologies into cash as soon as can be arranged. Thus in November 2002 the British press announced the birth of Genesis, a Holstein heifer who in fact had been cloned two years before from Zita, an American prizewinning dairy cow who produces 4,000 gallons of milk in a year. This is about twice what the milkiest cows produce in Britain, about four times the British national average, and twelve times more than a wild cow. In 2003 (after the text of this book is complete, but before publication) the US Food and Drug Agency is liable to grant a virtual carte blanche to cloned cattle. Milk from cloned cows could (we are told) be 'on the American breakfast table' by 2004. Milk from cloned cows will be chemically identical to milk from conventional cows (or at least, there is no reason to suppose otherwise). It will carry no special health risks. Cows find it difficult to carry the udders required to produce 2,000 gallons a year, but farming, we are increasingly told, is just 'a business like any other' and it is taken to be self-evident that everything must bow to the dollar. Chemical distinctiveness and special health risk are in general the only qualities that require special labelling, so we need not expect to be told that the breakfast milk comes from cloned cows. There will be no opportunity to object on aesthetic or moral grounds. Aesthetics and basic morality simply do not feature on the list of criteria that are considered worth taking into account. Many would argue still, even in this debased age, that a world stripped of aesthetics is not fit for human beings at all. Try telling that to a panel of modern politicians, financiers and scientists.

The Biotech of Molecules: Genetic Engineering and the Rest

The other group of biotechnologies involves the manipulation of molecules, and in particular of DNA, the stuff of which genes are made. The most powerful and far-reaching of these biotechnologies is properly called 'recombinant DNA technology' since it involves recombining pieces of DNA from different organisms. But it is colloquially known as 'genetic engineering'; and it is genetic engineering that gives rise to 'genetically modified organisms' or GMOs. Conventional breeders, both Mendelian and pre-Mendelian, can acquire new genes for their crops or livestock only by cross-breeding, and successful crossing is possible only between organisms of the same species or (sometimes) closely related species. But genetic engineers in principle can take genes from any organism and transfer them into any other: human into bacterium, cauliflower into cow, mushroom into cauliflower. Through genetic engineering, every organism becomes a part of every other organism's gene pool. More than this: genetic engineers in principle can remove genes, modify them, and then put them back. In principle, too, and sometimes in practice, they can even create brand-new genes in the laboratory, without precedent in nature, and introduce them into bacteria, or plants, or animals, or what you will. These are still early days, but the potential is astonishing.

Genetic engineering is now abetted by two further molecular technologies that greatly enhance its power. For one of the main problems has been to know which pieces of DNA are actually worth transferring – in other words, which pieces of DNA correspond to which genes. This is now being unravelled by the science of genomics, which aspires to describe the entire complement of DNA in an organism, and to show how which stretches of DNA function as which particular genes. Most famously, the human genome has now been at least roughly mapped; but similar work is well in progress on livestock including pigs and poultry, and a swelling catalogue of crops. In *Science* in 2002 (vol. 296, pp. 79–92 and 92–100) a team from Switzerland reported the draft gene sequence from the rice subspecies *japonica* (which is favoured in Japan) while a Chinese group reported similar studies on the subspecies *indica* (which is widely grown in China). The complete picture should be forthcoming by 2004. All the cereals are fairly closely related, so the rice genome is a good model for all of them; and studies have begun on maize, barley and wheat.

The second great subsidiary biotech is proteomics, which is the art and science of predicting what kind of protein a given chain of amino acids is liable to form, and how that protein will behave. The time will come when scientists will be able to specify what kind of protein they need to perform a particular function, then specify the sequence of amino acids in that protein that will produce the required result. Then, using the knowledge of the genetic code that is already to hand, they will be able to design and synthesize DNA of the kind that will produce the required amino acid sequence. This is power indeed.

DNA itself was first discovered in the late 1860s by a young Swiss physician called Johann Friedrich Miescher, first in the cells of pus and then in the gonads of trout. Orson Welles's comment in *The Third Man* that the Swiss in 500 years had contributed nothing to world culture but the cuckoo clock, is thus resoundingly untrue. (A Swiss scientist provided DDT, too, as we have seen.) Miescher showed that his new discovery is an acid, that it is rich in phosphorus and has very large molecules. He called it 'nuclein'. A student suggested the term 'nucleic acid'. Soon it became clear that DNA was compounded of many thousands of sub-units, which were termed 'nucleotides'. Each nucleotide had three components: a sugar, which was deoxyribose; a phosphate radical; and a 'base' – an organic (carbon-based) entity containing nitrogen. By the 1920s it was clear that there were two kinds of nucleic acid. In one of them – the one that Miescher first discovered – the sugar in each nucleotide was deoxyribose; so the whole molecule was called 'deoxyribose nucleic acid', or DNA. The other contained a slightly different sugar (effectively the same one with an extra oxygen), called ribose. Hence: ribose nucleic acid, or RNA.

But it wasn't clear until the 1940s that DNA is actually the stuff of which genes are made. The general reason for this is that living systems are extremely complex and nature does not come with labels attached, and all serious knowledge is hard to come by. But there were also specific reasons.

A succession of scientists in the early twentieth century showed that genes exert their effects by producing proteins: and proteins are (potentially) infinitely versatile molecules which, in effect, run the entire body. They form much of the substance of cells; some of them function as hormones; others as antibodies; and many – crucially – function as enzymes, which are the catalysts that control the whole body metabolism. As we have seen, it was known from the early twentieth century that genes are carried on the chromosomes. But the chromosomes have two main components: one is DNA and the other is protein. Which of them was the stuff of genes? For a long time DNA seemed a poor candidate. Its basic chemical structure seemed too simple: just a string of four different nucleotides. Surely such a (conceptually) simple molecule could not provide the code for the potential infinity of proteins, which in any case were compounded from twenty different amino acids. It seemed that the genes must themselves be made of protein. Thus the protein component of the chromosomes was taken to represent the genes, while the DNA was construed simply as the glue that held the chromosome together.

That DNA is the stuff of genes was finally made clear in the 1940s. It was known by then that bacteria pass on genetic information to each other in an ad hoc kind of way: essentially a form of sex. A bacterium that has received genetic information from another is said to be 'transformed': the same term that the genetic engineers employ, for they in effect make use of an analogous process. If the chemical make-up of the genetic messages that pass between bacteria could be discovered, then the true nature of genes would be revealed. In 1944 the Canadian Oswald Avery and his colleagues provided the answer. They showed that if the genetic message that passes between bacteria was attacked en route by enzymes that break down protein, then the message got through anyway. But if DNA was removed from the message, then nothing got through. Genes, therefore, were made not of protein, but of DNA.

So it was finally acknowledged that DNA is not just glue for sticking chromosomes together but is, in fact, the most important molecule in nature. Biologists on both sides of the Atlantic set out to unravel its structure and explain its modus operandi. Many contributed, but as all the world knows the race (for such it became) was finally won in Cambridge, England, by the English physicist Francis Crick and the American biologist James Watson, who in 1953, using wire and cardboard, cobbled together the first model of the instantly famous 'double helix'.

Thus was born 'molecular biology' (which is usually taken to mean the science of DNA) and 'molecular genetics' (in which genes are perceived not as abstract entities, but as lengths of DNA). In the twenty years after 1953 work proceeded extraordinarily quickly. It became clear almost immediately that RNA, the second form of nucleic acid, acts as the intermediary between DNA and proteins. As Francis Crick put the matter, 'DNA makes RNA makes protein'. Soon, too, Francis Crick and Sydney

Brenner showed how it is that DNA, with its relatively simple structure, could provide the code for proteins, which seem infinitely various. It transpires that the nucleotides in DNA operate in groups of three. Since there are four different nucleotides, sixty-four different combinations of three are possible (4 x 4 x 4). Each combination of three is known as a 'codon'. The sequence of codons in DNA corresponds to the sequence of amino acids in protein, and since there are sixty-four possible different codons and only twenty-four different amino acids, DNA can clearly carry all the information that's needed with ease, and indeed with plenty of spare capacity. In the same way the twenty-six letters of the alphabet, plus a few accents, code all the sounds needed to create all the hundreds of thousands of words in the English language and most of those in the thousands of other languages in the world as well. The mutations that Hermann Muller identified when he worked with Thomas Morgan result when some change is made in the order of nucleotides in the codon. In short, everything fell into place beautifully.

In the 1960s, too, it became clear that the DNA molecule itself is served by a whole coterie of enzymes (enzymes that were made in the first place by the DNA) which help the DNA to replicate and to repair itself, and to correct mistakes that occur during replication. So in 1972 Janet Mertz and Ron Davis, in California, used combinations of these enzymes to cut a piece of DNA out of a bacterium, and to insert that fragment into the DNA of another bacterium. This was an artificial procedure, of course; but Mertz and Davis made use of those molecular tools that nature has provided itself, for nature's own purposes. The transferred pieces of DNA were recombined with the host's own DNA by the host's own enzymes: hence, 'recombinant DNA'. It seemed that by this means, any gene from any organism could be transferred into any other, just as any bit of a Ford engine can be transferred into any other, and work accordingly. Hence the soubriquet, albeit in many ways misleading, 'genetic engineering'.

It was obvious from the outset that genetic engineering was one of those technologies which, like steam and electronics and nuclear power, had the potential to take humanity and hence the world at large into a new age, with new ways of thinking and new economic structures. Many in recent years have chosen to deny this, suggesting that genetic engineering is just one more technology among many, and is merely an extension of conventional breeding. But although genetic engineering may be (and so far in practice generally is) deployed simply in the low-key, ad hoc fashion that this disclaimer implies, it clearly has connotations that extend far beyond conventional breeding. Non-scientists feel this intuitively, and the more that scientists and industrialists deny the broader connotations the more they generate mistrust. As an ordinary citizen, after all, one is always somewhat wary of the power wielded by experts; but one becomes positively alarmed if the experts themselves seem unaware of the implications of their own work. In the same way, advocates of nuclear power generate no confidence at all when, in the face of Three Mile Island and Chernobyl, they continue to insist that nuclear power is perfectly safe. Don't they read the newspapers? one is inclined to ask.

In fact, many scientists did perceive right from the beginning that genetic engineering does have enormous implications and in a letter to *Science* on 24 July 1974 the American molecular biologist Paul Berg (who later received a Nobel Prize for Chemistry) warned of the inherent dangers, and went on to suggest a moratorium. His and his colleagues' caution seems exactly comparable with that of some at least of the early nuclear physicists. Such caution does raise public confidence, precisely because the experts are seen to be taking serious matters seriously. Today, many science ethics committees on both sides of the Atlantic continue to discuss the intricacies (and I personally have great respect for Britain's Agriculture & Environment Biotechnology Commission, led by Professor Malcolm Grant of Cambridge, with which I have had some dealings, and which, among other things, holds all its meetings in public). The existence of such committees and the continuing involvement of very good people is enough to demonstrate that there is still a great deal to be discussed. Nonetheless, American biotech companies in particular have ploughed ahead with a range of GM crops (notably soya and rapeseed) and leading politicians, including Britain's Tony Blair, certainly give the impression that they believe that GM, like nuclear power, is a fait accompli at least in principle. It's their confidence that's so disquieting. One feels (as Niels Bohr said in the early days of quantum mechanics) that anyone who thinks the matter is easy has not understood the problems. So what are the difficulties?

8

GMOs and the Corruption of Science

Genetic engineering can be seen as the highest of agricultural high tech, at least up to now. GMOs are extremely significant in their own right – although not necessarily in the ways that their advocates would have us believe; and they illustrate the problems of all science, and all technology, when applied to any human activity. The issues are practical (what is actually done? is it helpful? is it beneficial?); and also have to do with the philosophy of science (in particular, how certain can we be of our own knowledge?); and with politics (who is doing what to whom and why, and is it good?).

The first point – which some of its advocates continue worryingly to deny – is that genetic engineering really is a heavy-duty technology which represents a qualitative shift in humanity's ability to manipulate living systems; and the fact that it has produced such a qualitative shift is itself important, and suggests the need for fresh discussions. Of course, human beings have been consciously altering the genetic make-up of their fellow creatures for thousands of years (and we have shaped our own genetic outcome, too, through mate selection and arranged marriages), and so genetic engineering can properly be seen as an extension of an ancient and inevitable process. But it isn't *just* an extension of ancient craft, or of classical (Mendelian) applied genetics. It brings significant new powers to bear.

Until the 1970s, as for thousands of years previously, breeders could make changes in their fellow organisms only by crossing and selecting. True modern breeders produce many a mutation within their breeding lines both by radiation and by chemical 'mutagens', so as to increase the range of variation they have to work with. But these are random strokes. In general conventional breeders are still obliged to work within the boundaries imposed by nature. Cross-breeding is their principal agent of radical, controlled alteration, and this can be effected only within species, or between closely related species, by sexual means, generation by generation.

But with genetic engineering it becomes possible to transfer genes in one single stroke between any two organisms, related or not; or indeed, in time, to endow any one organism with genes from dozens or hundreds of other organisms, so that a wheat might be fitted with genes from bacteria and mushrooms and oak trees and cattle, and indeed from human beings. Or a plant might be given genes made from scratch in the laboratory that are not found in nature at all. Crop plants could be like genetic Christmas trees. Whether they started out as wheat or oats or rye would no longer be relevant; all that would count would be the qualities with which they were adorned. GMOs of the future need not even be recognizable as plants, animals or fungi. Flesh itself could be re-conceived. Future biotechnologists will create the designer organism.

All the great advances in biology in the twentieth century were preceded by learned statements from leading scientists solemnly asserting that whatever it was that was being mooted was 'biologically impossible'. When I studied biology formally in the 1960s we took it to be self-evident that it was and would always be biologically impossible to transfer genes between unrelated organisms, for example between animals and plants. Yet by the 1970s this was achieved. In the 1980s a leading German embryologist declared in the learned journal *Nature* that 'cloning of mammals by simple nuclear transfer is biologically impossible'. Yet this was the technique that produced Dolly barely more than a decade later. I first floated the idea of the 'designer organism' in print in the 1990s (though I don't claim I was the first to think of it) and was solemnly assured by a well-known professor of biology that I was fantasizing. Now (November 2002) comes news that Craig Venter, who helped to unravel the human genome (his team in America worked in parallel with Sir John Sulston's team in Britain) is intending to produce a completely novel bacterium, with an assemblage of genes unprecedented in nature; and 'designer organism' is precisely the term he has applied.

All in all, it should be clear as we enter the twenty-first century that the expression 'biologically impossible' has become obsolete. Everything is biologically possible provided only that it does not transgress the laws of physics – or at least that has become the only safe and sensible assumption. The only restraints apart from those bedrock laws will be those we impose upon ourselves, through the law and – much more importantly – through the sensibilities that in the end provide the only safe foundation for law. In principle, we can rethink flesh itself from scratch, creating living systems that go far beyond Craig Venter's initial forays. Such visions may not be realized in the twenty-first century, but scientists should still be at work in 500 years, or 1,000, or 10,000, and these are very early days in the history of biotechnology, or indeed in the history of science. It is not irresponsible to speculate, as some conservatives continue to insist. Now that we can see that anything must be considered possible, it is irresponsible not to. If this is not a qualitative shift in our relationship with nature it is hard to envisage what would be. Yet genetic engineers, industrialists and politicians continue to insist that there has been no conceptual shift.

Why does all this matter? Isn't it good for humanity to have such power? Isn't the whole point of scientific endeavour to acquire such power? Isn't it perverse to want to call a halt, or suggest any change of direction? Many argue this way; as if it was somehow our destiny to follow our scientific noses wherever they may lead; as if technological change was simply an adventure, not to say a lark. But actually, as suggested in earlier chapters, science should be seen primarily as an aesthetic pursuit, a means to appreciate nature. The idea that there's some compulsion always to turn the findings of science into new technologies, putatively to make life more comfortable and to generate wealth, is an add-on, and a pernicious one. We really ought to be prepared to leave some discoveries on the sidelines if they don't lead us where we think we ought to be going. This principle is already well established in other fields. Even war has rules. Moral and aesthetic restraints should be applied to all our endeavours. Powerful science doesn't have to be converted into powerful technologies, however tempting this might seem to some active players.

So what are the guiding principles in this case? Why are so many industrialists and politicians now urging us to press ahead with GMOs with all possible speed? What is the case for caution? How do we judge?

Risk–benefit analysis is necessary, of course: to consider all the possible advantages and conceivable dangers and see which comes out on top. Yet such analysis, though necessary, is not sufficient. Risk–benefit analysis is applicable to technical matters that in principle can be quantified; for instance, whether a GM crop is or is not nutritionally inferior (or superior) to one produced by conventional means (though in truth this is very hard to assess). But as with all technologies – and especially in agriculture – the issues are not simply technical. They are also aesthetic, social and ethical, and such matters are not easily quantified. If risk–benefit analysis is all that's applied, then the broader connotations tend to be ignored altogether – which has commonly been the case whenever new technologies have been introduced, in any context. We should come back to this. But let's see, first of all, where risk–benefit analysis can lead us. First, the putative advantages.

The Pros of Genetic Engineering

It's easy to be excited by GM crops. I was when I was first writing about them in the 1980s, when the idea of them was still new. You can begin simply by making a list of all the improvements that seem desirable in crops and livestock, and asking what genetic engineering might contribute. What deficiencies might it help to make good? What more, as yet undreamed-of possibilities might it raise? Such lists seem to write themselves, and they include items of world significance – potential life-savers for many communities worldwide – and others that may seem more whimsical but are enticing nonetheless.

For example, it is a prime requirement of plant breeding to provide crops that are more resistant to stresses of all kinds: drought, heat, flood, soil toxicity including salinity, pests, diseases. The need becomes ever greater as human numbers grow and more and more people are obliged to raise crops in marginal land; and as the climate changes, it seems that whatever is to be done must be done quickly. In general it takes a dozen years to produce a new variety of wheat, say. But the greenhouse effect could transform the climate of a continent in a season.

Important programmes are already in progress. For the people of the Sahel, to the south of the Sahara, the staple crop is sorghum, which is astonishingly resistant to drought and heat (the soil is often too hot to stand on) – but not quite resistant enough. Scientists from ICRISAT (the International Crops Research Institute for the Semi-Arid Tropics) searched the international gene banks for relatives of sorghum that could be crossed with sorghum to provide the genes needed for super-toughness, but found none that could do the trick. The last time I spoke to them they were wondering if such genes might be introduced from groundnuts, which are ridiculously tough. But groundnuts are legumes (relatives of beans) while sorghum, a cereal, is a grass. So the necessary genes could not be introduced by conventional breeding. Genetic engineering would be necessary. Here (if it can be made to work) is a prime example of the highest of technologies deployed to help the world's poorest people. For people in some of the harshest environments, such science could in principle be a godsend.

For less beleaguered peoples we might envisage truly frost-resistant strawberries – perhaps by introducing the genes that enable Antarctic fish and crustaceans to operate and indeed to move at speed at sub-zero temperatures. Such a feat has been attempted in various forms. Those same strawberries – or cucumbers, or kiwi fruit, or what you will – could be combined with Virginia creeper. Suburban houses or urban skyscrapers could drip with produce that could be harvested just by leaning out of the window, all year round. This is whimsy of course; though any move to make fresh fruit and vegetables more available in cities and suburbs is surely welcome.

Somewhat less whimsically, it should be possible to fit crop plants not with genes that kill insect pests (as is already in train) but with genes that produce the insects' own alarm pheromones, and warn them to stay away. Aphids produce such pheromones when attacked, and they are highly specific. If the warning is intended for aphids, then bees and butterflies would still feel welcome. Alternatively, hedging plants could be engineered as insect lures: fitted with pheromones that attract pests away from the crops. If such seductive plants were also crossed with something like Venus fly-trap, and so were insectivorous, they would dispose of the pests at the same time. Shades of the Pied Piper; or indeed of the giant Indonesian flower *Rafflesia*, which smells of rotting meat and so attracts flies, which it then digests. If those alluring, engineered plants were also succulent and non-toxic, they could (when well fed on putative pest insects) make marvellous fodder for goats and cattle. Here again could be a very high technology that would be of particular value to subsistence farmers in the world's poorest communities.

Anything, in short, that you might care to imagine could come about. Yet however valuable or amusing such enterprises might seem, they are bound to ring alarm bells. Any attempt to take liberties with nature is liable to misfire. Theory and some early experience of GMOs suggest that such misgivings are entirely justified. This does not mean that we should halt all genetic engineering in its tracks. It does mean that public mistrust of the unalloyed enthusiasm in high places is very much to the point.

The Drawbacks

Genetic engineering even at its simplest implies the ad hoc introduction of exotic genes into the genomes of established organisms; and this, in principle, immediately suggests a hierarchy of possible problems.

Most obviously, the newly introduced gene could disrupt the host genome in undesirable and quite unpredictable ways. The theoretical problem can readily be seen through an analogy. It's often said that the genetic code is 'digital', and so in a general way it is. Each gene and so, by implication, each functional length of DNA, corresponds to some specific 'bit' of information. We get closer to reality, though, if we compare genes to language (as in the title of Steve Jones's 1993 book: *The Language of the Genes*). Individual genes are then compared to words. But the meaning of individual words is not to be captured in their stripped-down, dictionary definitions. Anyone who tries to speak a foreign language out of a dictionary knows how droll the natives find such efforts. The meaning of words depends very much on their context – what words they are surrounded by. Behind the dictionary definitions of individual words lies the syntax of the language, and the actual use of it: the colloquialisms, the cross-references, the historical allusions, the puns. Genes work in this way too because genomes evolve, trailing their history behind them. They are not simply 'digital', but work to rules that are in part logical and in part a matter of historical accident. If genes are compared to words, then the genome of any particular creature as a whole should be compared to literature. Genetic engineering is not really engineering. It is more like gardening, where you plant and then stand back and watch; or, to pursue the present metaphor, it is more like editing. Every writer knows that the injudicious alteration of a single word can change the import of a text absolutely, and prays for a gentle and competent editor.

At present, after 100 years of formal Mendelian genetics and a few decades of genomics, we have some small insight into the function of a few genes in a few organisms (including a few human genes). For some organisms, in short, we have the beginnings of a dictionary. But the genome of an organism – any organism – might be compared, in literary terms, to some sacred, poetic text written in a language of which we have virtually no inkling: medieval Tibetan, or Linear B. Would you, or anybody who was halfway sane, undertake to edit such a text if all they had to guide them was a bad dictionary?

Whether the editing of genomes has evil consequences in practice depends very much on the organism in hand, and what is done to it. But we can easily envisage (and to some degree have already witnessed) undesirable outcomes. One problem with crop plants is that many of them have toxic wild ancestors (including virtually everything that isn't a cereal – cereals on the whole are remarkably benign). Some of the modern crops have presumably lost the genes that made their ancestors dangerous. But in others, the genes of toxicity may well remain. They might simply be repressed; for how a gene operates, and whether it operates at all, depends very much on the company it finds itself in. If a novel gene is dropped ad hoc into some modern crop it could well (in principle) reawaken undesirable genes that have been lurking; provide the very stimulus they need to enable them to be expressed. Thus an innocuous gene added to an innocuous crop could have a very undesirable outcome. This can be tested, of course. Rats can be recruited to sample the novel crop, in the time-honoured fashion. But once a novel gene is introduced into a crop-breeding programme, it could subsequently, by further breeding, be introduced into a wide range of varieties, each providing a different genetic background. In short, the possible ill effects of the initial introduction would not necessarily become apparent within one generation. Years might pass before toxic strains appeared. All this is theoretical. But it's the kind of theory that needs to be taken seriously.

With animals as with plants: it is not difficult to envisage applications of genetic engineering that could be very helpful. In principle, after all, animals might be produced that are innately resistant to, say, foot-and-mouth disease. This would be a huge bonus, not primarily to prevent epidemics like the one in Britain in 2001, which should have been preventable by other means, but to eliminate the infection from, say, the whole of Asia, where it is largely endemic. Foot-and-mouth was traditionally allowed to run its course in India (though now vaccination is common). But it can permanently depress the productivity of cattle (milk yield, ability to pull ploughs) and causes the animals, and their owners, a great deal of misery. If it were possible to produce foot-and-mouth resistant cattle by a breeding programme that began with the introduction of some exotic gene, this surely should not be dismissed out of hand.

But animals raise a new issue: that of welfare. Sometimes (as with putative foot-and-mouth resistance) genetic engineering could improve animals' lives. But often it would not. Within the next few centuries (which isn't so long) it would surely be possible to create, say, a cow that was reduced entirely to an udder (which could be fed artificially), or a sow as prolific as a termite queen – effectively a giant uterus. You may feel this is simply grotesque: alarmist nonsense. But in the spring of 2002 British television featured the featherless chicken, bred (so its creators assured us) to feel more comfortable in the tropics. Chickens are, of course, descended from Indian jungle fowl, and are tropical to begin with. (It is truly alarming that many modern biotechnologists, though highly competent technically, are not biologists at all. They may literally have no idea what the creatures they work with actually are. Such technologists are 'trained', but at no stage are they educated.) More to the point, chickens are sentient creatures, and we can reasonably infer that feathers play a large part in their lives, just as our own physical appurtenances do in ours. I have been to conferences, too, where commercial agricultural scientists seriously discussed the production of entirely brainless chickens that wouldn't mind what was done to them. Nothing stands between us and the cow-qua-udder or the sow-qua-uterus or the brainless, featherless chicken except human sense and sensibility; and there has long been abundant evidence that within some of the most commercially successful reaches of livestock farming both have long since been abandoned. Intensive turkeys have been reduced to globes of sticky white flesh. It is technically wrong to suggest that genetic engineering is simply an extension of traditional breeding but it is entirely correct to suggest that the creature reduced to mindless and shapeless flesh would merely be an extension of existing, hyper-intensive production. If we don't like the idea of it, then we ought to be saying so while we still have the sensitivity to respond.

More immediately, a gene dropped into an established genome can easily throw the entire organism off course. The result is a monster. If the organism in question is a bean, then perhaps this does not matter. It's one more failed plant for the bonfire. But if the creature in question is a pig, a sensitive being, then we really should take such outcomes seriously. Crude attempts were made to produce a fast-growing pig in the 1980s (the infamous 'Beltsville pig') by adding a gene that would produce more growth hormone. The resulting creature was so beleaguered, shapeless and unable to stand, that even its creators felt obliged to have mercy and put it down. In practice, too, the routine genetic engineering of animals is not practicable without employing the technology of cloning, of the kind that produced Dolly, and then Polly. Genes are introduced into cultured cells and entire new creatures are created from the genetically transformed cells. In fact, Dolly was not genetically transformed but her successor, Polly, was. Dolly herself was not produced without casualty, however. There were about 300 failures along the way and although most were very early abortions (the new embryos dying when they had just a few cells and were well short of sentience) there were also late abortions, neonatal deaths and deformities. In short, the creation of genetically engineered animals is bound to leave a trail of failed creatures that would be very miserable indeed.

I would not presume to criticize the research that led to Polly. She has been produced not for agriculture, but for medicine. She is fitted with a human gene that produces a blood-clotting factor (a protein), which she secretes into her milk. She will give rise to a dynasty of offspring who in principle will need no further manipulation, but will simply breed like any other sheep (except that they will be more cosseted). The protein factors they produce in their milk will be used to treat forms of haemophilia, and any procedure that reduces the use of human blood (especially now that Aids is with us) has to be welcomed. In this context, the reward certainly seems (to my mind) to justify the means. But as discussed in Chapter 4, there is very little reason to raise the world's output of meat at all, and absolutely no reason at all to do this by increasing the growth rate of animals that already grow far faster than common sense says is reasonable. The urge to produce GM animals in an agricultural context is purely commercial. Yet it isn't good commerce: we are not talking here about the satisfaction of serious need or reasonable demand, or a fair day's money for a fair day's work. In this book I have avoided using emotive and generally somewhat vacuous terms like 'greed'. But greed is the only motivation for any endeavour in genetic engineering that is intended simply or primarily to increase the growth rate of livestock.

All these issues spring from just one problem: that it is impossible to predict the influence of any one added gene on the genome as a whole. But GM crops in particular raise a further series of possible drawbacks – for each of which there is already some evidence: not much, for these are early days, but enough to show that they should be taken seriously. To begin with, a plant that has been fitted with a gene that kills pest insects (which is a common line of research these days) may finish up killing non-target insects. Thus, studies in the United States already show that pollen from GM maize adversely affects wild monarch butterflies. You might say (as some American agriculturalists do say) 'So what?' but if we think that our fellow creatures and the environment at large have any worth at all, then we ought to be alarmed. Even if monarchs per se are not considered important, they can be taken as warnings, like miners' canaries.

Then there are two theoretical mechanisms by which genes introduced into crops could find their way into wild plants. Firstly, some crops have wild relatives that grow locally, and they may cross-breed with these relatives. This is not true of maize grown in Britain (maize is very much an American plant), but it is true of rapeseed wherever it is grown. Rapeseed is a member of the botanical family Cruciferae and more specifically of the genus *Brassica*, and brassicas are notoriously promiscuous. William Cobbett made this point as long ago as 1822: 'This very year, I have some Swedish turnips . . . but [they are] a sort of mixture between that plant and rape . . . The best way is to get a dozen of fine turnip plants, perfect in all respects, and plant them in a situation where the smell of blossoms of nothing of the cabbage or rape or even of the charlock kind, can reach them.'* Old-style farmers knew all about the dangers of outbreeding between crops and wild plants.

Secondly, biologists are more and more aware of the phenomenon of 'horizontal transfer': transfer of genes between unrelated plants (or indeed unrelated organisms of any kind) by viruses. Nature herself, in short, practises genetic engineering. Of course, any gene might in principle be transferred from creature to creature by some wild vector. But it is at least sensible to inquire whether a gene that was introduced via a vector in the first place (by the genetic engineer) might not be more inclined than usual to hitch a ride on a wild vector. These questions need to be asked. If genes that confer toxicity to insects or super-resistance to drought or salinity or whatever did find their way into wild plants, then they could generate a new race of super-weeds (comparable, say, with the pesticide-resistant populations of insects that made such nonsense of many early control methods).

Or, of course, the engineered crop itself might escape and become a serious pest. Many domestic plants worldwide – including both crops and garden ornamentals – have escaped to become major pests, and plants pre-equipped to be super-resilient could be formidable opponents indeed. California is awash with Mediterranean escapes (and eucalyptus

* *Cottage Economy*, Oxford University Press, 1979, p. 89.

from Australia and kudzu from Japan), and the Mediterranean is full of American plants (including cacti). GM crops of the present generation include strains of rapeseed equipped with genes that make them resistant to herbicide. The farmer can plant them, and then spray the entire field, leaving the protected plants to grow on. In theory, this can have environmental advantages: for example, a farmer can allow weeds to grow for a time, confident in the knowledge that he can zap them when he needs to. But American farmers already complain that seeds from last year's rape crop, left behind in the fields, can make very resilient weeds the following year. By the same token, the super-resistant sorghum plants intended for the Sahel might spread to become a serious bane through all of dry Africa. In general, genetic engineering at best is more like gardening: you plant the gene in the genome, and then stand back and see what happens. Once we take the whole ecology into account, we might compare GM technology to genetic fireworks. Light the blue touch paper, and retire. But even that analogy does not quite convey the essence of the problem. Fireworks explode, and then they are done. Genetic fireworks might smoulder and fizz through wild populations for ever.

Of course, there are experiments and trials afoot in many countries to see if any of these conceivable disasters could come about. Industrialists and politicians alike have rounded upon the protesters who, particularly in Britain, have pulled up the trial crops and disrupted those trials. Is it not perverse (the GM advocates demand) to disrupt the very experiments that are designed to supply the missing information? Yet as things stand my sympathies lie with the protesters. Supporters of the status quo, eminent though many are perceived to be, just don't seem to have thought things through.

GM advocates are wont simply to argue that 'there is no evidence' that, say, GM crops are particularly toxic, or that GM maize kills more than a few marginal creatures of which we know little, and so on and so on. Yet this general defence of genetic engineering is answered by the most elementary adage in the philosophy of science: 'Absence of evidence is not evidence of absence'. On the other hand, the evidence that might be accumulated from the (largely interrupted) trials can never be sufficient. Absolute certainty is of course too much to ask for, but we ought surely to ask for the kinds of standards demanded in courts of law, to remove 'all reasonable doubt'. New technologies, however, are innately uncertain. Civil engineers illustrate the point with alarming regularity: it's truly remarkable how often modern buildings fall down.

A lovely new footbridge, taut and fine as a piano string, was built over the River Thames in central London in 2000 – the 'Millennium Bridge'. The design was new but the physics was of the kind that Isaac Newton first worked out 350 years ago, and the materials have been investigated to the last molecule, each delineated by a dozen coefficients. Yet when people set foot on it, which was supposed to be the idea, it swayed. Alarmingly. As Galileo commented in a somewhat different circumstance, 'And yet it moved'.

In civil engineering, every factor that seems halfway pertinent can be measured, and will typically have been studied for decades or indeed centuries. Giant programmes on vast computers work out every possible combination of circumstance. Still it may go wrong. We can partially explain these failures by the rules of chaos. Thus at any one time an effectively infinite number of factors may influence the outcome of any one series of events. The effect of any one factor is likely to be 'non-linear': that is, the relationship between cause and effect is not at all simple (not least because other factors influence the outcome). It is clear, however, that very small factors can have very large effects. The famous metaphor (demonstrated on computer simulations) is that of the 'butterfly effect': a butterfly that flaps its wings in Texas might (in principle) precipitate a storm in, say, Chile. Worse: it is impossible to know which of the many possible factors that could have an influence are liable to do so in any one instance. Worse still: it is theoretically impossible to know in advance what all the possible factors are that might come into play. Some matters of deep theory have become apparent only after particular pieces of novel technology have created the circumstances in which they become manifest. A famous example is that of the Comet airliners, once the pride of Britain, which, as they fell apart, first revealed the general phenomenon of metal fatigue.

If civil and mechanical engineers can get things wrong (regularly), when working with exhaustively explored physical principles and materials that have been measured every which way, how much more will we get things wrong in biology, where the complexities are multiplied by orders of magnitude, and – relative to that complexity – almost nothing is known? We drop novel genes into genomes, and exotic organisms into ecosystems, at our peril – our peril, and the world's. There is simply no way of knowing, a priori, what will happen.

Yet the most stunning point of all is again one of philosophy. We can envisage science as a way of lighting up dark areas of the universe – not physical regions, but phenomena that are not understood. Does the light of science at any one time illuminate the whole universe (apart from the odd shadow), like the floodlights in a football stadium? Or has it merely laid out a trail of torches across the fen? It is impossible to know. Formal science is still new, and it seems intuitively most likely that our knowledge so far is, to pursue this metaphor, little more than a parade of torches in the darkness. But we are standing in the little light there is, and cannot see the vast dark space between. So the line of bright spots looks to us like floodlighting. At any one time it seems to us that we understand much more than we really do. Clearly, though, it is logically impossible for us to gauge the extent of our own misunderstanding. We could see what we still have to find out only if we were already omniscient. In short, we can be sure at any one time that we don't know everything there is to know, and be reasonably certain that we don't know everything that may prove pertinent. Yet we can never know how much we don't know. Nature does not come with labels. The simplest insights are very hard won indeed. Nature, in short, cannot be second-guessed. All technologies are innately uncertain; and those that presume to trade in biological matters always run on a wing and a prayer. In short, genetic engineering seems to offer us unlimited power, and in principle it may do. But there is no corresponding omniscience, and never can be. If we do not find the combination of mounting power and inescapable ignorance innately terrifying it can only be because our senses have already been dulled.

Still, though, the advocates of genetic engineering point out that if humankind had always allowed itself to be put off by such philosophizing, we would still be living in caves. There can be no progress without some risk. Admittedly, the risk cannot always be accurately measured; but then, the immeasurability is part of the risk. All we can do, they say, is measure what can be measured before we begin – which is what the GM field trials are supposed to do. After that, we just have to stay alert to see if anything further goes wrong.

Even so, there are two large caveats. The first is that in nature very rare events may be important. Thus there may be only one chance in 100 billion that a gene that confers resistance to pests may escape from a crop into a wild weed; and one in 100 billion may seem very long odds indeed. But a plant may bear 100 seeds and a commercial field might contain 100 million plants and suddenly, one in 100 billion seems a fair bet. Yet ordinary field trials are most unlikely to detect such rarities. Secondly, disasters may be out of hand, or virtually so, before they are noticed at all. An example similar to the present context is the scourge of the possum in New Zealand. Possums are Australian marsupials: cute creatures, which the Australians love, favoured for their fur. New Zealand had no native terrestrial mammals apart from bats, and the European colonists introduced possums (and rabbits, rats, stoats, foxes, deer, sheep, and so on) to enrich the environment (or so they thought). But New Zealand did have a wonderfully rich native flora and fauna even without mammals, not least of reptiles and birds; and the possums have been laying the natives waste. In the dense forests on the steep hills they are impregnable. They were out of hand before anyone realized they were a problem. Similarly, blue whales may well have been over-hunted before the whalers acknowledged that they could be in danger, and they might yet prove to be in terminal decline. The world's climate may already be out of hand, even though the American leadership in particular has not yet deigned to acknowledge that it's an issue at all, or at least not one that America should modify its way of life for. And so on. Very rare events happen, and can be important; and they can be out of hand before the problem is realized. Every new technology is innately risky but sometimes the risks seem altogether out of proportion. When we contemplate dropping new organisms (or newly transformed organisms) into established ecosystems, the possible consquences are quite beyond prediction but could well be horrendous.

The more moderate protesters at this point tend to invoke the 'precautionary principle'. This is eminently sensible. In general, it acknowledges that new technologies are necessary, but urges that all possible attempts should be made to anticipate possible snags and, if in doubt, then whatever it is, don't do it. I suggest, however, that when the uncertainties are in principle very great (as in the case of GMOs they must be) and the dangers could be enormous and are potentially uncontrollable, then the usual form of the precautionary principle does not quite meet the case. We need a more stringent rule: that such technologies should not be floated at all unless we can see very clear advantages, not to say overwhelming necessity. The US Food and Drug Administration (FDA) already applies this principle. New drugs must be shown to be safe (insofar as this is possible), and must also be shown to be superior to whatever is on offer already. If this more rigorous principle was applied, how would the present generation of GMOs stand up?

The Realities of GM Crops

On 23 May 2002 Prime Minister Tony Blair made a speech to Britain's Royal Society, the most venerable of the world's scientific clubs. He effectively expressed the belief that without high technologies in general and genetic engineering in particular, the future population of the world cannot be fed. If this were really so, then the case for GMOs would be (almost) open and shut. Risky though they might be, if people must starve for lack of them, then the risks must be taken. Physicians apply the same principle. If the patient is dying, then heroic surgery and untried physic must be applied. To be fair to Mr Blair and other leading politicians, industrialists and some scientists, it seems they really do think that the world positively needs GMOs – crops, if not necessarily livestock – and so believe that the objectors are Luddite, self-indulgent, muddle-headed, gratuitously nostalgic, unrealistic and effete. In short, the defenders take the moral high ground. If their claim is correct – if we really do need GMOs to feed the pending 10 billion – then the high ground does indeed belong to them.

But actually, they are profoundly wrong. Mr Blair called for 'evidence' in his Royal Society speech, but the evidence has been misread. In general, defenders of the high-tech approach to agriculture argue and clearly believe that traditional methods have in general proved inadequate, and that farming worldwide and hence humanity have been rescued, effectively in the nick of time, by science and the technologies it provides. This is simply not the case. Specifically, advocates of genetic engineering clearly believe that GMOs have already enhanced output, and are poised to push the world's crops ahead by leaps and bounds, and that unless they do so, we will soon be in serious trouble. In reality, genetic engineering has so far contributed nothing that can truly be said to be of any significant use at all in feeding the world. Its contributions have purely to do with (putative) ease of husbandry, and hence with reduction of costs. There is no good reason to assume that genetic engineering will contribute anything that the world actually needs within the next half-century – in which time the world population will have stabilized, and the heat will be off. These are not simply Luddite arguments. Absolutely not do I suggest that the world does not need science or high tech as a whole. It is still possible, too, as outlined above, to find serious tasks for genetic engineering to address (including the hyper stress-resistant sorghum). But the cross-the-board enthusiasm for genetic engineering in high places reflects a profound misreading of history, and a complete lack of understanding of the nature of science.

On the broad front, history and cool assessment of the status quo do not show a general failing of traditional husbandry, with science effecting the rescue. Tragically, all too often we see the precise opposite.

Thus, as outlined in Chapter 2, the prodigious rise in human numbers ever since the Neolithic Revolution can be ascribed entirely to farming. As outlined in Chapters 6 and 7, science had very little impact on farming practice and virtually no impact at all on total food output before the late 1920s, when world population was around 2 billion. Thus, human numbers rose from 10 million to 2,000 million entirely on the back of traditional husbandry: on farming that was run entirely on the basis of craft, which means experience and common sense. Traditional craft alone effected a 200-fold increase.

Since the 1920s, human numbers have risen to 6,000 million. Can this further, threefold increase be ascribed to the impact of science? Not at all. For since the 1920s, the total area under cultivation has increased by at least 50 per cent, not least by further encroachment in North America and the former USSR. Traditional farming alone, then, with more space to work in, could have taken us to 3 billion, a 300-fold increase in human numbers since the Neolithic Revolution. Science might generously be credited with the further twofold increase. The arithmetic is crude but the point is clear enough. Agriculture is primarily craft. Science is the gilt on the gingerbread. Historically speaking, too, its impact is almost ridiculously new. The craft of agriculture has been at least 10,000 years in the making (probably with a 30,000-year prelude in the Upper Palaeolithic). Science has made a serious impact only within the past seventy years: much less than 1 per cent of the whole. The idea, implicit in Blair's speech (and in the general zeal), that all this traditional know-how should be replaced with all possible dispatch by spanking new, largely untried, agricultural high technology rushed almost without a by-your-leave from the laboratory to the field is, one might politely suggest, ludicrous. Ignorance, in fact, is the only excuse for it. To urge such a transformation knowingly would be wicked.

There is worse. First, if we ask what science has done for agriculture that really matters, then the list of unequivocal triumphs is remarkably short. Actually I can think of only one that truly deserves to be included.

The Green Revolution surely should be seen as a triumph of modern science. In the 1960s it really did look as if Malthus's predictions were coming true. In particular, yields of wheat and rice on the Indian subcontinent were not keeping up with numbers that were rising faster than at any time in history, and they seemed incapable of doing so. There was a succession of famines. The solution seemed all too obvious: the fields were not fertile enough to sustain high yields, and in particular lacked nitrogen. But the traditional varieties and landraces of wheat and rice did not provide more grain when they were heavily fertilized. Those ancient varieties, like those of most of the world, were tall, with long stems, and when they were given extra nitrogen they grew even taller and then fell over, or 'lodged', as farmers say. So breeders, in particular at CIMMYT in Mexico, the international research station for wheat and maize, introduced genes from Japanese wheat varieties that produced short stems, and so created a range of semi-dwarf varieties. Soon, scientists at the International Rice Research Institute in the Philippines (IRRI), one of CIMMYT's sister organizations, produced semi-dwarf rice. When these new, short-stemmed varieties were heavily fertilized, they did not grow too tall. They produced more grain. Within less than a decade both India and Pakistan had become grain exporters. The turnaround was astonishing. Norman Borlaug of CIMMYT, prime mover in the creation of semi-dwarf wheats, absolutely deserved his Nobel Prize. Semi-dwarf wheats soon had a comparable impact in Mexico and Turkey.

The Green Revolution has been much criticized on social and economic grounds. The new crops required greater inputs and so favoured richer farmers. Millions of poor farmers and their families went to the wall. This issue must be faced. Humanity in general has made a very poor fist this past few hundred years of introducing new technologies. Few countries, until very recent years, have thought hard enough about what will happen to the people who are inevitably displaced. This is what the British Luddite riots of the early nineteenth century were about: traditional weavers displaced by new machines, with nowhere to go. The Green Revolution recreated the same kind of misery on the grand scale. So, too, on an even grander scale, does the current worldwide obsession to reduce agricultural labour in general. But in 1960s Asia the new technology itself, and the science behind it, was necessary.

Yet, there are conditional clauses. First, the introduction of semi-dwarf wheat and rice can be seen as a one-off. The varieties developed over the previous 10,000 years had been bred to be tall: in general, with low (traditional) levels of fertility, big plants produce more grain, just as big oak trees produce more acorns. When soil fertility is raised, a different shape of plant is needed. But, as we have seen, wheat and rice are by far the most important crops, and now this shift has been made, that's the end of it. Now all the world's breeding programmes include a range of semi-dwarfing genes, which are found in the vast majority of modern varieties (and can be introduced to create semi-dwarf versions of traditional types). There is nothing in this to suggest a general opening of scientific floodgates, to transform all the world's crops to a comparable extent, or with comparable need. There is no other task in agriculture of comparable significance, and that task is now accomplished. Note, though, that all this was done in the decade before the term 'genetic engineering' even came into the world's vocabulary. The artificial nitrogen fertilizer that fired the Green Revolution was and is produced by First World War industrial chemistry (courtesy of Haber and Bosch), and the semi-dwarf cereals were created by conventional, mid-twentieth-century breeding (with a few bells and whistles). Yet, the importance of wheat and rice is such, and the need to provide maximum fertility is so fundamental, that this combination of vintage technologies can reasonably be said to account for at least 50 per cent of all that science has done truly to feed the world's people. We could make a strong case for suggesting that it accounts for 80 per cent of all that science has done for farming that is really worthwhile.

Neither is it true, even in the context of the Green Revolution, that science has ever simply displaced traditional agriculture. As Chapter 2 and the last two chapters show, innovators in every generation, whether farmers or traditional breeders or scientists, have always inherited a great deal from the generations before. The farmers of classical and biblical times were not primitives. They inherited highly developed crops and livestock, and a wide range of effective techniques, developed over the previous 8,000 years (at least). The great agricultural innovators of the eighteenth century, in England, mainland Europe and America, were building on tremendous accumulated know-how. The first bona fide scientists, including Mendel and Darwin, acknowledged how much they owed to their forebears – who for the most part were craftspeople. Mendel was the son of gardeners, in an obsessively agricultural community, and Darwin maintained a lifelong dialogue with pigeon-fanciers and breeders of fancy goldfish. Science has indeed been the gilt on the gingerbread. Always it has built on broad and deep foundations of craft; and without that craft, it would be founded on quicksand. Agricultural scientists who understand both science and farming, know this. But we are bringing up a new generation who have no such background, and who apparently believe that they are inventing the world afresh, and that all that has gone before is obsolete, and should be replaced as soon as possible by spanking new chemistry and GMOs. This is a very dangerous illusion. It is founded on the most profound ignorance, both of history and of science.

Is it really the case, though, that GMOs have made no significant contribution to world food security? Even if they have done little so far, won't they be needed in the future? If so, then the present zeal could be justified. There may be dangers but, as the enthusiasts suggest, we will just have to bite the bullet.

The startling truth is (at least I think it's startling, in view of the hype) that genetic engineering has contributed nothing of significance to world food security – that is, to issues that really matter – and is not likely to do so in the foreseeable future. As far as human survival goes, its contribution is precisely zilch. In reality, it is locked into and is designed to promote an economic strategy that is already proving pernicious, and in the longer term could well prove disastrous. The net contribution of genetic engineering to human well-being is negative.

The reasoning is simple. Wheat, rice and maize between them contribute half of all human food calories, and (directly or indirectly) most of the protein. All other crops and commodities – even including those that have such tremendous economic and social impact, such as beef – are marginal by comparison. Genetic engineering has contributed nothing truly worthwhile to wheat, rice, and maize and is unlikely to do so. For that reason alone, whatever else it may do is minor. Some of the things it might do – including the creation of super-tough sorghum – could be extremely valuable locally, and therefore would be worth doing. In practice, however, such projects are largely sidelined, and are dependent on international aid and trusts set up by philanthropists. The genetic engineering projects that have made an impact so far, and are destined to do so in the next few decades, are designed entirely to enrich and generally to strengthen the grip of particular biotech companies, and the corporations that increasingly employ the companies.

The impact on wheat, rice and maize has been nugatory, and is liable to remain so, partly for botanical reasons. Cereals are grasses; and grasses on the whole do not lend themselves readily to genetic engineering. Crucifers, such as rapeseed, are more easily transformed. So, too, are the Solanaceae, which include potato, tomato, capsicum and aubergine. But although these crops are important commercially none of them – even the potato – is in the same league as the three great cereals. All are minor and hence genetic engineering itself is minor – if, that is, we are talking about feeding the human species, and not simply about making money.

Of course, wheat, rice and maize need further improvement. IRRI scientists are seeking varieties of rice that on the one hand can withstand drought, and on the other resist excessive flooding. As the climate changes over the next few years, more versatile rice will surely be a premium. But the necessary qualities and the genes that support those qualities are all to be found within existing rice varieties, or among the eighteen known species of wild rice. Conventional cross-breeding programmes, already in train, will do what needs doing. If they do not, then there is no good reason to suppose that unrelated flood-resistant or drought-resistant plants can supply the genes that are necessary.

Engineers of rice will surely protest at this. In particular, not least at IRRI, scientists are developing strains of 'yellow rice': rice rich in carotene, which is the raw material from which the body synthesizes vitamin A. Deficiency of vitamin A among other things causes the cornea to become dry, leading to 'xerophthalmia' (Greek for 'dry eyes'). An estimated 3 million children are currently blinded by xerophthalmia, mainly though not exclusively in poor countries, where rice is commonly the staple. Vitamin A-rich rice therefore is surely a godsend, and only the effete, afflicted only by the long sight of genteel middle age, would presume to protest.

But carotene is one of the commonest molecules in nature. It is the yellow pigment found in yellow fruits such as mango and papaya but also – much less exotically and expensively – in green leaves of all kinds, including spinach. Traditional farming always included horticulture. The vegetable patch and the occasional fruit tree were and are standard; taken for granted, like chairs and tables. So long as people have horticulture, they have all the vitamin A they need. Obsessive monoculture, in which there is no room for local produce to feed the local people, is a modern aberration, another example of obsessive commercialism. It is in many ways pernicious, socially, economically, ecologically; and the blindness of children is only one of the consequent evils.

Xerophthalmia in poor people is presented to the world as yet another example of backwardness and technical inadequacy, cured by heroic high technology of the kind that only the West can provide. The absolute opposite is the case. In traditional societies doing traditional things it does not exist (or only in the strangest circumstances). It is caused by the imposition of Western economics. Yellow rice is the heroic, Western, high-tech solution to the disaster that heroic, Western commercial high tech has itself created. Regrettably (as discussed further in the next chapter), yellow rice is a symbol of much of what modern, heroic, Western high tech is about: expensive, much-hyped, but at bottom fatuous.

Finally, there is very little evidence indeed that traditional husbandry, when left alone, cannot feed people. Sub-Saharan Africa has become the world's symbol of disaster but there cannot be a country in Africa that could not feed its people several times over and which in the past has not done so. This does not mean that those countries could not benefit from a little well-directed science. Of course they could. Preferably, of course, the science should be generated indigenously; or, better still, science really could be globalized (as to some extent is already the case). But the idea that Africa's own traditional techniques are inferior, and should be overridden, is either wickedness or madness. It has nothing to with the facts of the case. Angola, for example, that poor beleaguered country on Africa's west coast, is two and a half times the size of France and has every kind of climate and terrain. Its farmers are wonderfully astute, and can grow anything. Angola has only 12 to 13 million people (the census has gone somewhat awry in the past few decades). It sometimes has trouble feeding itself but only because, for the past thirty years (until the late summer of 2002) it has been at civil war. Every now and again the farmers have had to leave their crops to avoid being shot. Traditional Angolan farming depends heavily on adroit exchange of seeds between different areas – constantly shifting the varieties around to create new combinations, to stay ahead of the pests and diseases, and get the best out of the local soils. This trade has continued but only with extreme difficulty and sometimes heroism, as the people (often the women) cross military lines to keep it going.

Cambodia provides an even more striking example. In the 1950s it could feed itself with superabundance. Then, during the 1960s, in its war with Vietnam, the US dropped more bombs on Cambodia than were dropped in all the actions of all sides during the Second World War. After the bombing stopped, Pol Pot became dictator. People like Pol Pot do not generally rise to power except when countries are on their beam-ends, and

Cambodia very definitely was. To the chaos that already existed he added more. He introduced a system of farming that was a grotesque parody of Mao's peasant-based policies of the 1950s, with many millions of people toiling on the land with no serious sense of direction. Not surprisingly it failed. Cambodia was traditionally a rice exporter. By the late 1970s, rice production had ceased all together. Now, helped not least by international agencies, Cambodia is putting itself back together again. It can be seen, by those with no sense of history, as yet another failed Third World country that needs help from outsiders to bring it into the twenty-first century. The truth is that it was doing very well indeed until the guts were knocked out of it. Perhaps it is simplistic to suggest that the US bombs were the sole cause of Cambodia's late-twentieth-century collapse. But they certainly made a difference.

It may seem astonishing that world leaders in politics, industry and science can dupe themselves in the way I am suggesting is the case; although less surprising that they often manage to dupe the rest of us, once they are themselves convinced. Yet history is littered with comparable aberrations. One might reasonably see the whole of the twentieth century as a series of grotesque errors in high places. More specifically, as I mentioned in the Introduction, in the 1970s textured vegetable proteins or TVPs (artificial meat made from soya beans, fungi or bacteria raised on oil) were seen as essential. TVPs have indeed found a role in the modern diet; but it's a minor one. We can reasonably doubt whether bean-based TVPs improve on the beans themselves (you have only to boil them up, after all), or on traditional bean extracts, notably tofu. But TVPs' advocates gave the impression – and clearly believed themselves – that without TVPs, the human race would soon be in dire nutritional straits. The nutritional theory of the day was still advocating the highest possible intake of protein; and, as was rightly pointed out, it was at least profligate to raise too much livestock.

But some nutritionists were already asking whether, in truth, people do need huge intakes of protein. If that was really the case, then how come humanity had survived at all? Why weren't most of us already dead? In the event, TVPs didn't catch on, although they have been secreted into pies. Yet there was a time in the 1970s when an entire Noah's ark of 'top experts' in science, economics and politics had convinced themselves that TVPs were world-savers. Within less than a decade it became obvious even to the zealots that this is not so, and never was. The present vogue for GMOs is surely similar: a South Sea Bubble. A lot of people are growing rich by it, and many 'top experts' have convinced themselves that GMOs are vital, and preach to the rest of us with evangelical fervour. This is precisely the kind of gaudy and clamorous bandwagon that politicians like Tony Blair feel obliged to climb aboard. Yet it is largely fatuous, just as TVPs were largely fatuous. The simple truth is that experts of all kinds, and politicians of a certain stamp, are wonderfully adept at chasing wild geese. This is one more reason for putting one's faith in the common sense and experience of humanity at large. This, after all, brought us the traditional agriculture on which the world as a whole truly depends; and it could still, if the top experts don't get in the way, take us into a long and agreeable future.

In Praise of Democracy

Much that is startling emerges from the debate on GMOs. Most striking of all is that the arguments and misgivings raised by people at large have been much more to the point than the assurances of many of the leading experts. Thus, from on high, has come the idea that the dangers of GMOs are very slight, whereas the truth is that they are unknown and in principle are unknowable, at least in detail. Many of the most influential people – including Mr Blair – clearly believe too that GMOs already play an essential role in feeding people, or at least will be vital over the next few decades; yet this is profoundly untrue. I have never heard anyone, whether on the debating platform or in the saloon bar, spell out all the issues as outlined above. But all the points are implicit in the kinds of arguments that have been made over and over, and that people feel in their bones to be the case. When dealing with new and potentially powerful technologies it is right in general to be extremely cautious; and not only cautious but to ask why we should be asked to take any risk at all. People at large have asked such questions, but officialdom has dismissed their fears – to the point where Tony Blair is known to be determined to push ahead with GMOs, come what may. People at large, too, have raised the most obvious point: the one that implies that the emperor has no clothes. If it is true, as the experts seem to be suggesting, that GMOs are so vital, how come land is currently being taken out of production? And how come we are not dead already?

Here is one very good reason (I suggest) for valuing democracy. Democracy suggests (at least in one of its versions) that public will and public instincts should prevail; and I am more and more convinced that public instincts and common sense are a much more reliable guide to strategy than the opinions of experts. The adage 'Experts know more and more about less and less' surely has literal truth. Notably, I seem to have heard dozens of homilies from scientists these past few years on the nature of risk: how it is easy to be fooled by appearances; how, in reality, it is hundreds of times more dangerous to drive on a motorway than, say, to eat GM soya; how necessary it is to invoke statistical 'proofs' of risk, which only experts understand; how necessary and sensible it is, therefore, to trust experts.

What's missing from such analyses is any realistic knowledge of the nature of risk in the world at large. When ordinary human beings assess risk we are not concerned only with the statistical chances of dying. Of course motorways are dangerous. So is skiing, or playing rugby. What matters are all the connotations. One of these is to ask what else we get out of our various adventures. Skiing is dangerous, sure, but I'm told by those who do it that it is also tremendous fun; there is nothing else quite like it, the enthusiasts say. The key issue is who is in charge, and who is making the choice. I don't ski as it happens but I do drive a car – but only when I want to. It's my choice; my risk. In the present world, in the context of food, we are being invited, not to say urged, to hand over key decisions of our lives to third parties, who appoint themselves as experts. Those experts will decide on our behalf whether GMOs are safe or not. They will do the trials, in field and laboratory, to assess the risks to other species. Even if those trials were adequate (which they clearly are not, and cannot be); even if the experts were fully abreast of all the relevant information (which is clearly not the case – certainly not when it comes to assessing the true importance of GMOs), we would still have cause for alarm. It simply is not comfortable to be in a position where other people, however expert they may think themselves, take important decisions on our behalf; and what we should be growing and eating is a very important decision indeed.

Yet I have often of late heard senior scientists (I won't embarrass them by naming names) drawing an analogy between the experts who create novel foods and novel crops, and, say, washing-machine repairmen. We rely on professional mechanics to fix machines that we ourselves don't understand. Why shouldn't we rely on professional scientists to produce novel crops which (especially if they are GMOs) are at least equally beyond our ken? Again it's a matter of who is in charge. The relationship between householder and the washing-machine repairman is comparable with that of patient and physician. The physician may do highly invasive things to the patient but – at least traditionally – the patient always remains in charge. The patient decides if he or she is ill in the first place. The patient invites the physician to take an interest. The physician (ideally) says what he or she thinks can be done, discusses the pros and cons of treatment, and then asks the patient if the patient wishes to proceed. If so, then the patient and the physician have a contract, and enter on the treatment together, with the physician committed to do his or her best, while the patient accepts the risks and is willing to take the responsibility. Similarly, mechanics do not fix washing machines except at the express invitation of the owner. But who asked the genetic engineers to fix our crops? Where is their invitation? Where is their contract? Where is their mandate?

At this point the experts are wont to argue that they need no special mandate. This is where the advocates of GM wheel out the argument that genetic engineering is simply an extension of conventional breeding; there has been no qualitative shift, they say. No one has ever seriously questioned the right of farmers or of traditional breeders to create new crops by conventional means. (People at large may not always like the new varieties, and may miss the old ones, but that's not the same thing at all.) The same tacit mandate that allows conventional breeders to produce new varieties must also, logically, extend to genetic engineers.

But genetic engineering does represent a qualitative shift. Conventional breeding does include a number of tricks (which I casually referred to earlier as 'bells and whistles') that to some extent presage genetic engineering. But the specific transfer of lengths of DNA between organisms that may not be related to each other is a distinct departure, with all the connotations outlined above. If democracy means anything at all, then a new procedure that affects our daily diet, could profoundly affect the environment at large, and has serious connotations for animal welfare and the structure and economics of agriculture and hence of society as a whole, requires a new mandate. None has been sought; and when people have protested, they (we) have been told that we are stupid, that we do not understand, and, like Victorian orphans in the workhouse, that we should eat up our GM porridge and be grateful. Leaving aside the obvious shortcomings of the self-appointed experts, the high-handedness is awesome.

For in the end, the nature and weight of the arguments that people at large have lined up against the new generation of GMOs is not the issue. In a democracy, public will should prevail; and if people at large say they don't want to eat GMOs for whatever reason, then that ought to be enough. Politicians are elected to carry out the wishes of their electorate. They are public servants. If the electorate don't want GMOs then that should be that. People may, with sound and honest argument, be persuaded that GMOs are good and necessary, and may then change their minds. This has happened often enough in the past. But the hectoring and the disdain issued *de haut en bas* in this present context, is intolerable. It would be intolerable even if the objections to GMOs were groundless. Yet the objections are far from groundless. It is very clear that the experts, yet again, have allowed their expertise to override common sense. Common sense is not the answer to all our problems, and of course it is part of the task of science to improve on common sense. Yet we abandon common sense at our peril. If we do choose to override what our instincts and experience tell us is the case, then we have to be very sure indeed that we know what we are doing, and why.

GMOs raise issues that are pertinent to all science. Science can enhance and enable everything that human beings aspire to do. In general, now that the world population is so high, it has become necessary; and I for one would hate to face the future without brilliant agricultural science (and would happily pay more taxes to support many of the research institutes I have been privileged to visit these past few decades). But science nonetheless, and the high technologies to which it gives rise, can just as easily be deployed perniciously. The Austrian philosopher Ivan Illich encapsulated the issue wonderfully in the 1960s with his notion of 'tools for conviviality'. Some technologies, he said, in some contexts, are liberating. They help to increase the autonomy of individuals and sometimes of entire societies, to help people get more out of life, and ensure that we are more in charge of own affairs and destinies. The examples he offered in the mid twentieth century were the telephone and the bicycle; both of which can transform people's lives, but both of which are simple enough to be produced and controlled locally. It is not necessary for those who adopt such technologies to sell their souls and their political freedom to foreign countries or to corporations who supply them. By contrast, many other technologies in other contexts are, in the end, agents of political and economic control. Illich saw conventional broadcasting in this pernicious light: a central voice that speaks to the population at large, and is deaf to the response. (Modern broadcasting, with its many channels and feedbacks, can in principle be more convivial.)

The technologies of agriculture illustrate Illich's arguments perfectly; and, as cannot be overstressed, nothing else is as remotely as important as agriculture. The argument is not with high technology per se. Many high technologies, including biotechnologies, could indeed be tools for conviviality. In the 1970s it was fashionable to argue that only 'low' or 'intermediate' technologies could truly be appropriate to poor societies. High technologies (it was felt) should remain in the domain of the rich. In truth, 'low' technologies have a great deal to offer. Thus, the French historian and philanthropist Jean Gimpel helped to introduce watermills to Nepal that followed the designs of medieval Cistercian monks. They have proved very effective, and are built by local people from local materials, with no dependency on outside agencies. Truly they are convivial, in Illich's sense. Yet we should not be too prescriptive. Sometimes – often – the highest technologies have the most to offer the poorest people. I visited a village primary school in India that was too poor to afford paper and books but could afford a television and took lessons from the satellite (while the children made notes with slates and pencils). India has its own space and electronics industry which, in this context, is cheaper and more friendly to the environment than traditional technologies. In agriculture, computer-controlled irrigation can make the best possible use of limited water. High-tech vaccines (perhaps produced with the help of genetic engineering) can increase protection against foot-and-mouth disease in harsh environments, and so on. The point is not whether the technologies are 'high', 'low' or 'intermediate', but who is in control of them. In practice, of course, it is generally easier for traditional societies to build and maintain medieval-style watermills than it is for them to create and control their own high tech. Yet biotech (in the broad sense) can be extremely friendly environmentally, precisely because it does not require the huge inputs of energy of traditional industry. It can also be small scale. For all these reasons, once established it can be cheap, and adapted to local needs. It can, then, be extremely convivial. All it really requires is talent, and there is no shortage of talent in poor countries; merely of opportunity.

So the argument is not with high tech per se. It is with the political and economic systems that ensure that high tech is designed not, primarily, to feed people, but to create systems of agriculture that in turn maximize disposable wealth. More generally, the quarrel is with technology that is not intended to increase autonomy, but to bring all societies, all people, into the same economic fold, more and more controlled by a few superpowers and a few corporations, each supporting the other. GMOs are worrying for biological reasons, as outlined above. Yet the greatest threat is political; they are helping to create a world in which the most important of all human enterprises, which traditionally has been the ultimate democratic pursuit, becomes the domain of big business, like everything else. More broadly, science as a whole should be the servant and the delight of humankind; but it is rapidly becoming the exclusive handmaiden of corporations, and is doing this with the eager assistance of politicians whom we elected to act in our best interests.

The research that produces science, and the high technologies that flow from it, are expensive. Science needs money, and it clearly flourishes best in a system that is designed primarily to create money. Science alone can supply the smart machines and the chemistry that replace human labour. In short, science doesn't have to be the handmaiden of big industry, and big commerce. But in practice, it slips very easily into such a role. If we want it to behave more decorously, then we have expressly to impose the necessary directives and restraints. That is what we have to do if we think enlightened agriculture is worth going for. If we are simply content to let things take their commercial course then science will become the exclusive property of big business, and big business (if not otherwise constrained) will create industrialized agriculture: simplified, mechanized, monocultural, denuded of human labour.

Industrialized agriculture can seem impressive (it clearly impresses politicians) but among the many drawbacks are the loss of active rural communities (and ways of life and cultural diversity), the despoliation of traditional landscape, the loss of wild species and – the greatest flaw of all – the fact that it is not sustainable. It is not sustainable for the simple reason that, in general, it is not designed to be. It no longer feels the need to march to the drum of biology. Science should be used to enhance good husbandry. When used in that capacity it is unquestionably a good thing. But the role of science in this new, industrialized agriculture is not to abet good husbandry, but to override it. Industrialized agriculture is flashy, it is profitable, it has scored many short-term victories, and it fills supermarkets like cornucopia; but taken all in all, it is not wise. Science, in short, which should be such a glorious pursuit, has been compromised. It is locked into a system which, if allowed to run along its present course, will surely be destructive. In such a guise, science becomes, in net, the enemy of humanity. The protests against so much modern science in general and GMOs in particular are not simply Luddite. For the most part the protesters are not anti-science, in any general way. There is simply the feeling that science has fallen into bad company; and that feeling is surely justified.

If agriculture was indeed 'just another industry', or 'just another business', as the modernizers maintain, the increasing concentration of power would be very worrying. Given that industrialized agriculture in many ways runs counter to the fundamental requirements of biology, and is designed expressly to override traditional husbandry, the trend is truly terrifying. GMOs alone are not responsible for the overall trend, but they are very much a part of it.

9

Of Cash and Values

We are moving absolutely and more rapidly than ever before from one kind of world into another. The transition is economic, political, social, cultural, ecological, and (if things continue as they are) global. The driver of change in the end is technology, which increasingly is the child of science. Transition itself – perpetual change – has become the *Zeitgeist.*

The transition is one way, like a weir or a transistor in historical time. There is no going back. Thousands of species and countless cultural variations, including hundreds of languages, have been lost over the past few centuries since the process fully began. We may in future resurrect a few extinct creatures through cloning and DNA technology, or reinstate a few dying languages in rudimentary form, but on the whole what's gone is gone. What's coming in place of all that history and diversity is not at all clear, but it would be naive indeed to suppose it will all be for the good, all of it better than what we had before.

Nobody is in charge of this transition, but some people are more in charge than others. These include senior politicians ('world leaders'), financiers, including private merchant banks, the World Bank and the International Monetary Fund (IMF); and industrial companies, and in particular the transnational corporations. Behind all of these lie intellectuals and technocrats: economists, scientists, technologists. These people meet in their own cabinets and boardrooms, at the UN, and at international meetings of which the latest at the time of writing was the World Summit on Sustainable Development at Johannesburg in August 2002. Symbol of the whole shift in the world's affairs and way of life, the body that comes closest to the status of organizing genius is the World Trade Organization (WTO), driven by the mission of a single, global market, in which everyone from General Motors to the smallest grower of tomatoes in West Africa will in principle be free to sell to everybody else in the world.

Yet if you talk to any individuals within these centres of influence, it's surprising how they tend to deny their own power, and hence their own responsibility. The most senior politicians will aver that they are merely a small part of their particular government, and that the government was elected, so they merely represent public will. The richest and apparently most powerful industrialists stress that their companies depend entirely on the good opinion and support of consumers. They too, therefore, are merely responding to public will. In short (the story has it) it's the people at large – all of us – who are really in charge. Nothing happens except through our collective will. People continue to starve, others are effectively poisoned (diabetes doesn't just happen) and thousands upon thousands are thrown off their land every week as the political, commercial and technical transition sweeps through the world's farms. But all this reflects, and is entirely dependent upon, public will, apparently. As voters and consumers all of us are presumed to be in ultimate control.

Yet the politicians, financiers, industrialists and technocrats who seem to be setting the pace do have a vision of their own. It is rarely spelled out but we can infer what it must be from what is actually happening. These leaders all speak of economic growth, which implies in a democracy that everyone should become steadily richer. No end-point is envisaged. The sky is apparently the limit. At the end of every rainbow, apparently, is the possibility of being President of the United States, or somewhat wealthier than Croesus. Along with this steadily rising wealth goes security, peace, good health and long life, and of course physical comfort: all the ingredients and trappings of general well-being. The vision, boiled down, is that all of us will, in the future, live like suburban Californians or downtown Berliners. No one to my knowledge has as yet directly promised the people of Angola, say, that this is what the future holds. But in South-East Asia, where such a vision would have seemed outlandish just thirty years ago, it seems in effect to be coming true, at least in for example, Beijing and Shanghai, and as quickly as can be arranged in Bangkok, Kuala Lumpur and Seoul. The dream is already out there. It is taken as a matter of faith that such a future is, in practice, available to everybody (provided of course that they toe the line, work hard, and effectively compete). It is taken to be self-evident, too, that such a future is desirable. Not only desirable but inevitable. It represents progress, and progress is good, and is our destiny. Urban or suburban comfort for all are the denouement of humankind (the implicit message has it), although an even more privileged few might aspire to the hacienda or the stately home, suitably guarded against the envious and the dispossessed. The 'developing' countries (dubbed 'the Third World' by French demographer Alfred Sauvy) must follow this path. This is what 'development' means. To resist is not only perverse, but wicked.

Yet for humanity as a whole such a vision is clearly unrealistic. Only a minority can ever be as rich as present-day Californians, because the world just doesn't have that many resources. Gandhi pointed out that the rich world was rich largely because the poor world was its empire: and who, he disingenuously inquired, will be the poor world's empire? Besides, many of those for whom the vision might be realistic, actually find it unpleasant. Many people are not especially materialist, and are content with basic comforts (among which good food is paramount), means of communication such as radios and telephones or hi-fi and Internet (which on the scale of life's commodities are cheap), and a few mementoes. Many have no desire whatever to compete with their fellows, beyond the bounds of friendly rivalry, and certainly not in any lifelong, life-and-death race for ostensible wealth. Many are content, in short, with what might be called good fellowship.

But those who resist are overridden, not because they are bad, or even because they are openly despised, but simply because, since they do not choose to compete, they are bound to lose out to those who do. Thus we are faced with a paradox: that those who are repelled by bullishness must become bullish themselves to defend their own affability. Adlai Stevenson addressed the same kind of paradox when he commented that we cannot afford to be tolerant of intolerance. But those who resist actively are put down, or at least they are when they seek to resist from a position of weakness. A proportion of those who protested on the streets when the WTO last met in Seattle and then Genoa, were indeed there only for the excitement, but many others were making serious points. But they weren't heard. It was far more convenient simply to treat all protesters as a general mob.

Much more to the point, though, is that the voices of opposition are largely misdirected, and that they do not cohere. The vision of global industrialization represented by the WTO has much to be said against it. In the context of agriculture, the most important human enterprise of all, it threatens to be disastrous, and irrecoverably so. But the vision of globalization is coherent, and can be made to sound good: freedom and opportunity for all, leading to wealth for all, or at least for those who work hardest and in the appropriate ways. The wealth that the ablest and most energetic acquire will then, it's supposed, 'trickle down' to the less forceful individuals. No doubt some supporters of the WTO are out-and-out cynics but the ones that I have met are sincere, and truly believe that they are on the side of good. The protesters, by contrast, tend to sound Luddite and reactionary, and are easily countered: 'unrealistic' and 'elitist' are common put-downs.

In particular, the protesters commonly attack the entire edifice of capitalism. Capitalism does of course provide the framework of the WTO and of the whole modern concept of 'development', and because the modern concept of development and the ambitions of the WTO are both highly questionable, capitalism itself emerges as the root of all evil. But in this way the protest is halted in its tracks. The principal and most coherent antithesis of capitalism is loosely called Communism or, more broadly, the centralized economy. It was given a reasonable crack of the whip in the twentieth century and in practice was not convincing. The USSR had some benign features (not the least of which was full employment) but the lack of democracy was an obvious flaw; and in the end it collapsed under the weight of its own muddle and corruption, after decades of perceived cruelty and deprivation that often seemed gratuitous. The Chinese regime was in many ways more subtle, yet still it was too harsh, and China is now becoming the new capitalist superpower almost as rapidly as can be organized (although not quite as rapidly as might be theoretically possible, not least because China for various reasons doesn't want simply to become a facsimile of America).

Capitalism, in wondrous contrast, clearly works. The Pacific countries that embraced it have leapt ahead. Contrast South Korea with the North, or Western Europe with Eastern Europe (although some of the East is catching up fast). Look and wonder at the wealth and power of the United States, the arch-exponents of capitalism. At this stage of history it really is unrealistic, even childish, to contemplate the downfall of capitalism. The protesters may argue that the WTO should modify its position and that the concept of development should be reassessed, but if this can be achieved only through the downfall of capitalism, then the protest is clearly absurd. Capitalism is here to stay, and capitalism conscientiously applied produces the kinds of changes that are now afoot worldwide. Romantics and others may not like everything that's happening but if indeed there are drawbacks, then they are the price that must be paid. Capitalism leads to material progress which improves human well-being and so is good; and, as its defenders commonly insist, you can't make an omelette without breaking eggs. Let that be an end to it.

So the protests tend to run into the sand. The great machine, the driver of change, continues. Yet the causes of disquiet are real. The vision of material growth for everyone really is unrealistic. Neither can it be taken for granted that as the rich grow richer their wealth will trickle down to the poor. It's perfectly possible for the rich to grow richer and the poor to grow poorer. Then again, the notion that rising material wealth necessarily correlates with rising well-being is naive in the extreme. For one thing, the means by which material wealth is created often destroy the very assets that make life worth living in the first place. It's amazing how much very rich people are prepared to pay to enjoy the simple pleasures that the world's allegedly poor people traditionally took for granted. The general showering of wealth from above has a smoothing effect, too, riding roughshod over landscapes and local customs. The current loss of cultural and biological diversity is surely the greatest of tragedies that can be conceived. What else is there?

But still it is a waste of everybody's time to resist capitalism in the round. It is childish, too, simply to resist the march of technologies, including the much-feared biotechnologies such as genetic engineering. Science and the new technologies can and must be developed and put to good use; and whatever is done in the future is in general bound to be done within a broadly capitalist framework. But still it is vital and ever more urgent to protest. Much is happening that is obviously bad, and the changes are far too precipitate, leaving no time to take stock; while many problems that really are important and urgent remain unaddressed. Species disappear, and the poor grow poorer. This is damaging to the extent where we must ask how much more the world can really endure, and most of the blame must be laid at the feet of capitalism in its present form. Yet the protests must be reframed and re-argued. This chapter is an attempt to do this.

In Praise of Capitalism

There are many good reasons, both negative and positive, for not seeking to 'overthrow' capitalism, as the old-style Communists used to put the matter. The negative reasons include the fact that there is no obvious off-the-shelf alternative (and we have to run our economies somehow), and that in any case capitalism would not go quietly, and the serious concerted effort that would be needed to get shot of it would require disruption on a scale that would wreck the world for ever. In short, capitalism has the quality that physicists call inertia, like a big ship on full ahead. It is here to stay. We have to live with it.

But there are positive reasons for keeping it too. Capitalism in many ways seems 'natural'. It grew out of the general notion of trade, and trade – the exchange of goods and favours – is surely older than humanity (as Matt Ridley has argued of late, for example in *The Origins of Virtue*). Trade is a key component of cooperation, and hence of societies that really do work in a cooperative fashion. Capitalism, defined broadly, merely puts trade on to a more formal basis. The fact that trade is natural does not of course make it morally right (a lot happens in nature that seems most regrettable), but the process of exchange for mutual benefit is certainly not obviously wrong and in general, things that come naturally are easier to work with and build upon than things that do not.

Trade at its simplest is barter: direct exchange between consenting individuals (if one does not consent, then the 'trade' is theft). The next logical step is the market. Everyone with goods to dispose of brings them to a central place, and everyone who wants anything comes to see what is on offer. In its pristine state, the market is a physical space: the village green; the quay. As things develop, the market can become more abstract, a general pool of goods and services where everyone can display what they have to offer, and from which everyone can draw what they need.

So far so good. Indeed the Scottish moralist and economist Adam Smith pointed out in his *Wealth of Nations* (1776) that the true advantages are even greater than they seem:

> by directing industry in such a manner as its produce may be of greatest value [the trader] intends only his own gain, and he is in this, as in many other cases, led by an invisible hand to promote an end which was no part of his intention. Nor is it always the worse for society that it was no part of it. By pursuing his own interest he frequently promotes that of the society more effectively than when he really intends to promote it. I have never known much good done by those who affected to trade for the public good.

Thus, according to Smith, the market is not merely a convenient way of distributing goods. A free market is a moral arbiter, and itself is socially cohesive. If every trader simply does his or her own thing, working in his or her own interests, and if every consumer does the same – in fact, if each simply tries to buy or sell for the best possible price – then an 'invisible hand' will ensure that a just society results. If any trader overcharges then – in a free market – consumers will go to some rival who charges less. But if any trader charges too little, just to beat down a competitor, then before long he or she must go out of business. If any traders cheat, surely they will soon be found out and shunned, and for a trader to be ignored is death. Thus, if all traders and consumers simply stay alert and look after their own interests, prices will find their own level and dishonesty will be kept to a minimum. All that matters, said Smith, is that there really should be competition – not just one or a few traders running a monopoly, or a cartel – and that the market should be free. Any restrictive practices, including tariffs (special charges on imported goods) are against the interests of the consumer. Thus, said Smith, the truly free market, fired by competition, is the engine of a just and productive society. It is also the essence of modern capitalism.

The late-eighteenth-century founders of modern America, including Thomas Jefferson, took Smith as their guide, just as the founders of Soviet Russia took Karl Marx. The first modern Americans also embraced John Locke, the seventeenth-century English philosopher who argued that human beings are a tabula rasa, a blank slate, who are written upon and moulded by their upbringing and environment. In short, all human beings are born with the capacity to become anything they want to be. Smith and Locke taken at their face value are a heady combination, especially in a vast and fertile country which (apart from the millions of people now known as 'native Americans') was perceived to be empty. The free market rewarded energy and enterprise, and yet seemed bound to ensure fairness, since people would not buy what they did not want, or what was overpriced. Locke and the wide open spaces seemed to encourage people to be bold, strike out, exert their individuality to the full. It was in many ways a wonderful vision. It certainly inspired Britons, for not all were enamoured of the monarchy which at that time was represented by George III (a significant agricultural reformer as it happens, known as 'Farmer George', but regrettably mad). The intellectuals, engineers, philosophers, biologists and clerics who formed the Lunar Society in the English Midlands, including Erasmus Darwin, Joseph Priestley and the potter and canal builder Josiah Wedgwood, were among the British enthusiasts for the new American vision. Those gentlemen were also good friends of Benjamin Franklin and loosely in touch with Thomas Jefferson.

In short, capitalism and personal freedom, which implied personal fulfilment, seem to go naturally together; and, according to Smith, led equally naturally to social justice. Here, surely, is a vision of democracy. Actually, it is only one of several possible versions of democracy, because many of the founders of the USA were vehemently against majority rule, which some see as democracy's central strut. Some of the early republicans felt that majority rule meant mob rule. The two opposing views of democracy are of course reflected in America's modern political parties. Broadly speaking, the Republicans see democracy as personal freedom, while the Democrats emphasize responsibility to society as a whole. The two have often fought and died side by side to defend democracy as a whole against various forms of perceived despotism, but they also fight with almost equal vehemence against each other.

Despite all this, the pristine American vision(s) of democracy remain attractive. In either version, they clearly have a lot to do with personal freedom and they reward energy and enterprise: key ingredients of personal fulfilment. Capitalism is their natural framework. As a moral vision, capitalism as conceived by Smith was as exciting as Communism was, as conceived by Lenin.

Furthermore, as is easily demonstrated by computer models (and as twentieth-century history seems to have borne out), the mechanism of the free market is innately more efficient than that of the Communist, centralized economy. Thus as the Cornell economist Bob Frank has argued (for example in *Passions within Reason* and *The Winner Take-All Society*), if a doughnut seller wants to set up a stall in an American campus he simply has to ask the university authorities, and away he goes. Whether he sinks or swims depends on the wishes of the students, who may or may not favour his wares. In a Communist state that is true to its Communist logic, the would-be trader would have to refer the matter back to some central committee. In short, the trading structure of the free market is, ideally, that of the neural net: the lines of communication between the participants are as short as they can logically be. In a true Communist system, the structure is more like the web of a garden spider, with everything referred to the centre and out again. The latter is demonstrably more complex, which inevitably means more bureaucratic. Tsarist Russia and Mandarin China were bedevilled by bureaucracy before their revolutions, and they were just as bad afterwards. Capitalism seems bound to be more efficient precisely because it seems designed to cut out all surplus bureaucracy. This, plus the emphasis on personal enterprise, makes a powerful engine indeed. All that, plus a claim to moral ascendancy which in its own way seems at least to match that of the more austere and cumbersome Communism.

Protesters tend to lose out, too, by carelessly attacking the tools and the trappings of capitalism: money, profits, accountancy in general. Of course, as I will argue, the modern shift of emphasis in all production and commerce from the task in hand, to the accumulation of money and profit, is more than regrettable, and in the context of agriculture it is rapidly proving disastrous. (John Maynard Keynes made the same point seventy years ago, with a great deal more authority.) Money and profit per se, however, are brilliant concepts, and clearly necessary.

There is money even in Communist, centralized societies. Trade by barter soon breaks down. If you have cows to sell and need clothes pegs, how do you arrange a fair trade? Should the clothes-peg seller make do with a horn or a hoof? With money, the universal medium, all such problems are instantly solved. If we didn't have money, we would need beads or cowry shells. Money is to all material things and human endeavours what language is to ideas; the abstract representation, much more easily manipulated than the thing it represents. Stock markets trade money as philosophers trade words. Banks hold cash as reservoirs store water, and move it to where it is needed. It is childish to complain about money per se or the apparatus of it.

Profit is the natural extension of money; and it leads to the greatest attribute of capitalism, which is investment. Profit is the money left over after the trader has paid his or her own costs. Because money is the universal medium, the currency, it can then be used to buy more goods and services to begin some fresh venture. Thus there is continuity. Last year's endeavours feed into this year's ambitions. Investment can be in anything: a new factory to create more goods; the children's education. Of course the system of profit and investment can go wrong if people are dishonest. This was recognized in the Old Testament, which speaks of excess profit as 'usury'. But if people work the system honestly, then in principle it provides the means by which anyone with a good idea can do anything. It is vibrant, it is liberating, and it enables societies to build on their earlier endeavours as no other system can do. Investment is the key; and investment depends upon profit, the surplus from previous efforts.

Modern agriculture is bedevilled by the over-emphasis on profit, as its critics rightly point out. Yet the defenders of the status quo, sometimes with genuine bemusement, protest in their turn that 'Profit is not a dirty word!' They are right too. The problem does not lie with profit per se.

The same broad principles apply to accountancy. Accountancy is a way of keeping the money tidy, making sure you know where it has all gone. Without good accountancy, there is waste: and waste in a world of finite resources, where goods are produced only by human labour, should be perceived as a sin (which is another powerful theme of both Testaments). To pretend to be casual with money, through sloppy accountancy, is indeed to be effete. Usually, it's other people's labour that is being wasted. In short, the moral and practical case for capitalism and all its appurtenances, money and profit and banks and accountancy and indeed stock markets, is powerful.

But although trade is older than history – there is evidence of Neanderthals trading – and the mechanisms and rules of cash and accountancy were clearly well advanced in Old Testament times, modern capitalism has arisen only in the past few centuries. Its recent rise, and its current world dominance, have not been brought about simply by refinements in trading, investing, banking and accountancy, which were already well advanced in Shakespeare's day. What has really mattered these past 400 years is the association of capital with science and technology. The three have operated together in a feedback loop. The combination is the most powerful engine of change in all of history.

The Feedback Loop: Capital, Science and High Technology

Science and modern capitalism have grown up together over the past 400 years. Some would say that the history of the past 400 years was dominated, until well into the twentieth century, by the fact of empire; and that the age of empires was then replaced by the cold war between the capitalist superpowers of the West and the Communist superpowers of the East. But the driver of all these processes has been capital and science working in harness.

Science and modern capitalism arose more or less separately, but both came of age in the seventeenth century. By then, world trade was well established, conducted through ocean-going sailing ships and backed by artillery, while craft-based mass manufacture first in cottages and then in bona fide factories was already coming on line. At the same time a succession of great theoreticians, notably Galileo and Newton, and of naturalists and physicians, including England's John Ray and William Harvey, were laying the foundations of recognizably modern science. They did not take nature for granted, as the old Greeks had tended to do, but observed it directly. They speculated on how it worked, and tested their ideas by experiments. They measured where possible (as the earlier alchemists, for instance, had not) and from their measurements derived coherent mathematical formulae, which they then tested again against observed reality. They began truly to conceive that the universe as a whole is orderly (perceiving the order to reflect the mind of God) and that its laws are universal: for instance that Newton's idea of gravity applies both to the behaviour of apples shed from their trees and to the movement of planets. They further believed that through diligent application of their God-given intellect they could understand those universal laws.

From the outset, these early 'scientists' (although in truth, the term was not coined until the nineteenth century) enjoyed a symbiotic relationship with the craft-based technologies that already abounded. Galileo's astronomy depended heavily on the new telescopes; Newton's theories of light emerged from experiments with glass prisms. On the whole, the early scientists did not disdain (as Plato, for example, had done) the idea that their natural philosophizing should be useful. Newton, for example, wanted to study the stars for their own sake (and/or God's) but also perceived that such studies would benefit the Navy's attempts to improve navigation, in turn to serve the interests of defence and trade. He was more than happy to work with Admiralty money, to the Admiralty's brief.

For scientific research is expensive. Even if it is carried out with simple instruments, it still requires time, and the people who do it need to keep body and soul together. Money was available in the seventeenth century, from trade and conquest, and the idea was well established that money should be invested – it should not simply sit around in coffers (another idea reflected in a New Testament parable). The new science was clearly worth investing in – not just for its own sake but also to improve navigation, ordnance, manufacture, medicine, or indeed everything. The new capitalism provided the money for the research. The research in turn produced new technologies – truly 'high' technologies since (in contrast, say, to medieval windmills or Roman aqueducts) these new technologies were products of science. These new technologies in turn generated more cash. This cash supported more research – and so on and so on and so on.

Thus was born the 'positive-feedback loop': capital provides finance for scientific research which produces high technologies that generate more cash. Switching the metaphor somewhat, we can see this loop as a dynamo, spinning ever faster; or as an auger, boring through all of life, drawing in what it needs, shoving aside whatever cannot be accommodated. Dynamo or auger: either way this feedback loop – capital–science–high tech–capital – has become the principal driver of the world's economies in the modern world. By harnessing money this dynamo draws upon all previous endeavour. By harnessing science it makes use of the most powerful agent of formal inquiry into the workings of the material world that perhaps is possible. The high technologies that are thrown off as the dynamo spins are expressly intended as agents of change. That is what they are for. The loop, in short, is the most powerful engine of technical, economic and hence social change the world has ever seen, or is ever liable to see.

So modern capitalism seems to have all the answers. People's desires become a market niche: and whenever there's a niche, someone will fill it. But anyone who tries to provide what people do not want must fail. Hence the market is bound to supply what people want, and only what people want. Unless people are crazy, then the things they want and are prepared to pay for must, to a significant extent, reflect their real needs. Thus the market emerges as an agent both of survival – supplying needs – and of democracy – satisfying desires. It is also bound to be efficient. After all, any participants who are not efficient are pushed aside by those who are. This is natural selection, the universal force. On the grander scale, the market ensures that all human endeavour, at least in net, is channelled to meet human needs and desires, as defined by the people themselves. The mechanisms of capital – money, accountancy, the concepts of profit and investment, banking and stock markets – ensure that the market works as fluently as possibly, responding instantly to the world's changing moods. The feedback loop, which brings science and its technologies into the fold, is the driver.

All in all it really is not surprising that modern capitalism – the market and money, the science and high technology that they so efficiently deploy – dominates the world. Neither is it at all surprising that many have found it so seductive. Notably, the 1980s saw the birth of 'monetarism':

effectively the belief that the best possible course for humanity is simply to ensure that the market is left as free as possible to do its thing. The market is bound to respond to our wants and needs, it is bound to be efficient and (following Adam Smith) it seems bound even to be reasonably honest, since consumers are so alert to dishonesty. Some modern politicians (such as Margaret Thatcher and Ronald Reagan) argued that morality itself can in effect be left to emerge as a by-product of the market; that is, what people are prepared to pay for is perceived to be 'right'. (In practice, of course, both Thatcher and Reagan, like all politicians, fed in their own moral add-ons, for example to do with 'family values'. But according to their core philosophy the general moral thrust should emerge from the market.) This view has pervaded current discussions on the rights and wrongs even of human cloning and of designer babies. Many advocates have argued that if people are prepared to pay for such procedures, then that is justification enough. I recently read some of the arguments that surrounded the first heart transplants in the late 1960s, and the beginnings of IVF (test-tube babies) in the 1970s and found that this particular argument did not feature. No one, in those days, was apparently prepared to argue in public that what people were prepared to pay for must by definition be acceptable. But now, apart from a few persistent taboos, this has become a standard ethical position. Thus to the obvious material power of the modern market is added moral authority too. Capitalism that has taken this final step, seeking to run all human affairs by the manipulation of money, might reasonably be called 'transformed capitalism', or 'hyper-capitalism'. Perhaps it is a mistake to coin such neologisms; but it does seem important to distinguish the kind of abstract transactions that take place now from the face-to-face trading in beef and linen and pots that Smith wrote about in the eighteenth century.

Money, Values and the Algorithm of Accountancy

The first drawback of capitalism began (as the most serious faults tend to do) as a virtue. Money is the essential medium of capitalism, as words are the medium of thought. But the money has to a large extent taken over, just as obsession with language has overtaken large tracts of philosophy. Profit should be a way of measuring success and, in the form of investment, the fuel for further endeavour. But in business as a whole, profit has largely become the main point of the exercise. Indeed, the nature of the exercise has become secondary. The most influential market in the world is not of motor cars, or food, or fine art, or even bombs, but of money itself: the stock market. The nature of the endeavours that gave rise to the money have ceased to be the prime consideration, or even indeed to be relevant. All are grist to the stock market's mill. Keynes warned that this might happen; and now it has.

If money really did reflect accurately the underlying realities of the world, and if the market really did run in the way that Adam Smith and the monetarists claim it does, then this shift of emphasis would not matter. But money does not necessarily reflect reality particularly closely. Things that are very important to human life, and particularly to the lives of other creatures, are often given no cash value at all. Other things are attributed huge cash values simply because they are perceived to be valuable; as auctioneers say, a thing is worth whatever people are prepared to pay for it, however perverse those people may be, and whatever their motives. If the planet Earth was infinite, and infinitely forgiving, it wouldn't matter that we attach so little cash value to so much that is obviously so vital: that we spend so little on the environment at large, for instance, or attempts to temper the climate. Neither would it matter that rich people are happy to spend several fortunes on bad art – except that all money these days is in the same pot, and in 1999 that pot was doling out less than a dollar a day to nearly half the people in sub-Saharan Africa. It is reasonable at least as a first approximation to try to run the world by manipulating money, because money is easy to manipulate. But it is deeply pernicious to do this when the values reflected in the money have so little to do with things that really matter; or at least not to recognize that this is the case. In farming, the mismatch between money and reality is all too obvious.

Ever since Cain and Abel (and doubtless before) all farming has been constrained and dominated by two economic desiderata: turnover and efficiency. Whatever farmers do, whatever they grow, they have to produce enough. A family that devotes all its efforts to just six chickens and an apple tree will starve, even if the resultant eggs and apples are show-stoppers. Farms also have to be efficient, efficiency being the ratio of input to output. Food energy is the ultimate measure of farm efficiency: it is theoretically possible, after all, to expend more energy in cultivating and harvesting a crop than the crop actually provides. This is what happens when a potato harvest succumbs to blight, or a suburban gardener spends an entire summer raising a row of tomatoes. Turnover and efficiency have a complex relationship, one to another. For instance, the biological efficiency of an individual animal is in general increased if it produces more. A cow that produces 1,000 gallons of milk in a year doesn't eat much more than one that produces only 700, so output relative to input is raised. But it can require a lot of effort (a high level of intensive husbandry) to get maximum output from an individual, and it often pays to keep more animals at a lower level of husbandry. It's a fine balancing act.

But within this general balance of turnover and efficiency there is a potentially endless catalogue of often conflicting details even on the simplest of farms. Should the farmer grow wheat or barley, or raise cattle or sheep, or all four – and if so in what ratios? Should he or she raise sorghum to eat, or tomatoes for cash (and then buy in sorghum)? Should the fields be cultivated by hand, or with animals, or is it worth buying a tractor? Or should the tractor be hired, and if so, should the driver be hired too, or should the farmer use home labour? On hobby farms on which, say, well-heeled German businessmen raise a few steers (hobby farms are particularly popular in Germany), the equations hardly matter. It's largely for fun anyway. When people farm for a living, the calculations make the difference between continuance and destitution. All successful farmers at least since Cain have been inveterate calculators. Much of the oldest known literature, on tablets from Mesopotamia, turned out not to be sonnets but accounts of farms. The account books of the great farming innovators of the eighteenth and nineteenth centuries are things of wonder – as every nail and hour of labour is recorded – and aesthetically pleasing, too, since the bookkeepers took such pride in their handwriting. George Eliot wonderfully captures the conscientiousness of these early-nineteenth-century farm managers in *Middlemarch*, in the character of Mr Garth.

Without the idea of money the calculations would be impossible at all but the most primitive level. How do you decide between high-cost cattle and low-cost sheep? Or the ox-plough versus the tractor? Or human labour that must be paid for day by day against a machine that requires a once-for-all outlay, plus maintenance? All the time, like must be compared with unlike. Without a universal medium of exchange, farmers could not even begin to make sense of their own enterprise. All successful farmers are accountants, in their heads. On the modern farm, the professional accountant is a key player. Of course.

But it's all too easy for the emphasis to shift. In the hyper-capitalist world money becomes not the tool but the motive of the exercise. The rules of accountancy begin to determine the overall strategy. Accountancy begins to emerge as what mathematicians call an 'algorithm': a rule of thumb that provides a sure-fire solution to a problem, even though the rule itself may not be understood. Thus the rules that we all learnt at school for doing long division (or at least, we used to learn them) are an algorithm. The guiding principles of the farm, as of all enterprises, become the adages of accountancy: 'the right-hand column'; 'the bottom line'. The farm is then organized specifically to maximize profit – measured exclusively in money. Input is reduced by cutting costs. The margin between input and output is raised by 'adding value' between production and sale. Efficiency means cash efficiency, which is directly reflected in profit; so profit is taken as the principal if not the sole index of efficiency. Efficiency is in general good (at least it seems preferable to inefficiency) and so (it seems to follow) profit is good too; and maximum profit, suggesting maximum efficiency, must be best of all. The modern economic climate is such that nobody can escape from this logic – apart from the hobby farmers who don't mind if they lose money, and the subsistence farmers of poor countries who are not yet caught up in the economic mainstream. For most modern professional farmers, profit now has to be the prime motive, and accountancy has to be the guiding principle. Those who try to buck the monetary rules are ousted by those who play by them. Those who play the financial game most astutely, or indeed most cynically, win. 'Farming is a business like any other' is the modern mantra, taught in agricultural colleges, and business, conceived in the hyper-capitalist manner, is about money.

Again, this would not matter if the rules of money reflected precisely the things that are really important: the requirements of biology, and human values. If the market, whirring energetically in the background, really did provide people with what they need and want, and really did create equity, and peace, and justice, and help to conserve the world at large, then we would have little cause for complaint. Demonstrably it does not, however. Again, farming is particularly compromised. Again, simple reasoning shows why.

Most obviously, neither the inputs nor the outputs of a farm are costed accurately or exhaustively, and neither can they be. The issues here are fourfold.

Firstly, it can be extremely difficult to cost even the material inputs, and often this simply is not done. Key ingredients are fresh air and fresh water. These are sometimes costed, and sometimes not. Many poor farmers have been (and are being) wiped out as their water supply is diverted or polluted. Sometimes this is compensated, sometimes it becomes an international *cause célèbre*, and sometimes it just happens. Too bad. Traditional farmers commonly rely on local forests for firewood and fencing. This may not be good environmentally, but the timber is a vital input nonetheless. Yet it is commonly left uncosted. Contrariwise, many commodities within traditional farms are recognized well enough by the farmers themselves, but not by the accountants who descend upon them as expert advisers. In village India, for example, the stalks of pigeon peas become fuel and fencing.

Vitally, too, human values remain uncosted. Of course it is often efficient in cash terms to replace people with machines. On traditional farms, labour is by far the most expensive input. The more the workforce can be cut, the greater the potential for profit. Thus Britain in the early 1950s had about 454,000 family farms, and now we have half that, and 23,000 of those that are left produce half of the food grown in Britain. The Government condones this, and indeed is part of the process. At the time of writing (November 2002) Britain's farming, like that of the US, still depends heavily on government subsidy, but as Graham Harvey comments in *The Killing of the Countryside*,* 'In a period of unprecedented public support almost a quarter of a million farms have gone out of business.' At the end of the Second World War British farms employed nearly a million people, but, says Harvey, 'by 1994 the number of farm jobs had fallen to 120,000, of which one-third were part time'. If a bank or supermarket chain sheds 100 workers, this in Britain becomes national news. Yet Britain's farms have lost an average of 350 workers every week for the past half-century without any significant comment outside the circles of farming itself.

Eric Schlosser notes in *Fast Food Nation* (Penguin Books, London, 2002) that when the US first came into being, Thomas Jefferson considered 'the hardy independent farmers [were] the bedrock of American democracy'. But now, says Schlosser, 'The United States has more prison inmates than full-time farmers.' It's cheap to cut the workforce, but human labour is not just a figure on a balance sheet. Work is one of the most important things that any human being does. The work we do

* Vintage, London, 1999, p. 72.

largely defines the kind of life we lead, and indeed the kind of people we are. The collective work of societies determines the mien of the entire society. The loss of work is often a huge personal tragedy. The Reverend Nick Read, based at the National Agricultural Centre in Coventry, looks after the social and psychological needs of Britain's farmers. He tells me that 10 per cent of those who remain are now taking antidepressants. Suicide has become a common occupational hazard as they are driven from the land, or fear they soon will be. It's at least reasonable to suggest that if more of America's people had the opportunity to work on the land that their ancestors so painfully carved out for them, fewer would be in prison. Yet I have heard defenders of the modern market economy declare that if people have left the land, it must be because they wanted to. To suggest that people are ever forced or obliged to do what they did not intend to do all along is, apparently, to be patronizing. Whatever happens in a market economy, the mantra has it, must reflect public will. Fair enough, I suppose. If people commit suicide, it must be because they want to.

Worldwide, United Nations demographers predict that by 2050 more people will be living in cities than now are living over the whole earth. There will be precious little for them to do there. Some cities in the modern world have grown on the backs of spanking new industries, from computers to tourism, but in many others the career options range only from guerrilla through prostitute to mugger. It seems tacitly to be supposed that in time every city can grow and develop as, for example, Bangalore in southern India has done as its computer industry has developed; that economies based on mugging and prostitution are simply an unfortunate phase, which is bound to end with more wealth. But to a large extent, the success of (some) new cities is illusory. Bangalore has always been well appointed and now is cashing in on a world market that has boomed as spectacularly as the railways did in the mid nineteenth century. But such booms cannot last for ever. Markets become saturated and consumers run out of buying power. It is not at all obvious that most of the world's poor cities ever acquire or generate wealth. Where from? How? There aren't that many resources in the world for them all to be like Berlin, or even Bangalore. The last thing they need is to grow. Traditionally, agriculture was the world's single greatest employer. Traditional rural communities are by no means always peaceable (another powerful theme of literature), but almost always they are better than the shanty town, and some rural societies have been among the most agreeable that have ever been devised. Very habitable villages can tick along with minimal infrastructure and capital investment. It's economical to shed labour – at least for individual farms, in the short term. Yet if farms did nothing at all except employ people, they would still be worthwhile. The food they produce might largely be seen as a bonus. For the world as a whole the frenetic urge to throw people off the land these past 150 years could well emerge as one of the greatest of human tragedies. But then: the simple algorithm of accountancy, simplistically applied, does not cost tragedy.

Thirdly, at the other end of the line, the products and the consequences of farming are not properly costed either. At least, the commodities that finish up in the supermarkets are costed; the prices are totted up. But agriculture, as emphasized throughout this book, stands at the core of all human activity. It affects everything we do. The cost of maintaining people who are unemployed, or are in prison, is not put on to the farming bill. More obviously, modern hyper-intensive livestock units are highly 'efficient' in cash terms. The capital costs are enormous but the running costs are low (since there is only one worker per thousand pigs or per hundred dairy cattle or per half million broiler chickens) and the yield of meat or milk or eggs per creature is stupendous. But so is the output of ordure, which may be far too copious and full of additives (such as copper, a common ingredient of pig feed) to be put on to the land. Often, the cost of disposal has again been paid from the public purse. The farm's accounts remain untroubled.

The frenetic desire to cut costs in the name of cash efficiency is dangerous. As outlined in Chapter 5, Britain's epidemics of BSE and of foot-and-mouth disease resulted entirely from Britain's cut-price farming policy. But again, society as a whole picks up the bill, while agricultural ministers and advisers continue to congratulate themselves on their wondrous efficiency.

Finally, of course, accountants (and modern economists and politicians in general) do not consider the long term. At least, they do: but 'the long term' is commonly construed to be thirty years. Some, acknowledged as visionaries, speak of the next fifty years, but in general it's recognized that detailed projections break down after the first few decades. The premiss of this book is that if we are serious about our species, *Homo sapiens*, and our fellow creatures, then we have to think in biological terms; and to a biologist, even a thousand years is modest. Ten thousand years have passed since the Neolithic Revolution, and it's at least reasonable to think in terms of the next ten thousand; or indeed, the next million. Self-evidently, agriculture geared entirely to the cutting of costs and the maximization of output, which does not properly cost its own inputs and consequences and has no intention of doing so, and is locked into a worldwide competition which ensures that those who play the game most bullishly will prevail, does not have its eye on the long-term future. Indeed, if we truly consider the long term, then present policies must be seen as the prime threat to all humankind. Yet politicians continue to congratulate themselves on their boldness ('bold' has of late become Tony Blair's favourite word), and as a cure for the ills which even they are forced to acknowledge, they recommend even more of the same. As Mark Antony commented in a slightly different context (*Julius Caesar* III. ii. 110), 'Judgement, thou art fled to brutish beasts, and men have lost their reason.'

All this follows from the present emphasis on money, as opposed to the endeavours that lie behind the money. But there is another great flaw in the capitalist vision, or at least the modern, monetarist version of it, the hyper-capitalist extrapolation of Adam Smith. The market itself, the universal stage and trading post, is all too easily thrown off course. Absolutely not is it the natural underwriter of justice and democracy that Smith envisaged.

The Market Freed: The Rise of the Corporations and Globalization

Theory suggests that the market would work beautifully, the smoothly oiled machine that Adam Smith envisaged, only if certain conditions were fulfilled. There would have to be an infinite number of producers and consumers, all well matched in strength, and all aware of each other but nonetheless acting independently. Then, and only then, can there truly be perfect competition, and free choice. Once the number of traders or consumers is restricted, or clusters of them start to cooperate, options become limited and competition is compromised.

Of course, there can never be an infinite number of traders or consumers, and people do talk to each other, so they are bound to begin cooperating. Even so, traditional markets do show Smith-like qualities. In many a town in the Third World you still find markets where dozens of more or less identical traders sell more or less identical piles of vegetables and spices and plastic shoes. In many old cities, too, as in Istanbul, you still find entire streets devoted to single commodities: fruit, cooking pots, electrical goods. If you want a kettle, you know what part of town to go to. Traditionally, it was all very amiable.

For various reasons, however, this pristine, innocent state is not stable, even though it may persist in any one place for centuries. For one thing, if any one trader is more bullish or lucky than the others, then he or she may start to prevail, and oust their fellows. In traditional societies where everyone knows everybody, the more competent may stem their own energy to give the others a chance, acknowledging that it is better to be a successful member of a flourishing society than the only kid left on the block. But in more modern, flexible societies, where local communities have largely broken down, the successful feel no compunction to hold back, and indeed are liable to be swamped if they do. In a modern society, then, natural selection ensures that the few who are most energetic (or ruthless) supplant the rest; and, of course, as all evolutionary biologists acknowledge, luck plays a large part too.

Then again, different traders may choose to cooperate. Cooperation is generally perceived to be good (it certainly seems preferable to out-and-out conflict), but it has its sinister side too. In nature, creatures commonly cooperate not primarily through feelings of altruism but to gain strength: working in groups, not to say gangs, they can compete more successfully with the rest. Thus lions commonly operate in pairs (usually of brothers) as they contrive to oust some ageing patriarch from his pride of females. The growth of technology further encourages cooperation among traders, because machines can be helpful but tend to be expensive, so it pays to join forces.

Cooperation may take the form of 'horizontal integration', in which different grocers (say) join forces (like the Quakers Joseph Huntley and George Palmer who teamed up in the early nineteenth century to create what became the biscuit giant). Or we see 'vertical integration', in which particular farmers (say) agree to deliver only to particular butchers, who in turn are linked to particular supermarkets, and so on. There is no theoretical end to such integration, except what may in the end be imposed by ad hoc legislation, including laws to restrict monopoly. The cooperators form cooperatives, or cartels, or companies, that may grow into corporations and eventually into transnationals that bestride the world.

Fine though it may be in principle, all such cooperation immediately shifts the balance of power within the market, and subverts Smith's vision both of perfect competition and of perfect, free consumer choice. This may be achieved within the law but people who succeed in trade also acquire power and in closed societies (such is human nature) they may soon begin to bend the law to their own purposes. The distortion achieved by vertical integration in early-twentieth-century small-town America is reflected in Sinclair Lewis's novel of 1919, *Main Street*:

Then, on the corner below her husband's office, she [Carol, the book's heroine] heard a farmer holding forth:

'Sure. Course I was beaten. The shipper and the grocers here wouldn't pay us a decent price for our potatoes even though folks in the cities were howling for 'em. So we says, well, we'll get a truck and ship 'em right down to Minneapolis. But the commission merchants there were in cahoots with the local shipper here; they said they wouldn't pay us a cent more than he would, not even if they was nearer to the market. Well, we found we could get higher prices in Chicago, but when we tried to get freight cars to ship there, the railroads wouldn't let us have 'em – even though they had cars standing empty right there in the yards. There you got it – good market, and these towns keeping us from it. Gus, that's the way these towns work all the time. They pay us what they want to for our wheat, but we pay what they want us to for their clothes. Stowbody and Dawson foreclose every mortgage they can, and put in tenant farmers . . . the lawyers sting us, the machinery-dealers hate to carry us over the bad years, and then their daughters put on swell dresses and look at us as if we were a bunch of hoboes. Man, I'd like to burn this town!'

Kennicott [Carol's husband] observed, '. . . They ought to run that fellow out of town!'

Here, the most powerful traders collaborated to give both the primary producers (the farmers) and the consumers a rough deal; and there wasn't a thing that either could do about it. In a perfect, Smith-like world, the farmers could sell their produce elsewhere and the consumers could buy more cheaply, but in reality, in the midst of the American Midwest, both were in thrall to the middlemen. The middlemen, of course, make a virtue of their ability to acquire the goods cheaply; and again, at first sight, this seems reasonable – in line with Smith's ideals. But again, this principle is fair and sensible only when the seller and the buyer have equal power, and equal options. If the middleman has a choice of producers, while the producers have no choice of middleman, then the deal can be extremely inequitable. As John Ruskin put the matter in *Unto This Last* (1862):

whenever we buy . . . cheap goods – goods offered at a price which we know cannot be remunerative for the labour involved in them . . . remember we are stealing somebody's labour. Don't let us mince the matter. I say, in plain Saxon, STEALING – taking from him the proper reward of his work, and putting it into our own pocket.*

I do not for one second want to suggest that any modern trader in food of the kind that deal openly in the high streets behaves in ways that are unlawful. However, the modern TV ads in which processors and supermarkets boast of the deals they do with farmers ostensibly on the consumers' behalf, I find chilling. It is truly dreadful that the survival of an entire family of peach-growers should depend on a nod from 'the man from Del Monte'. Where is freedom here, of the kind that Smith envisaged? In another ad shown locally in my corner of England a supermarket tells us that it scours the world for 'the best deals'. This means the cheapest. This is obviously bad for the producers but in the end, too, it is bad in all ways for the whole world. Up to a point, cheap production reflects commendable efficiency. After that point, it is unjust; and it implies cut-price husbandry which, we as have seen all too clearly this past few years, endangers all of us.

Neither is it true (as the politicians and the industrialists insist) that cost-cutting necessarily leads to cheap food. Profit is the ultimate driver, and profit derives not simply from cutting costs but from maximizing the difference between production costs and selling price. The whole thrust of modern food processing and retailing is to 'add value'. This is achieved partly by processing and packaging but mainly by nudging the whole of agriculture towards livestock, as outlined in Chapter 4. To be sure, some foods may sometimes be sold cheaply. In recent years in Britain milk and lamb in particular have sometimes sold for less than the cost of their production. But nothing comes free. Such anomalies occur either for public relations purposes, or to squeeze out smaller traders; or they reflect the fact that Britain's agriculture, like America's, is highly subsidized. Overall, although the husbandry is cut to the bone, and despite some spectacular bargains, food in modern shops is not cheap. Huge industries have grown rich by 'adding value', which means by making it more expensive.

As the modern economy has unfolded, the whole supply chain, from

* Penguin Classics, 1985, p. 133.

field to table, has become more and more integrated, horizontally and vertically. Elected governments have been key players in this. As Schlosser comments in *Fast Food Nation* (p. 8), 'During the 1980s, large multinationals – such as Cargill, ConAgra, and IBP – were allowed to dominate one commodity market after another.' By the 1970s it already seemed that the food processors would soon control the entire supply chain, and in the US that is still largely the case. In Europe, however, the balance has largely shifted to the retailers, so that in Britain, 88 per cent of all food is sold through supermarket chains, with the top five (Tesco, Safeway, Sainsbury, Asda and Somerfield) accounting for the lion's share.

Most decisively, the big corporations that run the world's food supply chains control information. Every despot knows that the first thing to take control of is the broadcasting station and the newspapers. Whoever controls information to a large extent controls what people think, because thought depends to a large extent on available data. Children, with no experience of their own to guide them, are particularly vulnerable. All parents know how their children are bombarded with exhortations to eat this, that, or the other. The food that is pushed most forcefully is the kind that is processed and packaged to the nth degree. The mark-up is maximal; and children grow up with the idea that food is not food until and unless it is suitably processed. (A small relative of mine recently rejected home-made chips. He wanted 'real' chips, with crinkly edges.) Insofar as prices are kept down, it is by the use of industrial chemistry – additives that prolong shelf-life; while others (notably sodium) exaggerate flavour. Even more perniciously, perhaps, a leading British professor of nutrition commented recently that it is difficult these days to identify nutritional research that is not in some way sponsored by the food industry. Game, set, and match.

Despite all this, I still do not want specifically to argue that the corporatization of the food industry is intrinsically bad. It would be hard to criticize if, for example, the food companies had the wisdom and morality of Solomon and if they were indeed, as they may sometimes claim, improving the output of traditional farms and protecting us from its hazards, and from those they claim are imposed by traditional butchers, bakers and grocers. Even if that were the case, however (and of course I don't think it is), we have clearly come a very long way from the free market as envisaged by Adam Smith. By the same token, the innate virtue that Smith claimed for the market – the invisible hand that guarantees the best possible outcome – has gone missing. The pristine conditions, of perfect competition leading to perfect choice, no longer exist. The claim of the monetarists, that by extending Smith's philosophy into the modern age they are thereby ensuring justice, is simply not defensible.

Yet the integration of small traders into big corporations is not the end of the matter. Although the market is no longer free in the way that Smith envisaged (and in practice probably never was), the current aim, indeed the obsession of world leaders such as Tony Blair, is to bring all the markets of the world into one. Globalization is the mantra; and the World Trade Organization is the controlling genius. The global market can indeed be seen as the denouement of hyper-capitalism. It is the logical outcome that we are bound to arrive at if we follow the current version of free-market economics.

I do not want to argue that the idea of globalization is entirely bad. For some commodities (like computers or ships or cars) it might even be good. When a friend of mine's computer recently broke down in England he called the repair line in Ireland and was put through immediately to Silicon Valley in Bangalore, India, and, he says, it was all very efficient, and despite the distances involved, the whole transaction had a family feel to it. Neither do I doubt (having spent time there) that the people of Bangalore and of India in general benefit from their involvement in the computer industry or that they bring a lot to it (there are 3 million BSc's in India).

But neither do I doubt, not for one second, that for agriculture, the economic drift of the past few centuries or so, culminating in the modern, bullish model of capitalism enacted on the global scale in a global market, is a disaster. The early signs of disaster are already with us; the current starvation in Argentina, discussed later, is demonstration enough. In the longer term – not simply in the next 10,000 years, but within the next century – the transformation of world agriculture that is now being enacted could begin to threaten the survival of our entire species, or at least of tolerable ways of life.

In addition, the idea of the global market is beginning to take over, effectively to hijack, the whole idea of Third World 'development'. That idea was suspect enough to begin with, but the added imposition of the global market makes it doubly so. To be sure, there is no a priori reason why the countries of Africa, say, should not soon be trading on the world market in electronics and cars, just as India, China, Korea and Malaysia are doing already. There is no shortage of appropriate talent in Africa. But for the immediate term, and perhaps for ever, the countries of Africa and most of the Third World will stake their claim in the global market not with computers and engines, but with agricultural produce; and the consequences of globalization on such economies could be so vile and far-reaching as to beggar belief. This, however, is what some of the world's most influential governments, including Britain's, are now urging.

The Idea of Development

'Development' at best is an equivocal concept. It is all too clear that some countries are rich while most are poor. The richest 20 per cent consume 58 per cent of the world's energy and account for 86 per cent of private consumption. In modern diplomacy, the rich countries are said to be 'developed' while the poor are presumed to be 'developing'. The poorest of all are called 'LDCs' – 'least-developed countries'.

But there are inescapable imperialist overtones in the idea and even in the vocabulary of development. The term 'developed' implies that some desired and desirable target has been reached: that the US or Britain, say, represents some denouement of humankind, or even its destiny. 'Developing' implies that the poor countries not only are tending to become more like us, but should indeed aspire and be encouraged to do so. There are many implications here, including the ancient notion that what rich people (which mainly means the Western countries) do is right and good, while what poor people do is probably bad or at least is primitive. It was with this conviction that armies of Catholic, Nonconformist and muscular Anglican missionaries imposed themselves so enthusiastically on the rest of the world over the past few centuries, justifying the muskets and cannon that had gone before them. Yet there are many reasons why the 'developed' countries should doubt their own superiority, which include their obvious problems at home, the obvious subtleties of other cultures that have largely been overridden if not wiped out, and indeed the assertion in America's own Declaration of Independence of 1776, 'that all men are created equal'.

On the other hand, it seems at least reasonable to suggest that poverty is a bad thing, and that something should be done about it. Thus over the past few years the general idea of 'development' has been conflated with and largely superseded by what is called the 'war on poverty'.

In 2000, at their Millennium General Assembly, the 189 member states of the United Nations pledged to halve extreme poverty by 2015. Thus in 1999, 47 per cent of the people in sub-Saharan Africa lived on less than a dollar a day, while in South Asia the proportion was 37 per cent. The figures were much the same in 1990: 48 per cent were on a dollar a day in sub-Saharan Africa, and 44 per cent in South Asia. So many people live on the streets in Bombay that this way of life has almost become respectable. Many families have their legally apportioned patch of pavement, just as westerners occupy their legally acknowledged surburban houses (and the Bombay children are at least as polite as their suburban counterparts). Dignity is maintained. But many more live on streets and in shanties worldwide, with no rights at all; not even the right to life. With such extreme privation, it is difficult even to function as a human being. Every day is spent seeking food, staying out of trouble in its many forms, and coping with disease. There is no scope for the dreams and speculations that make human beings human. Of course people must be lifted out of this state. 'Halving poverty' at first sight certainly seems self-evidently good.

Even so, there's a carelessness in the present policy; assumptions and connotations that don't seem to have been thought through. Thus it seems to be taken for granted that as personal wealth increases, general well-being must rise commensurately. This is surely true of people living on their beam-ends: a roof over the head and a square meal cost money, and are infinitely better than none at all. Once the basics are taken care of, however, it clearly is not true that more personal wealth means more well-being. Context is all. Life in a Greek village on ten dollars a day with a diet of olives, rough wine, excellent bread, sardines and goat cheese is surely more agreeable than in a US trailer-park on thirty dollars, with noise, oil-spills and soggy pizzas. If the war on poverty simply means an increase in personal spending power, then it may achieve little; unless we assume that the power to buy more hamburgers denotes a general rise in status.

It also seems to be assumed that a general rise in a country's wealth must benefit the people as a whole. Yet this is even less true. John Ruskin again summarized the point, in *The Veins of Wealth* (1862):

> It is impossible to conclude, of any given mass of acquired wealth, merely by the fact of its existence, whether it signifies good or evil to the nation in the midst of which it exists. Its real value depends on the moral sign attached to it, just as sternly as that of a mathematical quantity depends on the algebraical sign attached to it.*

Most obviously, wealth does not simply trickle down from the rich to

* Penguin Classics, 1985, p. 187.

the poor, as seems so blandly to be assumed. It often happens that the rich grow richer while the poor grow poorer. Thus in 1960 the richest 20 per cent of people in the world had thirty times more wealth than the poorest 20 per cent. By 1990 the ratio of the richest band to the poorest was 60 to 1; and by 1997 had reached 74 to 1. Worse is that the real incomes of many of the poorest are falling in real terms. Thus Schlosser points out in *Fast Food Nation* that 'Adjusted for inflation, the hourly wage of the average US worker peaked in 1973 and then steadily declined for the next twenty-five years.' In the US too, over the past twenty-five years, 'the inflation-adjusted value of the minimum wage declined by about 40 per cent'. On the other hand, there is no shortage of extremely rich people even in the poorest African countries. Angola, for example, with a mere 13 million people, is the second largest producer of oil in sub-Saharan Africa, which in 2001 was worth $3.31 billion dollars. It also has diamonds. Yet in wealth per head of population it ranks 146th out of 162 countries in the United Nations Development Programme's latest Human Development Index. Sudan received about a million dollars a day for its oil in 2002, which is precisely the sum the government was spending on the civil war that has sputtered on for decades.

It matters too, of course, how the money is raised. Sometimes – often – the means of making money destroys much that is good in a society. It costs a great deal of money to turn a village into a small town. The former may be the more agreeable yet any such transition is perceived as 'progress', and therefore good. To pick up on Ruskin's point, the only certainty is that injections of money into any society must be disruptive. In the short term at least, some people will benefit more than others. Existing hierarchies may be reinforced, or turned on their heads. Typically, the pattern of labour changes. Of course, if well deployed, extra money can do good. Sometimes it is the sine qua non. Surprisingly often, however, the added wealth creates more problems than it solves. Politicians acknowledge this: at least, a standard excuse for not spending money on awkward problems is that they can't be solved 'simply by throwing money at them'. This is true; and it applies absolutely to the general 'war on poverty'.

In short, no one can doubt that widespread extreme poverty is one of the horrors of our age and we cannot consider ourselves civilized unless we try to do something about it. Even here, though, we must tread carefully; not rush in, as the old-style missionaries so often did, where angels would have feared to tread. It is naive in the extreme to assume that additional wealth will necessarily do good. It's important, when generating wealth, not to destroy the good things that were there before. Most of all, it's important to recognize that poverty has many causes. Perhaps there have been instances in the history of the world where poor people have been poor simply because their technology or their politics were inadequate, and in such cases (perhaps) complete transformation was necessary. But it is very difficult to find a modern case where extreme poverty reflects innate inadequacy. A remarkable proportion of poor societies have simply been knocked off course by other people's politics, and all they ever really needed was a little peace. Angola was engaged in civil war for thirty years, until the ceasefire in 2002. Cambodia has become one of the modern disaster areas, but only in the aftermath of US bombing.

The official brief of the World Summit in Johannesburg in August 2002 was 'sustainable development'. This seems to mean something rather sophisticated: the idea that poor countries should indeed 'develop' but only in ways that could be continued into the indefinite future. But many observers from NGOs (non-governmental organizations) – who tend to be among the most alert participants at summits – report that Western governments (including Britain's) effectively refused to discuss the broad concept of sustainable development at all. They wanted to focus on the 'war on poverty'. Apparently they took to be self-evident that if poverty is somehow 'solved' (halved by 2015, for instance) then everything else would fall into place. In similar vein, the Western powers were also at great pains to ensure that no conclusions from Johannesburg would conflict with the present (Doha) round of agreements on world trade.

Simplistic though it obviously is, such a stance might be defensible if the questions that it raises were being conscientiously addressed. For example, is it in fact necessary to increase a country's total wealth at all, or do the problems lie merely with the distribution of wealth? If the latter, then more wealth might make matters worse – the rich growing richer, and the poor poorer. If wealth is to be increased, then how should this be done? After all, the manner of creating wealth can make life worse, at least for many people in the short term; and there is no guarantee that those who miss out in the short term will ever recover. Then of course there is the central question of all politics, which Lenin summarized as 'Who? Whom?' Who has the right to do what, to whom? It seems good, in principle, for rich countries to seek to help poor countries – but only if the poor countries invite them to do so, and remain in charge of their own destinies.

Of course, many politicians and diplomats in rich Western countries acknowledge all of these caveats. Many will be irritated by the previous few paragraphs, and invoke adages that involve grandmothers and the sucking of eggs. Nevertheless, the rich world is behaving in ways which suggest that all the subtleties have been put to one side, or have never been recognized at all. In particular, Britain's present government speaks and acts as if it believes that wealth per se can end poverty, and that the alleviation of poverty per se will do all that is required. Worse, it clearly believes that the best, easiest, and indeed the only 'realistic' way for poor countries to develop is to join the WTO and trade in the global markets. The only commodities that most poor countries have to offer are those of agriculture. In other words, the concept of 'development', which has always had some unfortunate connotations, has now been compromised even further. Now, Third World countries are being invited to 'develop' by 'globalizing' their own agriculture.

Globalization of agriculture is bad for agriculture as a whole, and for the long-term prospects of humanity, and hence for the world itself. For Third World countries, it seems to offer a one-way ticket to utter destitution.

The Absolute Destructiveness of the Global Farm

In March 2002 a group of farmers from Andhra Pradesh in southern India travelled to England to ask Britain's government not to invest £65 million in their country. Among other things, the investment was intended to 'modernize' their farming: more machines, more chemistry, more biotech, up to and including genetic engineering. The farmers delivered a petition to Britain's Secretary of State for International Development, Clare Short, in London, and were granted an audience. I met up with the farmers along the way. Their case was simple. The agriculture of Andhra Pradesh as it stands is already doing a good job. All the farms in all the world could doubtless be improved, but what the farms of Andhra Pradesh clearly do not need is the kind of transformation that Britain's hand-out will promote – a shift from the traditional structure into a more Western, industralized mode. Among other things, the Indian farmers pointed out, this would put at least 20 million people out of work and in India there is even less for disenfranchised people to do in the cities than there is in Britain. Ms Short replied that her dealings were with the Andhra Pradeshi government, which was elected, and therefore she was clearly acting democratically. The fact that the electorate itself was requesting something different apparently had nothing to do with the case.

Ms Short's attitude to the farming and the farmers of Andhra Pradesh typifies that of Britain's government towards traditional farming as a whole. It also typifies that of the Western world in general, and that of some (but by no means all) governments in the Third World. Those modern governments perceive traditional farming to be inadequate. Even if it does seem to be doing a reasonable job here and there, it is nonetheless perceived to be old-fashioned. It does not partake of the modern industrial methods (in which labour is replaced by machinery, chemistry and biotech), and it is taken to be self-evident that industrialization is necessary. Tony Blair recently announced in Africa that he has a 'vision' for Africa which, beyond doubt, includes the 'modernization' of its agriculture.

Thus in the gung-ho, Western, modern approach to development, as represented by Mr Blair and Ms Short, there are two strands of thought. The first is the traditional concept of development: that Third World countries are in effect destined to become more like those of the developed world, and that to achieve this they are bound in a general way to re-enact our history. That is, they must shift from a primarily agricultural economy (of the kind that Europe and America once had) into an industrial economy – or, as in modern Britain, into what Mrs Thatcher called a service economy. The industrialization of Third World agriculture is seen to be part of this essential and inescapable process. This way of thinking has a pedigree of several centuries, and has effectively been consolidated by the United Nations since the Second World War.

The second strand of thought is more specific, and more recent. It is rooted in monetarist hyper-capitalism and in the idea of the global market. It says that the global market is in general a good thing; but also, much more to the point, that it is in the best interests of poor countries to join in with the global market, and with the WTO that runs it, with all possible speed. Indeed it goes further. Politicians like Mr Blair effectively argue that for Third World countries, the global market is a sine qua non: the only realistic option. Agriculture in turn is perceived, as is the modern way, simply as a business like any other. It follows that the Third World's agriculture must be thrown with all possible dispatch into the global financial pot. For countries whose economy is still overwhelmingly agricultural, this would be a commitment indeed. It means that countries that are already poor should put their entire economies, their entire futures, on this one option. Indeed it conjures visions of desperate punters preparing to stake their life savings on an outsider, recommended as a cert by some racecourse tipster.

Yet there is more to it than that. Firstly, there are the general objections to the industrialized, globalized approach to agriculture as a whole, already outlined: that it is absolutely antithetical to the principles of enlightened agriculture. In a nutshell, industrialized, globalized agribusiness is not designed primarily to provide plentiful, safe, nutritious food that meets the highest standards of gastronomy, or to provide employment in an age when employment is at a premium, or to protect landscapes and other species; and what it is not designed to do, it is not likely to do, and in fact spectacularly fails to do. For humanity and the world as a whole, in the long term, the present trend can be seen to be disastrous. But for the Third World the problems will be more immediate, and have additional implications. We are asked to believe that the farmers of Angola, say, will gain wealth for themselves and their country by putting their produce on to the global market where they will compete with farmers from America and Europe (and Australia and New Zealand, and increasingly from China). They will be able to compete successfully (the mantra has it) because they can produce food more cheaply, and provide food of a kind that cannot readily be produced at all in America or Europe.

But if we ask why such countries will (perhaps) be able to produce food more cheaply, then the flaws in this simple scheme immediately become apparent. First, it seems to be assumed that the poor countries can produce food easily because they tend to be tropical and plants need warmth and sunshine. Indeed they do; but they also need water and various forms of fertility which in poor tropical countries are typically lacking. Although most poor countries (I am inclined to say 'all', but that would be a guess) could easily be self-reliant in food, they will not in general find it easy to out-produce and undercut the well-appointed farmers of Spain or Holland on the world market (although the Spanish and the Dutch are right to be worried; the fight will damage them, too).

Second, of course, the poor countries are supposed to undercut the rich because their labour is cheap. But here we have the ultimate catch-22. The poor countries are urged to join the global market so as to get rich.

But they will be able to compete in the global market only if they stay poor and continue to work for rock-bottom wages. This is wonderfully obvious, and yet the advocates of globalization as the route to salvation grow ever more confident that this is the way ahead.

Suppose, though, a country like Angola does decide to go down the modern road, and put its agriculture up for global tender. Suppose it does put down a few thousand hectares to oranges and a few thousand more to French beans (as Kenya has done). At that point it would come up against realities. The first is that they are not, of course, the only players in the field. They may aspire to sell in Britain's supermarkets (say), but so do the British, the French, the Italians, the Australians and New Zealanders (who still regard Britain as their traditional market), the Argentinians, the Californians, and so on. Enthusiasts for the global market are wont to suggest that for poor countries it is not only the best if not the only option, but is virgin territory, crying out to be occupied. Yet there cannot be a market in the world where the consumers have cash to spend (as in Britain) that is not already filled half a dozen times over. Nobody gets in except by fighting.

Even to enter the fight the contenders must first meet the stringent standards imposed by Western supermarkets. The fruit must be free of visible blemishes, and raised to precise specifications of colour, shape, texture, juiciness, and so on. These standards depend on the whims of the supermarkets themselves. They are not necessarily (or usually) those of nutrition or gastronomy. They have to do with immediate visual appeal, shelf-life, and so on. Of course, farmers in Third World countries include many of the most capable in the world (they have to be) so, given the resources, they could grow whatever the Western supermarkets want them to as well as any Californian or Dutchman. But they can't do this as they are. They would first have to industrialize, which means borrowing Western money (nothing is for nothing) and importing Western technology. If they join the global free-for-all, in short, Third World farmers will not be able simply to put their present-day, traditionally grown produce on to a plane bound for England or the United States. The point is not that their produce is inferior (absolutely not). But the criteria of the Western market are as they are, and those who do not meet them don't get in. It isn't a matter of being good; simply of conforming. To meet the Western criteria, absolute transformation would be needed. But then, of course, you never know when the criteria might change. In any case, part of the purpose of the standards is to enable the retailers to fend off potential surpluses: at any time they can simply tell the growers that their produce doesn't come up to scratch, and so leave the growers to bear the costs.

But suppose the farmers of Angola did manage to produce oranges, French beans or whatever which met all the Western criteria; would they then be able to sell freely in the markets of the US or of London or Paris? What a hope. One grand and overwhelming point is, of course, that Western governments (in the EC, and especially in America) subsidize their agriculture on a scale that is out of the reach, generally by orders of magnitude, of poor countries. This is absolutely against the spirit of the allegedly free global market, as the Third World countries themselves complain. Even if the playing field were more or less level, however – even if the food trade were just the free-for-all that it is supposed to be – it would still be very difficult indeed for poor countries to break in.

At the same meeting where I met the Andhra Pradeshi farmers I also spoke to small farmers from Wisconsin, for small farmers the world over have common cause. One of them, John Kinsman, produces milk from a small dairy herd at La Valle, where enormous herds (of many hundred beasts) are the norm. If he sold directly to the consumers, he says, he could charge twice the standard 'farm-gate' price, and yet sell his milk more cheaply in the shops. But he can't get his milk into the shops. The shelf space is contracted to the corporations. His colleague, Jim Goodman of Wonewoc, pointed out that although there are laws in America to prevent such restrictive trading, they cannot in practice be enforced. Laws may be made centrally but they are enforced locally, and there isn't a trading law in the history of the world that has not in practice been bent in favour of vested interests. Sinclair Lewis made the point nearly a century ago in *Main Street*, and nothing substantive has changed. The might of the United Nations or the WTO will not in practice be brought to bear upon Midwestern trading posts. But it's what happens in those outposts that really matters. The European Union began more than half a century ago (as the Common Market), primarily as a free-trade area. Still, and notoriously, English lamb has often failed to make the journey across France, while English fish has rotted on French quaysides. If free trade can't be made to work between neighbouring countries of equal political clout that are supposed to be cultural, economic and military allies, and if small American farmers cannot sell their produce even to their neighbours, what chance would a consignment of bright, shining, perfectly spherical Angolan oranges have in Sacramento or Madrid? If it wasn't tragic; if it weren't that some of the world's most powerful people are eager to make it happen, this whole concept would be a joke.

Then again, the WTO agreements are in principle based on reciprocity. Poor countries gain the right to trade freely in the lands of the rich, but the rich farmers have the same rights to trade in the poor countries. Coca-Cola machines are currently appearing all over India. The traditional tea-sellers are going out of business, to join all the other jobless.

The globalization of agriculture has two very obvious implications, both of which are most unpleasant. Firstly, we can envisage that some big Western company will acquire the right to grow, say, a few thousand hectares of kiwi fruit in some hypothetical country in Africa – which, after all, has every kind of climate, and in principle can grow anything. The company will of course take the lion's share of the profits because that's why they're there. Local dignitaries will take their cut. The company will employ a few local people (or even a few hundred), who will be photographed with flashing smiles for the company brochures. If market forces are allowed to operate, however, those workers will be employed only on a daily basis, laid off in the close season, and at the best of times paid astonishingly little (the kinds of practices that are still common worldwide in all kinds of industries). They will have nothing resembling security and certainly no pension.

In addition the land that would now be growing food for export would previously have been farmed by hundreds or thousands of individual farmers and their families: not rich, of course, but reasonably secure (if left alone) because they were good at what they did and their strategy was geared not to maximum output, but to security. Some of those farmers will now be working as labourers on the new company estate. Most will be out of work. In times of drought, with no jobs and no land, they will starve. The food grown on the estates will continue to be exported (for in any case, people can't live on kiwi fruit). Western governments will send 'aid' (lorryloads of Western wheat and maize, which helps the donors to dispose profitably of embarrassing surpluses). Western schoolchildren will raise money for their less fortunate brothers and sisters in distant lands and be told that they are doing good.

There are hundreds upon hundreds of precedents for just this scenario. It has been enacted in every country in the world. It has happened in Britain and the United States, steadily and inexorably, but with occasional spectacular bursts, including the Highland clearances of the late eighteenth and early nineteenth centuries, when Scottish crofters were cleared away en masse to make way for big estates of sheep; and in the American prairie, as so graphically documented by John Steinbeck in *The Grapes of Wrath*. Examples from the Third World are on television every week, or would be if the channels did not so quickly grow tired of them.

As traditional farms are expropriated and famine follows in their wake, Western ministers will wring their hands, and say how cruel the world is, and hold up the new company estate with its smiling workers as a model to us all which, they will urge, should be followed as rapidly as possible by everybody, everywhere – as it would be, if only the awful backsliders (effete, nostalgic, out of touch, unrealistic) did not keep complaining. If they live in Britain, those ministers will be rewarded in due time with extremely large, inflation-linked pensions, make speeches in the House of Lords, and die in the certainty that they have fought the good fight. So too, of course, will the company directors. The road to hell is not simply paved with good intentions, but is concreted and metalled, and officially declared open by the world's most powerful leaders.

Even at best, the globalization of agriculture implies a global dogfight, as foul and bloody as any that might be staged in remote rustic barns or urban warehouses. In traditional markets traders recognize that they are all in the same boat and help each other out. But in the modern world small farmers are obliged to compete with the corporations (which, among other things, make it difficult for them even to sell their produce to their own neighbours) and with farmers from every other country in the world. All the world's farmers are being thrown into the ring like old-style prize-fighters and invited to fight to the death.

For whose benefit? Consumers and voters are being told that it's all for our benefit, for this gladiatorial contest is bringing down the price of food and hence the cost of living, and so we will all be better off. But the point of the modern food supply chain is not to produce cheap food, but to increase the gap between the 'farm-gate' price and the final selling price; in other words, to increase profit. Not simply to increase, either, but to maximize, since in this ultra-competitive world those who are most profitable must prevail, since they have most power to reinvest. Thus with the frenetic reduction of costs goes the equally frenetic 'adding of value'. John Kinsman's point applies: without the middlemen, the food could be a lot cheaper. Even more to the point, in a morally acceptable world we should be asking *why* so many people are poor (even in the richest countries), and attacking the causes at their roots. One of the most obvious reasons, of course, is that many of the poorest people no longer have access to the land where their forebears once worked. In any case, the solution to poverty is not simply to be cruel to livestock or to invite farmers to fight to the death to reduce the cost of production even further. Among other things, cheap husbandry is dangerous, and it's the taxpayers who pick up the bills for the epidemics, and sometimes die in them.

Farmers on the other hand are told that the new battle is for their benefit. At least, the ones who are able and willing to play the game will do very well indeed and the rest – well, who cares? On a plane recently I met a young woman who bought avocados from Mediterranean farmers on behalf of a supermarket. Some producers, she said, met the company's requirements better than others; those who fell short were being 'weeded out'. This is a chill phrase indeed: seventh-generation farmers whose activities support entire communities are being 'weeded out'. Those who are weeded are not necessarily bad farmers. By the standards of enlightenment, they might for all one knows be among the best: the most humane, the most environmentally conscious. But they don't conform to the commercial fashion of the moment, and so they must be packed off to Malaga and the Costa Brava, to wait on tourists and sell lottery tickets. Many will remain unemployed, of course. In benign societies they will be supported up to a point at taxpayers' expense; and in less benign societies, who knows? The process will, though, produce a few agribusiness tycoons, as indeed it has already done; and they will take government ministers to lunch and each will convince the other that everything is fine (if only it weren't for the clamorous small farmers, and the poor in general, who are such a nuisance).

Yet in all but the shortest term, this dogfight will not be to anyone's benefit, except the organizers'. That is the way with dogfights. The combatants do not win (they finish up dead), and neither do the punters (who finish up fleeced). In twenty years' time, if the present generation of agri-industrialists and their political backers have their way, the agriculture of Europe, the US, New Zealand, India, Cambodia, Mozambique, Angola, Uruguay, and indeed absolutely everywhere, will be dominated by a few corporations. To a large extent, this is already the condition in the US and Britain. France, Germany, Italy and Spain still have strong traditional agricultural lobbies, and are still holding out against such transformation (and if you want to eat well, then these are still the places to go). For the most part, the French stand on agriculture is nationalist and self-centred. By no means do they occupy the moral high ground. But their position is infinitely preferable to Blair's vision of globalized industrialized agribusiness. *Faute de mieux*, in the short term the French farmers are worth supporting.

Adam Smith's vision of a self-adjusting market, of traders competing with each other, and in balance with consumers, will retreat further and further into the mists of quaintness – although he will doubtless be dusted down at intervals for public relations purposes, as a respectable eighteenth-century Scots intellectual and moralist who argued that the market guarantees justice. The corporations will sell whatever they find most convenient, charge what they like, and increasingly control all sources of information. Governments and international bodies will sometimes get hot under the collar and invoke laws of monopoly, and may even send the occasional out-and-out crook to an open prison, but the die will be well and truly cast.

Whatever resources are needed to keep up the flow of high-value goods will be pressed into service. Environmentalists will continue to catalogue the shortcomings: the loss of soil, water, and other species. Politicians will doubtless continue to attend summits, at their electorate's expense, and contrive to convince the rest of the world that whatever is happening is not their fault. Recent history does not suggest that the world as a whole takes much notice of environmentalist warnings. America in particular is openly defying the Kyoto agreement on carbon dioxide emissions, and America in this context is equivalent to most of the rest put together. Scientists, more and more, will do what the corporations want them to. The corporations, after all, will be their only employers. The loop will be complete.

But if the world's politicians who are supporting the industrialization and globalization of world agriculture are not wicked, then what motivates them?

The Mistakes Behind the Grand Mistake

Modern agricultural policy at the most influential levels of national governments and the WTO is rooted in a series of profound misconceptions. Most fundamental is the notion that agriculture is 'just a business like any other', which implies that we can farm well – indeed, farm as well as is theoretically possible – just by following the rules of money: cutting costs, maximizing turnover, adding value, maximizing profit.

Common sense and a growing catalogue of experience, of which BSE and foot-and-mouth just happen to be conspicuous examples, proclaim that such a notion is ludicrous. Yet the idea prevails. Anyone who suggests that our ideals should not be those of maximum profit, or that the simple rules of bullish commerce might lead us in wrong directions, is held to be unrealistic. Embedded in this is the notion that commercial competition means cheap production (which it does), and that this leads to cheap food (which in reality it does not), and that this compensates poor people for their poverty. Once spelled out clearly, we can see how threadbare the thinking is, logically, historically and morally.

Agribusiness is driven by the feedback loop, the dynamo, of capital–science–high tech–capital. Agriculture benefits from more science, after all, as all human activity might; and agribusiness is the most efficient way to pay for it. In this little chain of logic lies another trail of deep misconceptions which, I believe, reflect the educational status of most of the world's leaders who are versed neither in science, nor philosophy, nor agricultural history, and who select the advisers who tell them what they want to hear.

For in this viewpoint first of all lies huge misunderstanding of what science is, and what it can do, and of its limitations. Many (particularly in high places) seem to believe that science tells us all we really need to know; or that if it does not, then it soon will. I discussed the lie of this in the last two chapters. Science is not, and cannot be, omniscient; and, vitally, it does not directly address matters of human value – justice, happiness – but these, in the end, are what matter most.

Despite these obvious limitations, we see in modern agribusiness policy a terrifying faith in the omniscience and omnipotence of science. We have seen this most starkly of late in the context of GMOs – yet the world's biotech industries are planning to impose GMOs on the world as a whole just as rapidly as this can be arranged. Governments, including or especially Tony Blair's, are smoothing the path, holding the door open. These politicians have been elected, but choose in the name of what they think of as modernity to override the doubts of their electorate. In religion, the term 'enthusiast' is often used pejoratively, to describe the wild-eyed zealots, the fundamentalists and extreme ascetics. Blair, in the context of modern biotech, is just such an enthusiast.

Of course, history does seem to justify a general, broad-brush belief in science and the high tech that it produces, and so of the economic system that allows them both to flourish. After all, the countries that have ridden the dynamo of capital–science–high tech–capital most adroitly this past century, decisively dominate the modern world. The USSR failed in the end not because it was cruel but because it failed to get its own dynamo whirring. In the end, though, the triumph of societies that run the engine of change most efficiently depends on military power. Yet it is at least naive to assume that the approach that brings success in war, and hence in political dominance, must succeed in all details and in all fields, and in particular, in the long term, in farming. War, to some extent, really is a business like any other. But farming is not.

In truth, as outlined above, farming has grown up over the past 10,000 years (plus a few more tens of thousands of years of prelude) as a craft. As a craft, it has succeeded brilliantly. It demonstrably allowed human numbers to increase 200 times, and probably nearer 300 times. Agricultural science could not have come into being except on the back of that craft, and has made a significant difference only in the past seventy years – seventy out of 10,000. Agriculture has been, and fundamentally it remains, a craft industry, and we forget that at our peril. But many of today's engineers and biotechnologists, and the agribusiness people who employ them, have forgotten, or never knew this. Those who seek to grow the same variety of the same crop from mountain-range to mountain-range, or to keep pigs by the million in multi-storey feedlots, or to populate the world with GMOs, can have no sense whatever of agricultural history.

In short, the modern enthusiasts have failed completely to learn the lesson that has been so starkly and horribly pressed home this past half-century: that science is brilliant indeed when it is used to abet good husbandry, of the kind that has evolved through craft; but can be very damaging when it is deployed to override the fundamental principles of good husbandry. Yet increasingly, science is deployed expressly to flout the tenets of good husbandry because good husbandry is perceived to be expensive. The ill effects of the new methods, when good husbandry is overridden, are paid for out of taxes, or by the misery of disenfranchised farmers. The examples of BSE and foot-and-mouth stare us in the face. They have come about entirely because the fundamental tenets of good husbandry were put aside, in the interests of cutting costs. But the enthusiasts, including Mr Blair's government which brought us foot-and-mouth, cannot see what is so obvious, any more than the mad-eyed religious zealots of the desert could see their own absurdity. The mad-eyed zealots, though, were condemned by their more sober contemporaries. The modern agricultural strategists, clamouring for more and more high tech at all costs and in all contexts, somehow continue as the world's leaders.

For it is taken for granted that traditional craft agriculture, left to itself, cannot deliver. But look in detail at almost all the countries that have suffered famine this past half-century and you do not find failure of craft farming. You find civil war, and mad despots, and foreign politicking. You find, in fact, that very few Third World countries have escaped being screwed up in half a dozen different ways. As James Meek put the matter in the *London Review of Books*:

It's not that there haven't been any wars since 1945. There have been about three hundred. It's true that people haven't fought and died everywhere in the world. They've fought and died only in Indonesia, Greece, Iran, Vietnam, India, Bolivia, Pakistan, China, Paraguay, Yemen, Madagascar, Israel, Colombia, Costa Rica, Korea, Egypt, Jordan, Lebanon, Syria, Burma, Malaysia, the Philippines, Thailand, Tunisia, Kenya, Taiwan, Morocco, Guatemala, Algeria, Cameroon, Hungary, Haiti, Rwanda, Sudan, Oman, Honduras, Nicaragua, Mauritania, Cuba, Venezuela, Iraq, Zaire, Laos, Burundi, Guinea-Bissau, Somalia, France, Cyprus, Zambia, Gabon, the US, Uganda, Tanzania, Brazil, the Dominican Republic, Peru, Namibia, Chad, Czechoslovakia, Spain, the Soviet Union, Britain, El Salvador, Cambodia, Italy, Sri Lanka, Bangladesh, Chile, Turkey, Ethiopia, Portugal, Mozambique, South Africa, Libya, Afghanistan, Jamaica, Ghana, Ecuador, Zimbabwe, Burkina Faso, Mali, Panama, Romania, Senegal, Kuwait, Armenia, Azerbaijan, Niger, Croatia, Georgia, Bhutan, Djibouti, Moldova, Sierra Leone, Bosnia, Tajikistan, the Congo, Russia, Mexico, Nepal, Albania, Yugoslavia, Eritrea, Macedonia, and Palestine. One crude tally puts the number of dead at well over twenty million.*

In short, all the evidence is that when traditional agriculture is given a chance – no civil war, no vicious dictatorship and, we might add, no expropriation of fields to keep foreign fruit-canners happy – almost all countries can feed themselves very well indeed. Certainly, those of the Third World can do so since most, after all, are tropical, and have a wide variety of climates. People would not be living in those countries at all if they were not habitable (except when they are refugees from somewhere else). The causes of famine, in short, are almost entirely political. Amartya

* 6 September 2001, p. 28.

Sen makes the general point in *Development as Freedom* (Oxford University Press, Oxford, 1999): 'No substantial famine has ever occurred in a democratic country – no matter how poor.' (Professor Sen is Master of Trinity College, Cambridge, and won the Nobel Prize in Economic Science in 1998.) Growing food is what people in Third World countries do. They don't need it 'cheap', not least because in its traditional form their agriculture does not partake of the world's cash economy at all, so the concept of 'cheap' does not arise. Most of the people in those countries are farmers, and very good farmers at that, and all farmers need is the freedom to get at their own land. This is the freedom that modern Western intervention, ironically including some at least of the 'war on poverty', tends to deny to them. Decidedly, it is not effete or nostalgic to ask that the craft of farming should be given a chance. This is where hope really lies. The 10,000 years of its evolution are the true foundation for the future.

Neither are the serious critics of the modern way necessarily Luddite. I'm certainly not. I have spent a lot of my life writing about agricultural science, and extolling its merits. All human activities can in principle benefit from a little know-how and a few more technical tricks, of the kind that science can provide. As discussed in the next chapter, the more go-ahead organic farmers are extremely keen on science, both to help sustain fertility and to enhance methods of biological pest control.

The huge mistake, however, is to assume that science necessarily implies industrialization. Historically, in the West, science and high tech have indeed led to industrialization. This is what they were geared up to do: primarily, to replace labour with machines and chemistry. So the kind of agricultural science that is now around is, for the most part, the kind that is expressly intended to industrialize; and if you want access to that science, you have to invite the companies that have developed it to come on board and run your agriculture for you (or else pay them a royalty). But the kind of science that now is provided by big industrial companies is not the only kind there is. Or at least, it is not the only kind there could be. A truly enlightened agricultural policy would produce science of its own, geared to the business of feeding people without cruelty or injustice, and without destroying everything else. Traditional farming could certainly benefit from more science. Of course it could. Indeed, this is necessary. But in overall structure, in general approach and technique, and in its social arrangement, traditional agriculture worldwide for the most part is brilliant. It has, after all, evolved by natural selection to fit in with its own environments. Sure, it may need tweaking here and there. But the idea that it should be swept away wholesale to make way for the particular kinds of science that are now being developed by corporations in the cause of industrialization and profit is madness. No other word quite expresses how very misguided the prevailing policies are.

The final gross folly is to impose this general trend on to the Third World. The rich world can handle the mess its new agriculture is creating – or at least, up to a point. The problems of the cities in Britain and the US where too many people have too little to do that is worth doing, and small matters like BSE and foot-and-mouth, aren't exactly encouraging, but at least the US and Britain still survive. We both have many badly fed people but we don't have mass famine, and we are not bankrupt. In poor countries, though, for which agriculture is the biggest employer, and where people eat what their local fields produce, the drawbacks of modern industrial and globalized farming could surely prove fatal.

The role of farming in Third World communities, as in most human communities through all of history (and much of prehistory) is quite different from in the modern West. Traditional people live much closer to the land. In Angola, 70 per cent of the whole population work on the land. In Rwanda it is 90 per cent. This kind of figure is more typical of the world as a whole, and of all societies in the past, than the 1 per cent or less that now work on the land in modern Britain or the US. I do not suggest that 70 per cent or 90 per cent is the ideal figure. When so many work on the land it suggests a shortage elsewhere: not enough drains, hospitals or schools, not enough builders, engineers, nurses, doctors or teachers. Still, though, common sense and all history suggest that for most countries agriculture should be a major and probably the main employer (and sometimes the main employer by far). If people do not farm, then they have to do something else (or else take 200 days' enforced holiday each year, like the plebs in Ancient Rome). South-East Asia and India are getting stuck into electronics, but although there is currently a boom, there's a limit to the number of computers people will want just as, in the alleged service economy of modern Britain, there's a limit to the number of times that most people want their hair cut. But there are always useful things to be done on well-run farms. Husbandry can always be improved with more people to take care of it. The trails of grey, skeletal people with their begging bowls in modern Africa are, for the most part, the families of ex-farmers. Seventy per cent of people on the land may indeed be too many. But the driving notion of the West –

that farm labour should be reduced as closely as possible to zero – is in all ways madness; all except for the fact that in the immediate term, it is profitable. But as a long-term solution to all the world's problems, it is clearly nonsensical.

Then again, it must be in the interests of poor countries in a competitive world to be able to feed themselves. As I argue in Chapter 11, self-reliance in food remains a good basic principle for everybody, and for Third World countries it must be their prime asset. All those countries are innately vulnerable, and always will be. In particular, they will never compete militarily on the world stage, and in the end it's military power that underwrites political clout. The modern notion that these poorest of countries should commit their entire economies to the vagaries and vicissitudes of the global market in which they would be very small players indeed and have almost no ability to control events (no matter what it might say on their trading contracts) ought to be laughable, were it not that this is now the world's prevailing policy. But if people are self-reliant in food, then at least they need not starve. That, as accountants like to say, is truly the bottom line.

If further evidence is needed that the modern drive to industrialize and globalize the world's agriculture is misguided, then it is being supplied as I write (November 2002) by the tragedy of Argentina. Argentina was never a Third World country. At the start of the twentieth century Argentina was perceived as the brave new world. Poor and entrepreneurial Europeans flocked there to make their fortunes, notably from the wheat and beef that could be raised in superabundance on the pampas. By the 1980s it was flourishing. But then the International Monetary Fund took a hand. Argentina was not Third World, but it wasn't First World either. But it was persuaded to join the global market and, among other things, to industrialize its agriculture. The arguments, in essence, were exactly those that are now directed at the world at large.

Then its economy collapsed. In general, markets do not work as smoothly as their protagonists claim. Free markets are not under perfect control (by definition) and they are not perfectly honest, and dishonesty can make all the difference. Argentina's farms could feed the native population ten times over but they are not set up to do so. The crops are for export. They are spoken for. People starve in the countryside surrounded by produce they have no access to. A contract is a contract. At least if countries are self-reliant they can spare themselves this final misery. Most of the countries in the world that are now being urged to toe the economic line as Argentina did are in a far weaker position than ever Argentina was. Yet the mantra has it that globalization and industrialization must be good for them as for the world as a whole, and are the way ahead.

In summary, the traditional forms of capitalism envisaged by Adam Smith have been transformed over the past few decades into what might be called 'hyper-capitalism'. Hyper-capitalism has four components: first the assiduous translation of everything – all goods, endeavours, values – into money. Then (abetted by technology, in turn aided by science) all endeavour is industrialized. At the same time, the industrialized endeavours are increasingly, and inexorably, brought under the control of a small number of corporations – as few as the world's various monopoly restrictions will allow. Finally, driven by the same relentless logic, all this monetarized, industrialized, and corporatized endeavour is conducted on the global scale. Thus the modern world is run not simply by capitalism, which takes many benign forms, but by the particular hyper-capitalist model that we might call MICG – monetarized, industrialized, corporatized and globalized. For some industries (like cars and computers) the MICG model might serve us well. For the infinite diversity of agriculture with all its environmental, social, aesthetic and cultural connotations, it is at least too crude, and in reality threatens to be a disaster that threatens all humanity, and indeed the fabric of the world itself. Yet the MICG model is the one taught at modern colleges of agriculture, and espoused by the world's most influential governments, including that of the US (of course), but also of Britain under Tony Blair.

Clearly, the world needs an alternative philosophy. The last section of this book attempts to outline what this implies. It begins by looking at two modern movements that are growing in influence, both of which claim they already have the solutions that we need: vegetarianism, and organic farming. Do they provide the answer?

Postscript: A Meeting in London

I presented some of the ideas in this chapter to a meeting in London in May 2003. The audience told me not to be patronizing. If African countries (say) wanted to follow the lead of the West – establish urban industries, and indeed build modern versions of Paris or Prague – then that was entirely up to them. It wasn't up to people like me – well heeled, ageing white Europeans – to tell them how to conduct their affairs.

I agree absolutely. I have no desire whatsoever to tell anybody else how to live. But the countries of the present Third World are currently being urged by Western governments and corporations, in effect if not expressly, to industrialize their agriculture as quickly as possible. This is presented as a necessary component of modernity and as a prerequisite for a future economy based on urban industry. Industrialized agriculture means agriculture with minimal labour; and the idea (insofar as it is made explicit at all) seems to be that the present farm workers should instead work in factories, or banks, or whatever.

Yet there is no historical precedent for such a shift. In Britain, the first off the mark, bona fide urban industrialism began in the seventeenth century and was leading the economy by the nineteenth. But through those early industrial centuries Britain's farming was as labour-intensive as ever, and was geared primarily to national self-reliance in food. Thus, Britain's urban industrial economy was built on the back of a robust but traditional agrarian economy. The same is true of the US – or France, Germany, or anywhere else. No modern Western country cut its agrarian labour force until its urban industries were well established. The shift from the countryside to the cities was often painful (*vide* English Victorian literature) but it did *not* cause mass, nationwide unemployment. In Britain, for at least two centuries, the growing urban industries merely creamed off the surplus labour from the burgeoning countryside.

For Third World countries to industrialize their agriculture *before* they have urban industries to take up the workers would be to precipitate social disaster of immeasurable horror, and ensure that those countries would never be in a position to establish stable, autonomous economies. In India, precipitate industrialization of farming would put 500 million people out of work. It is hard to imagine any other industry that could absorb so many. Even if such industries could be envisaged, the experience of the West says that it would be well to wait till those alternative industries are in place. Depopulation of the countryside is not a prerequisite. The opposite is the case.

If 'developing countries' choose to follow the path that Western governments and corporations are currently urging, then that of course is up to them. At present, however, they are being plied with misinformation (and I have seen it being dispensed). If it is patronizing to suggest that it is bad to misinform, then so be it.

IV

Enlightened Agriculture

Do not wander far and wide but return into yourself. Deep within man there dwells the truth. (*Noli foras ire, in te ipsum redi. In interiore homine habitat veritas.*)

St Augustine, quoted by Pope John Paul II in *Faith and Reason* (Catholic Truth Society, 1998)

10

Alternatives off the Shelf: Vegetarians and Organic Farmers

The industrial kind of farming that now prevails – high on capital, chemistry and machines, low on labour – leaves much to be desired. It focuses on commodities that are not necessarily needed in large amounts (notably meat), but which happen to be particularly profitable. It claims 'efficiency', but its efficiencies are measured in cash and depend entirely on how the many inputs and the consequences are costed (or left uncosted, as the case may be). If biological or social criteria are taken as criteria of efficiency (as in principle they could and surely should be), then the industrialized approach is often revealed to be very bad indeed: for example, it is not efficient, in the long term, to squander existing soil through over-zealous cultivation; and it is socially very inefficient indeed to make productive people unemployed, which accountant-driven agriculture does above all else. Industrialized food production claims to give people what they want and hence to be democratic, but in fact it depends heavily upon bullish marketing and other commercial devices (and the support of governments) to control production, manoeuvre public tastes, and commandeer the outlets of sale. Obviously, too (the *coup de grâce*), industrial farming is not sustainable.

Yet the supporters of the status quo have one final defence. There is, they claim, no alternative. Huge piles of food are needed; and only the methods of industry, extended as far as can be achieved, can provide what is and will be needed. So we might not like everything that now is done in the name of food production, but there is no other way. At least, it is easy to propose alternatives – prettier, perhaps kinder. But these alternatives are not 'realistic'. They cannot deliver what is really needed.

Of course, in principle, there is an infinity of possible alternatives. The world is not a tabula rasa, but it can be very flexible in the short term (over periods of decades, or centuries, or even millennia); so in the short term we can get away with almost anything that does not break the laws of physics. Yet whatever is done has to be able to feed all of humanity, now and as it will become. We cannot in reality indulge every caprice that might occur to us. So of all the alternatives we might dream of, are there any that can really deliver what's needed, now and in the future? In the end, are the industrialists right? Their methods may not always be pretty, and sometimes may seem harsh, but are they in truth the only realists?

Throughout this book I have intimated that there certainly is a realistic alternative – far more realistic than the industrialized route: and that is to conform expressly to the demands of biology (the physiology of human beings, animals and plants, and the ecology of the earth as a whole), and to shape farming policies according to human values of aesthetics and morality (kindness, justice, peace of mind). Such a design I call 'enlightened'; and I will describe what is entailed formally and briefly in the next (and last) two chapters.

Already, though, among the many alternatives proposed, two in particular are up and running, and gaining strength by the month. One is the vegetarian movement, which is focused on consumers – what we eat: and the other is organic farming, which focuses on the means of production. Why not simply opt for one or other of these – or indeed combine the two, as is often done, and propose a system based on organic vegetarianism?

In truth, both vegetarianism and organic farming have a tremendous amount to offer. We can learn enormously from both, and incorporate much of their philosophy, lore and technique. Both have kept important fires burning while the powers that be, the axis of Western governments, big business and much of the best-financed agricultural science has been taking the world on its present wild adventure.

What Does Vegetarianism Have to Offer?

Vegetarians are divided into two schools: the vegans, who eschew all animal foods (and leather shoes and bone buttons) and the lacto-ovo vegetarians, who allow themselves eggs and milk, and do tend to wear leather. Vegetarians of both types have a variable attitude to fish. Some people who are otherwise vegans do allow themselves fish (which presumably are thereby elevated or demoted to honorary vegetables) and so, of course, do many lacto-ovos. Various religious groups have from time to time been vegan or lacto-ovo fish-eaters, and some of these groups have interpreted 'fish' broadly, to include anything that lives in water, at least some of the time. Thus, and notably, Japanese Buddhists have often eaten whales, and Christian monks in medieval Austria made free with the beavers which abounded in medieval Europe (and have now been reintroduced). Beavers have scaly tails which (so modern theologians suggest) would have allowed them to slip through the regulations laid down in Leviticus, which forbid the consumption of aquatic creatures that lack scales. By dint of their rough tails (in truth no different from a rat's, only flatter), beavers became honorary fish.

Clearly, through the ages, people have become vegetarian for different reasons. People in areas where animals are hard to come by (or are too small to be worth catching) have often had vegetarianism thrust upon them. In any case, as the American anthropologist Marvin Harris has emphasized, dietary restrictions that are ostensibly religious in origin (including those of Leviticus) often have economic roots that run even deeper; so that people living in a region where animal food was too rare, dangerous or small to catch, have tended to invent taboos expressly to forbid what might be tempting, but is not really worth doing.

In the modern world, however, vegetarians of all kinds have typically been guided by considerations of nutrition and personal health; of animal welfare; and by concern for the environment – friendliness to wildlife, and an eye to the long-term future. Vegetarianism in all its forms certainly can be helpful on all of these fronts, but not quite so decisively as vegetarians are wont to claim; and it raises a few problems of its own.

Nutritionally, vegetarianism of all shades has obvious theoretical advantages. Most obviously, vegetarian diets should be low in fat in general, and vegan diets should be especially low in saturated fat, which in general implies animal fat. All vegetarian diets should, however, be high in dietary fibre. Protein should be more than adequate, despite the fears of the mid-twentieth-century nutritionists. The Royal College of Physicians in their report of 1976 referred to epidemiological studies which suggested that monks on essentially vegan diets suffered less coronary heart disease than the general populace. Britain's Vegetarian Society can provide many comparable statistics. Vegetarians of all kinds, too, should escape the hazards of food-borne animal infections, including BSE, so kindly introduced through the cut-price policies of the British food industry.

But there are theoretical shortcomings, too, which sometimes become apparent. A vegetarian diet with too much unrefined plant material may to be too dilute in energy to provide enough for daily needs: a special problem among some lactating women in societies where cereal-based gruels provide most of the provender. Vegetarian and especially vegan diets may sometimes lack various essential micronutrients, notably metals such as zinc, iron and calcium; and the fibre may inhibit uptake of vitamin D, leading to deficiency which manifests as rickets. The body can also synthesize vitamin D for itself, when exposed to the sun. However, Asian children living in the English Midlands in the 1970s were reported to be in danger of rickets. Their high-plant diet was low in calcium and vitamin D, and the fibre inhibited uptake of what there was; and this was combined with a cloudy climate and traditional dress that covered all but the face and hands. It doesn't take much to deal with the problem (no serious inroads on treasured ways of life are required), but it does have to be recognized.

In other words, to thrive on a vegetarian and especially a vegan diet, you have to be astute and conscientious, or to be lucky. You need access to a wide variety of plant materials, including nuts (rich not least in oils; although some people are of course allergic to nuts). The modern health-food shop and/or high-street delicatessen can be a great standby. Without such advantages, shortfall is always possible. In the early nineteenth century the archetypal Romantic, Percy Bysshe Shelley, took up vegetarianism, and became even more poetically languid than usual. One of his firmest but most unlikely friends was the robust, bucolic satirical novelist Thomas Love Peacock, who recommended 'three mutton chops with pepper'. Peacock surely hit the nail on the head. At the other end of the scale, I have a (not untypical) book of lacto-ovo vegetarian recipes that rely heavily for their flavour on cheese and eggs – and are in fact high in saturated fat and cholesterol. Thus the lacto-ovos can easily squander their apparent nutritional advantage, and sometimes clearly do.

The welfare case for vegetarianism seems open and shut – especially for veganism, which makes no use of animals at all. Yet the matter is not as straightforward as it seems. For a start, lacto-ovo vegetarianism has many dubious welfare features. Most obviously, it does make use, sometimes very extensively, of eggs and dairy products. Modern laying hens are extremely hard-pressed. They begin laying at a few months of age and in general are allowed only one year in which to produce 300 eggs or so – about twenty times more they would do in the wild.

They are customarily kept four (or even six) to a cage, with no peace, no opportunity to stretch their wings, no perches; and they must lay their eggs on the wire floor, sloped so that they roll into a gutter for easy collection. The sound in such places is not the gentle cluck of contented poultry but an unceasing alarm call. Commonly, out of boredom and frustration, the birds peck each other. Many finish up bald, and some are pecked to death. It is hard to envisage a more hideous existence. Of course, hens don't have to be kept like this, and vegetarians who care about welfare commonly choose the free-range option (although legal definitions of free-range tend to involve only theoretical access to fresh air, and in general are niggardly in the extreme). But then, many people who are not vegetarians are equally concerned about the welfare of laying hens. The consumers' vegetarianism does not of itself improve their lot. Indeed, in principle it might make matters even worse, since people who eat no meat are liable to rely more heavily on eggs, and so provide an excuse for even more intensive production.

Then again, although the vegetarians do not eat the flesh of the chickens, they have their blood on their hands nonetheless. A wild hen might live for a decade; but the modern battery hen is killed after a year. In general, half the eggs that hatch produce male chicks, which are no use in the laying flock (and in general are of the wrong breed to raise for meat), so they are killed instantly. A favoured modern method is to drop them off the end of a conveyor belt through whirling blades, like the rotors of a helicopter, which chop them in the blink of an eye into small pieces. This is said to be humane, but it is certainly not aesthetically attractive. Various techniques are now on line to tip the birth ratio in favour of females. Even so, the fact is inescapable: vegetarians who eat eggs have almost as much blood on their hands as those of us who also eat chicken. The main difference is that in a vegetarian world the birds that the egg producers kill – the male chicks and the still-young hens – would simply be thrown away, rather than consumed. Such waste is intrinsically profligate, and as such might be thought sinful. Of course, the surplus males and the superannuated hens could be kept in rest-homes to live out their natural lives, but the profligacy of this is obvious, and gratuitous profligacy is bad whatever the context.

Those who drink milk are similarly indicted. Cows do not produce milk unless they first have a calf. In traditional systems, the calves of specialist dairy breeds were raised for a few weeks or months and then sold for veal. The veal trade has come under much attack of late, especially in Britain, simply because the food trade as a whole – the retailers and then the farmers – have again turned the screws too hard for the usual opportunist, commercial reasons. Traditionally, veal was just young beef: the animals were weaned, fed on a fairly normal diet, and then slaughtered while still young. But the vogue grew for white veal. The animals were kept confined in 'crates' so they did not exercise and strengthen their muscles, and fed on a low-roughage, high-cereal diet which left them craving fibre (so that they commonly licked and chewed their own hair) and anaemic (which whitened the meat). This was obviously cruel. In Britain it is now illegal to raise veal calves in crates, but this is still done elsewhere, and in any case the die has been cast. The veal trade has been sullied. Thus does greed and fashion that is merely effete destroy systems that were otherwise sound. In most modern systems, dairy cows are crossed with bulls from beef breeds (or, more usually, inseminated by AI with beef-breed semen). They still have a calf, and so lactate, but the calf itself will be sturdier than the specialist dairy offspring, and so can be raised for beef (typically, these days, an animal slaughtered at over eighteen months).

But again the question arises: what do the lacto-ovo vegetarians think happens to the calves? A few of the calves are kept on to become dairy cows in their own right. These future milking animals are produced by crossing their mothers (generally, the best of the herd) with bulls of a dairy breed to produce a pure-bred dairy calf. Again, though, only the young females are retained; the pure-bred dairy male calves are surplus to requirements. But since cows typically remain in the herd for at least five lactations, and have a calf every year, only a few of those calves are needed as herd replacements. The rest must be disposed of, which means killed. They can either be raised for meat or (like male chicks), simply thrown away (or sold for pet food). Again, to throw them away seems like a sinful waste: the world cannot afford such profligacy. Again, some vegetarians have proposed providing rest-homes for superfluous calves to live out their lives, but cattle easily live for ten years, and simple arithmetic shows that the surplus cattle even from the British dairy herd would soon fill all of mainland Europe. Not only is this obvious nonsense; it would also make life impossible for all other animals, which surely defeats its welfare purpose.

Old cows, too, like old hens, have to be killed. The world is not big enough to keep them in retirement homes. In practice they are typically fattened in a last flourish of heavy feeding and then put in pies, burgers or pet food. This makes sense: again it would be profligate simply to squander the flesh of an old cow. But – as is usual in modern farming – the final fattening of the superannuated cow is often done cynically and cruelly, with the animals packed into 'feedlots' with no regard to their social sensibilities. Again, modern intensive livestock farmers, quick-marching to the drum of commerce, generally contrive to turn otherwise sensible practice into something disgusting. The best response to this, though, is not to become vegetarian (which does not necessarily help), but to clean up animal husbandry in general. Various pressure groups are working on this, though always in the face of po-faced government reluctance.

Fish raise welfare problems too. Traditionally, those countries that at least have some laws pertaining to animal welfare have not seen fit to extend them to the cold-blooded types. The assumption has been that fish don't mind being hooked, or suffocated in nets, or beaten clumsily about the head when landed, and if they do, so what? But more and more research suggests that fish are far more intelligent and sensitive than their unchanging expressions suggest, so vegetarians (or near-vegetarians) who allow themselves fish are not absolved from welfare issues. Even the fishing industry now acknowledges that many major fisheries and some entire commercial species are now under threat (including even various cod and skate), and fish farming now provides 50 per cent of all the fish that human beings consume (as opposed to turn into fishmeal for fertilizers or animal feed, or give to pets, or whatever). 'Fish' in this context is defined loosely, to include prawns and other invertebrates, but the figure is astonishing nonetheless. Intensive fish farming has not yet received as much attention from campaigners as intensive pigs or poultry have done, but the issues surely must be faced sooner or later.

What of the environment as a whole? Vegetarians point out that a hectare of wheat or pulses, say, produces about ten times more protein or calories as would the same area dedicated to beef or sheep. Therefore, they say, agriculture devoted entirely to plants must be far more economical on space, and so in principle should leave far more land spare for wilderness, and the wildlife that lives in it. If the 10 billion people who will be with us by 2050 were all vegan, then the world population would indeed be 10 billion. But if mid twenty-first century people all eat as much meat as is projected, then (as discussed in Chapter 2) the effective population – people plus livestock – will be 14 billion.

There is a great deal in this argument. Farming would indeed be more sustainable, and wildlife-friendly, if humanity in general ate less meat. Even so, the case is nothing like so clear-cut as the vegetarians suggest. As already outlined in this book, and further discussed in the next chapter, the traditional role of livestock is to clear up odds and ends ('tail corn', leftovers, surpluses in general). Farmers should always aim for a small crop surplus for insurance, in case the weather fails and reduces the crop. In most years, then, they will indeed produce more than is strictly needed, and without animals to mop up the excess, this surplus will be wasted. In addition, cattle and sheep in particular (sometimes supplemented with goats and camels) can always produce at least a little meat or milk from hillsides, uplands, wetlands and semi-deserts where crops cannot be grown at all. Finally, too, animals provide an essential source of fertility in the form of manure; and as noted in Chapter 2, US cereal production waned somewhat in the 1920s and '30s until more livestock were introduced. In other words, agriculture that is plant-orientated but also includes some livestock here and there, in the end uses landscape more efficiently than agriculture that is geared exclusively to crops. Thus, agriculture with a judicious mix of plants and livestock (mostly plants, but at least some livestock) should occupy less space than an all-crop agriculture, and so (in principle) would leave even more wilderness for other creatures to enjoy.

Fish are relevant in this context too. Fish farming as at present practised can be extremely damaging in several ways. Farms for fish (defined loosely) are often established in environmentally sensitive areas. Thus many a hectare of mangrove has been flooded to provide lagoons for prawns. Disease and parasites build up in fish farms and escape to the wild; and so do the drugs and pesticides that are often used to suppress the disease. Farmed fish, bred for captivity – or even genetically engineered, as in salmon and trout – escape to the wild, and threaten wild stocks with what might be called 'genetic pollution'. These are technical problems that could be solved with money: it is in principle easy enough to isolate fish farms from wild sea and rivers (just as in modern Dubai, human tourists enjoying a swim are in general kept away from the open sea). But everything in this modern world is controlled by money, and so long as farmed fish is intended simply to be cheap, the dangers must surely continue.

The final indictment, however, is extrapolated from the moral philosophy of Immanuel Kant, who was one of the greatest of moralists.

He argued that no ethical principle is really acceptable unless it could in principle be recommended to the whole world. Vegetarianism, and especially veganism, clearly cannot be recommended with a clear conscience to absolutely everybody, even if more than a minority would accept it. For many people in many difficult environments, including high latitudes and semi-deserts, vegetarianism is hardly an option at all. If we all were determined to be vegans nonetheless, the necessary agriculture would in practice take up more room than is strictly necessary (albeit less than it does now). Vegetarianism does not, then, and cannot, provide what we ought to be looking for: a system of farming that would suit all of humanity (and our fellow creatures) for all time.

Taken all in all, though, vegetarians clearly do have a great deal to teach the world at large. They have demonstrated that an all-plant diet, or one that is very nearly so, can indeed keep people in excellent health. They have also shown that vegetarian food can be gastronomically superb. Forty years ago, at least in Britain, vegetarian food was typically dire. Restaurants at best tended to offer salads (and a typical English salad is lettuce) or the traditional English meat-and-two-veg, without the meat. Now there is far more imagination and general eclecticism – including a small wave of South Indian restaurants offering such delights as dosas, patras, and many variations on a theme of dhal. In short, vegetarians have shown us that a diet that consists entirely or almost entirely of plants need hold few fears, even for those of us brought up in the meat-obsessed middle decades of the twentieth century. Despite the conditional clauses, too, as outlined above, vegetarians have kept the home-fires burning both of animal welfare and of environmental concern. Overall, then, vegetarianism serves as a kind of beacon, a reference point, an on-line demonstration of essential principles. Yet the search for the system that would serve all our needs for ever must continue.

Organic Farming

The organic movement has ramifications that run far beyond food, deep into the philosophy of science, morality, aesthetics and religion; and partly for those very reasons, it has all too often been dismissed by the orthodox and the strait-laced, as 'muck 'n mystery'. But it deserves to be taken very seriously. Enlightened agriculture, expressly designed to serve all humanity for all time (and our fellow creatures), should surely incorporate and build upon many of the techniques and the broad ideas of the organic movement. Yet organic farming as it now stands does not quite provide what the world needs. Like vegetarianism, it can be seen more as a demonstration of what can be achieved outside the simplistic industrial approach; and also as a stepping stone to truly enlightened agriculture. This may sound a little lofty, even arrogant. But that's the way it is.

In truth, it isn't easy to pin down exactly what 'organic farming' means, or implies. It ought to be easy enough to establish a universal definition for purposes of law, but even this has so far proved difficult. The international organization IFOAM (the International Federation of Organic Agricultural Movements; pronounced 'if-oh-am'), based in Rome, has established universal principles, but for the time being organic farmers seem primarily to be guided by criteria laid down by their own country. Britain's organic farmers, for instance, largely take their lead from the Soil Association.

Most of the rules in most countries relate to what cannot be done. In general, organic farmers may not make use of artificial fertilizers – which for the most part means nitrogenous fertilizers produced by fixation of atmospheric nitrogen by the Haber–Bosch process. Nitrogen fertility must come from the mineral content of the soil itself, from manures, or from nitrogen fixation by bacteria in the roots of plants (notably of legumes) or by cyanobacteria (as in paddy fields). Organic farmers must also eschew 'artificial' pesticides and herbicides, including the organophosphate pesticides and the organochlorines such as DDT and dieldrin. Various high technologies are also forbidden including, notably, GMOs. Farmers cannot declare their crops 'organic' until their fields have received no artificial fertilizers for at least four years, and unless there is a suitable buffer zone between their farms and any conventional farms that do use the proscribed pesticides, or grow GMOs. In short, the rules are clear enough; but there are many complications and caveats, practical and theoretical. Once the farmers are certified 'organic', however, their produce attracts a premium.

On the practical front, it is hard for conventional farmers to go organic. Yields are typically less on organic farms. In fact, yield per hectare may be just as great; but in general it is easier to boost yields with artificials, and the organic farm as a whole is liable to have smaller fields with more hedgerows, which reduces the overall area that is intensively managed. There are compensations for this, of course. On the non-financial side, organic farms typically are very pleasant places to be, with more birds and wild flowers, and requiring more intricate and benign husbandry. But farmers cannot live by job satisfaction alone, and because the yields tend to be lower, the premium on the produce is needed to balance the books. But in the four years' interregnum, before the farmer is officially recognized as organic, he or she has to make do with smaller output, without premium. Spacing is a problem, too. For instance, the Maltese government has been toying with the idea of making its agriculture entirely organic. While it makes up its mind, many individual farmers are choosing to move in that direction, not least because they sell so much of their produce to European tourists, more and more of whom prefer organic produce. But the holdings on Malta are very small; many tiny farms, cheek-by-jowl over the hillsides. Because of the need for buffer zones, no one farmer can go organic unless his neighbours do so too. So in practice, the shift probably has to be made hillside by hillside. But farmers tend to be individualistic, and none more so than rural Mediterraneans. So a concerted move is necessary, but is extremely difficult to bring about.

The greater problems, however, are conceptual. Critics of the pedantic kind point out that the term 'organic' is somewhat confused and confusing, not least because it has several quite different connotations. To a chemist, 'organic' simply means 'carbon-based'. The organochlorine and organophosphate pesticides are, by this definition, 'organic'. Yet they are emphatically banned. On the other hand, the organic rules do allow some pesticides to be used that are non-organic, such as Bordeaux mixture, based on copper. Copper in more than minute doses is highly toxic. So too are some of the organic compounds that are allowed, such as nicotine. In close detail, in short, some of the rules have a strangely arbitrary, and indeed a muddled feel to them.

By the same token, it would be possible in principle to use genetic engineering in extremely benign ways: for instance, to produce crops that repel insects by producing flight pheromones, one sniff of which prompts invading pests to turn tail. If such pheromones were produced only by the non-consumable parts of the plant (for example, by the leaves of wheat) then there should be no hazards whatever, even theoretical, for consumers. Pheromones tend to be highly species-specific, too, so only the target insects would be influenced. Bees would ignore signals aimed at aphids. This would be an advanced exercise in biological pest control, lying some time in the future (although much of the groundwork has already been done). But organic farmers cannot even entertain such possibilities, because of the blanket ban on genetic engineering. Thus organic farming can seem gratuitously technophobic; a tendency that exasperates many who would otherwise be among its allies. Some great twentieth-century agricultural scientists, such as Sir George Stapledon and the Nobel laureate Sir John Boyd Orr, in many ways inspired the organic farming movement and yet did not formally sign up to it. In particular, Stapledon and Orr (and others) advocated judicious use of artificial fertilizers, expressly banned in organic circles. More generally, they were clearly put off by what they felt was simply technophobia.

Organic farmers have also demonstrated that fields can be kept extremely fertile by organic means alone, and stay in better 'heart' (that is, with a good open structure and resistant to compacting) when farmed organically, and that pests can be kept at bay by subtle husbandry, so why rely on toxic and potentially hazardous industrial chemicals? Many enthusiasts further maintain that crops and animals raised by organic methods are superior nutritionally and gastronomically to those raised conventionally. As a bonus, too, organically grown crops are free of all pesticide residues. In practice, such residues are probably low on the list of life's hazards (as advocates of conventional methods are wont to emphasize), but most thinking people concede that given a choice – pesticide residues versus no pesticide residues – then the latter is surely preferable. Organic farmers show that there can be a choice. Most of the time, most crops can do without pesticides, if other aspects of husbandry are in order.

Even more to the point, though, is the grand strategy and ethos of organic farming. Organic farming tends to be defined for legal purposes in negative terms (the list of technologies that may not be used). But this is unfortunate. Its real message lies in positive propositions that are very much in tune with the requirements of enlightened agriculture, as outlined in earlier chapters and summarized in Chapter 11. Above all, organic farmers emphasize the principles of good husbandry. They believe (rightly) that good husbandry is intricate, and therefore requires a lot of people; and also perceive that when many people work on the land, some of the most enviable human communities result. A countryside full of small farmers is far to be preferred (surely?) to a teeming city packed with out-of-work ex-small farmers and their families. Organic farmers emphasize the need to be kind to farm animals, and that this can best be achieved by devising husbandry that conforms to their physiology and psychology: for example, small herds of cows in meadows, as opposed to huge anonymous herds effectively force-fed on customized ryegrass.

Organic farmers perceive that farming as a whole cannot be sustained in the long term unless it marches to the drum of ecology, and they further observe that industrial technologies (both artificial fertilizers and pesticides) have often upset the ecology in many ways, not least by pollution. Some organic farmers are good philosophers of science, too, and know full well that science cannot be omniscient, and that the consequences of applying any one technology cannot be predicted in detail. Therefore in general they urge humility; one of the grandest and most necessary of concepts in all dealings with nature. More specifically, organic farmers point out that regular heavy doses of artificial fertilizers can deplete the soil microbes, and rightly point out that those microbes have a tremendous influence in the growth of plants but that the extent of that influence is largely unknown. *Faute de mieux* it is wise to respect the adage that nature knows best, for the sound evolutionary reason that the creatures that compose its many ecosystems have all, to a large extent, adapted to each other and to a significant extent benefit from each other's presence (even though there are antagonisms too). Thus for reasons both ethical (in the broad sense) and evolutionary it seems wise to encourage the soil flora to flourish.

In spirit and in essence, then, the ideals of organic farming are very much in line with the broad requirements of enlightened agriculture. The shortlist of intentions and ambitions – good, kind, sustainable husbandry; friendliness to wildlife and the perpetration of true rural communities – are precisely in tune. Sadly, though, the organic movement all too easily becomes a bandwagon, like anything else that becomes popular. The wagon appeared in literal form in my local market town the other day when a truck the size of a small warship thrust its importunate but precarious way along the high street bearing a logo three feet high proclaiming 'Fresh Organic Produce'. No doubt: but such quantities, dished out from some central clearing-house, are not in the spirit of the thing. Thus it is perfectly possible to conform to the letter of the organic law and yet impose the overall structure of mass production and distribution, which in many ways seem to abnegate organic principles. But the mass producers are powerful people who can and do lean on the law-makers, and by a tweak here and a twist there can compromise the fine ambitions entirely. The legal definition of 'free-range' – effectively not much more than a theoretical glimpse of sunlight – shows how smart lawyers and relentless commerce can pervert the best of intentions. As discussed in the final chapter, one of the grand tasks for the present and future world is to devise ways of ensuring that big ideas that catch on do not automatically give rise to pyramidal, centralized systems of organization, geared to profit at the expense of the task in hand. Good initiatives should be big, even universal: but they should be perpetuated by many different people doing, effectively, their own thing. The desirable structure for agriculture is not the power pyramid, but the neural net. In food and agriculture, conglomeration is almost always compromising.

How Organic Farming Came About

As Philip Cornford excellently describes in *The Origins of the Organic Moment* (Floris Books, Edinburgh, 2001), the organic movement grew from the late nineteenth century onwards, first in Britain, then in particular in Germany, Austria and the United States. It arose in parallel with the new science and industrial form of agriculture which now (despite its novelty) is called 'conventional' – and to a large extent is a reaction to it.

Like all great movements (science, socialism, Christianity), the organic movement has mixed origins. Its British pioneers included several agriculturalists with strong Far Eastern connections, anxious both to learn from the obvious abilities of traditional Asian farmers but also to build on them. The first was Robert Elliot, who wrote *Agricultural Changes* in 1898, based not least on his experiences in India, and (says Cornford) 'anticipated many of the organic movement's concerns'. The American Franklin H. King was similarly impressed by his own visit to Korea, China and Japan in 1907: his book, *Farmers of Forty Centuries*, published posthumously, was another powerful spur to the organic movement. Arthur Howard, educated as a chemist at Cambridge and then at Wye (agricultural) College in Kent, pioneered composting techniques in Indore, India. This was the famous 'Indore system', which established the formal approach to composting now reflected both on organic farms and in many millions of suburban gardens. (The trick is to balance carbon-rich, structurally tough materials like straw with nitrogen-rich but usually flimsier materials like grass mowings – creating an open, airy but moist structure where aerobic bacteria can thrive.) Major-General Sir Robert McCarrison was yet another great India hand, who focused on diet – particularly that of the Hunza people of the North-West frontier who ate simply, whatever nature provided, and were astonishingly healthy.

Some of the key players supported the broad aims of the organic movement but took issue with the details. Sir George Stapledon and Sir John Boyd Orr were leading British agriculturalists who both directed research institutes. Stapledon argued that good husbandry must respond to the particular needs of landscape, while Boyd Orr stressed that farm output should be geared to good nutrition – fundamental requirements that have been comprehensively ignored or at least overridden by modern governments, wedded as they are to single-minded commerce. Both philosophies inspired the organic movement – but neither Stapledon nor Boyd Orr, agricultural scientists both, eschewed the use of artificial fertilizers.

The organic movement has had a powerful spiritual history: notably through Rudolf Steiner, founder of the Anthroposophy movement, who among other things advocated planting according to the phases of the moon. The Catholic and High Anglican influence has been powerful: Hilaire Belloc, G. K. Chesterton, T. S. Eliot. Eliot published the last three of his *Four Quartets* in the *New English Weekly*, which promoted organic ideas; and, as an editor at Faber & Faber, he established that publisher's substantial list on (largely organic) farming. Politically, the organic movement has not always been well served. Some of its early protagonists were alarmingly right-wing: the poet and critic Edmund Blunden was a supporter of General Franco in Spain and a Nazi sympathizer; Arthur Bryant, the historian, was lucky to escape internment in the Second World War; Jorian Jenks, who later edited the organic magazine *Mother Earth*, was active in the British Union of Fascists in the later 1940s; and Henry Williamson (best known as the author of *Tarka the Otter*) also admired Hitler and was arrested in 1940. Thus the organic movement has partaken of the Germanic rustic 'romanticism' that inspired the Nazi party itself and was reflected in Goebbels' propaganda films. But of course, the organic movement lends itself at least equally to the Left. The modern stereotype, after all, is not the jackboot, but beards and sandals. The organic movement is not, innately, any kind of party political movement.

The movement acquired formal status in the mid twentieth century. A key player in Britain was Lawrence Hills, who founded the Henry Doubleday Research Association in 1954, which remains one of the world's most important organic research stations. His counterpart in the United States was the millionaire businessman Jerome I. Rodale, who among other things established *The Organic Farmer* in 1942 and now (posthumously) lends his name to a publishing company. Most importantly, perhaps, at the end of the Second World War, Lady Eve Balfour, who had studied agriculture at Reading University (and incidentally was the niece of a British prime minister, Sir Arthur Balfour) founded Britain's Soil Association, which can be seen as the parent organization of the modern movement, and in Britain at least sets the standards by which organic farmers are assessed.

In short, the organic movement is well established intellectually, spiritually and in its methods, and has become a serious commercial player. What should we make of it?

What Does the Organic Movement Really Have to Offer?

The proper answer to the question posed in this sub-head is the one that Mao Tse-tung proffered when asked to comment on the effects of the French Revolution: 'It's too early to tell.' I am sure, though (and so are many professional, orthodox agriculturalists), that its ideas will prove in many ways to be of supreme importance, and that they deserve to be explored in depth.

To begin with, organic farmers make many different claims: that food grown organically is superior gastronomically, with a better flavour and texture; that in various ways it is more nutritious; and that organic farming is kinder to livestock and to the environment as a whole, and has many social advantages – not least in fostering working rural communities. If these claims are true, then of course they are important; and at first sight, it should be easy to show whether they are true or not. In practice, however, few critical studies have been carried out, and those that have serve largely to illustrate how difficult it can be to provide certainties in areas such as this.

On the matter of flavour, it is easy enough to set up tasting panels. The judges are asked to compare vegetables and fruit and meat raised organically with similar produce raised conventionally. Of course the tests are 'blind': the judges do not know in advance which is which. In practice, however, such tests immediately raise the general problem of 'confounding variables'. Thus the varieties of vegetables and fruits favoured by organic farmers are typically different from those grown through normal commerce. So are any perceived differences due to cultivation, or to the breeding?

With meat, the problems run deeper. Pork, for instance, is conventionally produced from hybrid pigs grown rapidly in intensive units. Organic farmers commonly raise traditional breeds of pig – Gloucester Old Spot, Berkshire, Tamworth. The organic animals are generally free-range and so have more exercise, and have a less concentrated diet (often with a high intake of grass), and so take longer to reach slaughter weight. Organically raised pork generally does taste very different from, and is infinitely superior to, the conventional kind. (I live in an area where it is possible to make the comparisons, and have long since given up on conventional, supermarket pork.) Again, though, we have to ask, does the difference result from the husbandry or the difference in breed? Berkshires, after all, were bred more for their flavour than for their rate of growth. Does the difference result from the more varied diet of the organic pigs? Or their greater age at slaughter? Both make a difference, because flavour in large part reflects biochemical complexity, and this to a significant extent reflects diet and age. Old animals tend to be tastier than young ones (though of course if they are too old they are tough and sometimes unpleasant).

If 'conventional' pigs were *always* hybrids, raised intensively in double-quick time, and organic pigs were always from named breeds raised slowly on grass, then these variables would not matter. 'Organic' pork would imply grass-fed Berkshires or Old Spots munching out on windfall apples. But it is perfectly possible to raise Berkshires or Old Spots on conventional feed, or on grass that has been fertilized artificially, and so would not qualify as organic: and it is possible to raise hybrids intensively (or at least semi-intensively) on organically raised feed. No one who likes meat and has the opportunity to make comparisons can doubt that the slowly grown named breed is superior, or that free-range is generally kinder than intensive husbandry. But organic farmers need to show that it's the organic side of their husbandry that really makes the difference – not the other factors that in practice tend to go along with the organic approach. Where I live, in Oxfordshire, a surprising and encouraging number of local farmers do raise named breeds (Angus and Dexters among the cattle; Berkshires, Tamworths and Old Spots among the pigs) free-range, and the meat is indeed excellent. But none of those farmers that I know is actually organic, or claims to be; and it is not a hundred per cent obvious that if they were – if their grass never saw a whiff of industrial nitrogen – it would make any noticeable difference.

The nutritional claims can and ideally should be studied from two angles: firstly to analyse the produce itself, to see if the organic kind differs significantly and consistently from the conventional kind; and then to look at the people who eat it, to see if those on organic diets are, or are not, healthier.

The first kind of study should be easy enough. Analytical chemistry has been developed over nearly two centuries and is now well advanced. Even so, present results are not clear-cut. Organic gardeners claim that their produce is richer in minerals, not least because organically grown vegetables tend to be deep-rooted, and this in turn is partly because they are given less water, so that they need to root more deeply. Thus they draw up minerals from deeper in the soil. Present studies are not definitive, however. Again the confounding variables continue to confound. To what extent is the perceived difference between organic and conventionally grown vegetables due to difference in variety? If it is due to husbandry, is this because of the organic treatments per se or the differences in watering regime? If there are measurable differences, how universal are they? Are organic vegetables always more mineral-rich, or only the outstanding ones?

To demonstrate direct effects on health is far harder, as outlined in Chapter 3. Such studies are far easier on rats – but rats are the wrong species. With people, it has proved notoriously difficult to discover beyond all possible doubt whether a diet high in saturated fat does or does not predispose to coronary heart disease. One problem is to find suitable groups to compare; and always the confounding variables are there to confound. Another is that human beings are long-lived, and the effects of a high-fat diet may take decades to unfold. It is impossible to control people's diets over many years unless they are prisoners (in which case more confounding variables come into play). In the late twentieth century a study in Britain to find out whether folic acid in pregnancy does or does not reduce the likelihood of spina bifida was run for several years and in the end was not conclusive. It was in effect called off because the clinicians already felt in their bones that folic acid obviously is beneficial and should be recommended, even though the evidence was not statistically cast-iron. They felt that to delay such a recommendation until the evidence was beyond all possible doubt was simply inhumane. This of course was good sense, but it is not of the kind that provides scientific certainty.

Issues like this – fat and coronary heart disease, folic acid and spina bifida – are conceptually straightforward. Fat and folic acid are easily measured, and the pathologies in question are only too obvious. Still, though, for practical reasons, they cannot be pinned down beyond all doubt. But food as a whole contains many thousands of ingredients, and again as noted in Chapter 3, we are far from knowing what is really important. Besides, common sense (and much evidence) suggests that no food component should be considered in isolation. The ratio of each component to all the others matters too. Thus (to take some very obvious examples) protein must be accompanied by fat and/or carbohydrate, or the protein itself will be burnt to provide energy. Some components of plants may inhibit uptake of calcium. And so on. Most of the interactions are clearly not understood and (as is always true of all science) it is impossible to know how much is yet to be found out. To compound the difficulty, 'health' and 'well-being' obviously imply a great deal more than a simple absence of gross pathology. An organic diet over thirty years might add a year to your life, and a year is worth having; but it would be well-nigh impossible to demonstrate such an effect. 'Well-being' implies an appetite for brisk walks into the ninth decade, a spring in the step and a cheery outlook. But if we cannot even show beyond all possible doubt that folic acid protects against spina bifida, how can we hope to show with anything approaching certainty that organic diets in general improve well-being in general?

A key question for philosophy, though, is how we should behave in the face of such uncertainty. The sensible line (it seems to me) is to do all the science that can reasonably be done, but then to admit that in some areas (as in much nutrition and a great deal of medicine of all kinds) statistical certainty of the kind demanded by editors of scientific journals is impossible. Thus when all the science is done we have to fall back on experience and common sense (which can be improved upon, but should never be abandoned). Here, too, is where experts who are truly expert are invaluable. Thus on the matter of fat and coronary heart disease, Britain's Royal College of Physicians offered advice on what they called 'the balance of probabilities'. The 'balance of probabilities' in most dietary and medical matters is the nearest we can come to truth; and only the best opinions of serious clinicians can tell us what the 'balance of probabilities' actually amounts to.

For my part, on the one hand I despair of those enthusiasts who insist that organic food *must* be better for us just because it is 'natural' or because some guru says it is. We must improve on guesswork and dogma. We have to do such science as can be done. On the other hand, I positively despise those scientists who write off the claims for organic diets on the oft-heard grounds that 'there is no evidence'. There are at least two kinds of answer to this that spring from science itself, and at least two more that are matters of the most elementary philosophy. Firstly, 'there is no evidence' for many of the claims of the organic movement because, for the most part, the necessary studies simply have not been done. It is hard to fund science of any kind these days. It is even harder to fund science of an unimpeachably independent kind, that is not supported by industry with a specific commercial target in mind. It would be extremely hard to fund an inquiry on the relative merits, over time, of organic versus non-organic potatoes, say, on people in Oxfordshire or New York or wherever. Even if funding was forthcoming, it would still be extremely hard to design studies that gave clear answers, and even harder to carry them out.

The first philosophic riposte is that scientists can ask only those questions that it occurs to them to ask, and which they feel they have a reasonable chance of answering. But of course, there is no way of knowing a priori which questions ought be asked because, logically, you can know this only when you are already omniscient. When scientists do have a cogent question to ask then in some cases, as in many issues of human nutrition, they often cannot for practical reasons carry out the necessary experiments. The second philosophic response to those who dismiss organic claims for lack of evidence is the most elementary observation in the philosophy of science, which is that 'absence of evidence is not evidence of absence'.

In short, it certainly is not demonstrated beyond all possible doubt that organic food is better for us. Such demonstration, probably, will always be elusive. But common sense, a respect for 'the balance of probabilities', and a great many straws in the wind (including the new science of nutraceuticals and the evolutionary thinking that derives from it), suggest that its claims should be taken seriously. There is clearly much more to diet than orthodox nutritional theory has so far investigated, and the general idea that diets should be biochemically diverse, and that organic husbandry promotes such diversity, is surely worth pursuing. We certainly should not be talked out of this idea by industrial scientists with school fees to pay or by finger-wagging politicians hooked on some vague ideal of 'progress', or indeed by bad philosophers. The lead of good physicians seems unimprovable: to use all the science there is but to acknowledge its limitations. Don't claim certainty where there is none; don't dismiss out of hand what cannot be satisfactorily investigated.

Most fundamental of all, however, are the claims and philosophy of Eve Balfour. She argued that the living components of the soil, and in particular the micro-organisms, significantly affect the growth of plants and also their nutritional value. 'Take care of the soil, and the crops will take care of themselves' is the adage. But again we must ask, what is the evidence? On the one hand the roots of most plants (including forest trees) form symbiotic relationships with soil fungi of the kind known as mycorrhiza whereby the plant provides the fungus with organic molecules and the fungus greatly extends the absorptive power of the roots (though there may be much more to the relationship than this). Nitrogen fixation, too, is carried out by many soil microbes other than the well-known rhizobia that live in the roots of legumes. On the other hand, many plants these days are grown hydroponically, in solutions of nutrients and without any soil at all; and those that I have seen (for example in Israel under glass, and China under screens) look very healthy indeed. So what, all in all, is the contribution of soil organisms to plant growth and human and animal nutrition and health? Are hydroponically grown vegetables really inferior, and if so how, and do the perceived deficiencies matter? Again, as so often in biology, our knowledge of soil microbes on the one hand seems vast – entire institutions are devoted to nitrogen fixation, for example – and on the other seems almost nugatory. Thus, studies of DNA in soil suggest that only about one in 10,000 or possibly even one in 100,000 of the microbial species living in soil have yet been named and identified. Thus, most of the DNA that is found in soil clearly derives from bacteria, but usually from species that have never been isolated and directly observed. How much can we claim to know if we don't even know to the fifth order of magnitude what is actually out there?

Pest control is a key issue, too. Organic farmers claim with some justice that they kept the idea of biological pest control alive when most of orthodox agriculture had bought the idea that pests must be zapped with industrial pesticide. Organic farmers seek primarily to control pests by many tricks of husbandry that range from crop rotation and intercropping to the encouragement and introduction of predators (such as spiders which have broad dietary prefences and parasitic wasps that attack specific pest species, notably of aphids). No one doubts that biological pest control is of huge significance, and in principle is preferable to chemical control.

To be sure, conventional farming seems to have some advantages over organic farming as now carried out. Notably, it delivers (largely) predictable quantities of crops on time, to tight specifications (on colour, texture, size, and what you will). Though it is getting better apparently by the month, organic produce has often seemed knobbly and small and sometimes ravished by pests. Yet, the organic farmers point out, the comparison is largely unfair. Scientifically speaking, we are now comparing a system that has been cosseted at every turn with one that has been largely neglected. Conventional farming has benefited this past 160 years from intensive scientific research, while organic farming has mostly been left to enthusiasts with little support of any kind. If the problems of organic farming had received the same scientific attention, how much more might have been achieved? Of course, the idea has grown up that the marriage of science and industry is absolute and all-embracing; that is, that science is concerned *only* with the problems of industry, and hence of industrial farming. Some politicians clearly believe this. In truth, though, science can in principle turn its attention to any problem we choose to confront it with (even if it cannot answer them all). The problems of organic farming have not, for the most part, been presented to it.

On the other hand (always in this debate we seem to have to say, 'on the other hand'), the rules that define organic farming for legal purposes seem to preclude much of the scientific research that could be extremely useful to it. Thus it is obviously true as Rachel Carson so ably described that the organochlorine and organophosphate pesticides, crudely and greedily applied, can do immense damage. Yet there are some situations even now (for example, in the local control of some disease-carrying mosquitoes) where DDT does the cleanest job; and surely there are some in agriculture too. Many who care deeply about the environment feel (as Britain's Sir Kenneth Mellanby argued in the 1960s – and Rachel Carson herself acknowledged) that even the organochlorines could sometimes do what needs doing. Contrariwise, some of the agents approved by the rules of organic farming, including copper, are known to be nasty. More to the point, the organochlorine pesticides are second-generation chemistry. We can envisage fifth-generation chemicals based upon them (or perhaps upon quite different molecules) that are highly specific and not persistent at all, and would oxidize away into the most innocuous vapours as soon as their work is done. But we are not going to get to the fifth generation if we preclude the research that could lead to it. Similarly, as already suggested, we can envisage ways in which genetic engineering could promote highly specific and ultimately benign control of pests, not killing them at all but merely shooing them away. But again, we won't get to that point if the entire discipline is banned out of hand.

Many agriculturalists, too, stress the advantages of minimum cultivation – which effectively means no ploughing. This is in line with organic thinking, or should be, since ploughing is known to reduce soil organisms. But minimum cultivation is most easily achieved by zapping weeds with selective herbicides. The present ones may not be satisfactory, but again it is possible to envisage more precise and benign chemistry which again cannot be realized if the research that can lead there is cut off at source as a matter of principle. Neutral observers might be forgiven for thinking that organic farmers have a down on chemistry in general. But the general point again is not to condemn particular technologies or sciences a priori, but simply to ensure that all technologies and sciences are geared to the greater requirements of humanity – to the ecology of the earth, and to human values. Chemistry can be humane too.

Finally, we should apply the Kant test. Organic farming has much to commend it, of course, but could it in conscience be recommended to all the world? I find it hard to see how. Notably, half of the nitrogen now applied to the world's crops is 'artificial', produced by the Haber–Bosch process. Of course organic farmers have alternative sources of fertility, but these have their drawbacks too. Manure can be polluting. I have read many an article recommending seaweed as a source of fertility, but what does the theft of seaweed do to the ecology of seashores? More broadly, could organic farmers really double their input of nitrogen, as they would need to do to maintain present agricultural output if artificials were banned? Could they double it again in the next fifty years, as world population doubles? Nobody knows (it's amazing what is not known), but the odds are surely against.

But then, what in the end is the objection to artificial fertilizers? Indeed there are some. If they are applied exclusively, then over time the organic content of the soil may go down, although not as dramatically as sometimes suggested, since a significant proportion of any crop is in the roots, and much of the root (or all of it, in the case of cereals) is left in the ground to rot when the crop is harvested. Loss of organic material is in general undesirable, not least because of the loss of soil texture, and its reduced ability to hold water. However, application of artificials per se need not destroy soil structure or seriously compromise the soil flora. Artificials can undoubtedly boost yields. To be sure, high yield is not the only desirable goal (in many circumstances, reliability of yield is more important, for example), but if yield is lower farming must then occupy more space, spreading into wilderness and into marginal land that should not be cultivated at all. Some object to the quantity of fossil fuel used to produce artificial nitrogen, but the amount is much less than 1 per cent of the world's total use – a small price to pay for half of agriculture's fertility. When there is no oil left at all it would be easy to run all the Haber–Bosch units we need on solar power (or on biofuels, if necessary raised on the farms that use the fertilizer). To be sure, organic farmers often suggest that extra inputs of artificial fertilizer will cause more run-off and hence more pollution of groundwater. But the key point is husbandry. If artificial fertilizer is applied 'by numbers' (as has often been the case with modern industrial farming), simply according to the calendar, then there can be problems indeed. If the ground is too wet or too cold when the fertilizer is applied the crops cannot absorb the added nitrogen – which then simply drains away. But if the farmer is alert, and gears his regime of fertilizers to the perceived needs of the crop (and measures the amount of nitrogen in the soil before applying more) then the growing plants will take up almost all of what's applied before it has a chance to disappear. By the same token, if organic farmers simply apply manure to cold, wet, naked fields (as has often been done), then this too can be horrendously polluting. Whatever the technology, there's no escape from the demands of good husbandry. As with pest control, an integrated strategy might achieve the best combination of yield and sustainability: a judicious mix of artificials and organics, with each taking the strain off the other.

By the same token, present-day biotechnology, including genetic engineering, is very expensive and highly esoteric, and is conducted by governments and big companies. But it uses little energy, and in principle could be carried out on a small scale in the countries that need it most, like Tanzania and Bangladesh, by the people themselves. At present genetic engineering manifests in the developing world as an agent of Western influence, but it doesn't have to be like that. Excellent scientists pop up everywhere, given half a chance: and the advantages of science in principle can belong to everybody (or at least of sciences like biotechnology that do not require, say, significant inputs of plutonium). GMOs are currently deployed for dubious economic and political purposes but the science that has given rise to them should not be banished out of hand. If the overall farming strategy was different (if agriculture was indeed enlightened), then genetic engineering could do real service.

All in all, then, organic farming has a huge amount to offer. In strategy and ambition, and notably in its emphasis on good husbandry, it provides much of the model for the enlightened agriculture that the world needs. It is not innately or inevitably technophobic, and we should be grateful to it for keeping so many threads of good and necessary research alive while most orthodox agricultural science was engaged elsewhere. Yet its restrictions do seem more arbitrary than is necessary, matters of dogma rather than of common sense, and they do seem to close the door to many possible approaches that could be extremely helpful. On the other hand, organic farming as it now stands provides a useful label. Organic farmers are on the whole deeply committed to the general ideas of enlightenment, and because they follow discrete sets of rules and are duly accredited, they are easy to identify. Hence, if we support organic farmers, then we know we are supporting enlightenment. But still it is not sensible to close the door to technologies that could further the grand cause simply as a matter of dogma or for legalistic convenience. Furthermore, as many farmers already demonstrate, it is possible to pursue excellence in husbandry (as excellent as it is necessary or reasonable to require) without being committedly organic. Farmers who raise Berkshire pigs free-range are surely doing as good a job as it is reasonable to ask for, even if the grass is sometimes egged on with artificial nitrogen.

So the organic movement is important and deserves our gratitude and support. But in the end its rules are too arbitrary. True enlightenment is at least as rigorous in its morality, but more accommodating in technique.

II

Biology, Morality, Aesthetics: The Meaning of Enlightened Agriculture

Enlightened agriculture on the one hand is rooted in the physical realities of landscape, climate, living creatures and humanity itself – in other words, it marches to the drum of biology; and on the other it is guided by the most fundamental human values, of morality and aesthetics. Like the eighteenth-century Enlightenment itself (at least in some interpretations), it is an exercise in sense and sensibility (whereas, I am inclined to say, the monetarized, industrialized, corporatized and globalized – MICG – agriculture that we have now has little or no claim to either). The principles of enlightened agriculture could in theory be applied in any economic or political context, provided only that the economics and the politics were adjusted to the needs of farming, and not – as has usually been the case – the other way around.

At the moment, the world as a whole is dominated by various manifestations of capitalism, some more attractive than others; and the most powerful nations and coalitions at least acknowledge the desirability of democracy, and the fact that they do so is perhaps the single most encouraging feature of modern life. Enlightened agriculture requires us to reverse some of the most bullish trends in modern politics – notably the globalization of agriculture, and the general cultural hegemony of a few Western powers – but it can probably be accommodated more easily in a broadly capitalist and avowedly democratic system than in any other. Capitalism at its best makes a virtue of its own flexibility, and when it is not corrupted it encourages variety and individual ingenuity; and the morality and ideology of democracy surely encapsulates some of the most fundamental of human values. It is possible to envisage societies that are not exactly democratic and yet are benign (as in a monastery, for example) but on the whole, enlightened agriculture in a non-democratic environment would not (one feels) be truly enlightened. (In monasteries, of course, people always have the option of leaving.)

In short, the world has got much of what's needed – excellent craft (basic husbandry and broad knowledge), a base of good science, an economic system that is flexible enough in theory to accommodate enlightened farming (and which at times has made a virtue of its flexibility), and an underlying ideology (democracy) that is broadly sympathetic. So the things that need to be done ought to be doable. In the next chapter I want to ask how we might get from where we are to where we ought to be. In this chapter I simply want to ask: What is the precise agenda of 'enlightened agriculture'? What is it trying to achieve? And what, if its targets were achieved, would it look like?

The Agenda for Enlightened Agriculture

The first requirement is to produce good food – where good means nutritious, safe, and also appetizing: and 'appetizing' is not as defined by me, or by cordon bleu chefs, or any other single pundit or group, but by people at large. In other words, the food provided by truly enlightened agriculture should support all the great cuisines of the world. There is nothing indulgent in this: as outlined in earlier chapters, people who are true gourmets, who really care about food, are much easier to feed than those who insist on hamburgers or, in the cause of austerity, will eat only lightly grilled sardines with a little salad. As noted in Chapter 3, this is one of the world's great serendipities (albeit one that the modern systems choose to override). Gourmets are not necessarily rich, and are certainly not food snobs of the kind who simply demand what they know to be fashionable. The people of the Peruvian highlands who can distinguish a hundred different kinds of potato and care about the difference are gourmets, or those of South Asia who know great rice when they taste it and who (as in India) prepare magical dosas (rice-flour pancakes) by the side of the road. The entire traditional rural populations of Italy and France were or are gourmets, as are those of Turkey and Morocco, who can conjure the most wonderful meals from wheat and mint and fragments of goat.

Of course, truly enlightened agriculture must provide enough for everybody, and do so unfalteringly. 'Food security' is the requirement; famine is not acceptable. Furthermore, farms must provide enough for everybody who is liable to be born on this earth until the end of accountable time – a maximum of 10 billion people at any one time (so the United Nations demographers now suggest) for the next 10,000 years, or indeed for the next million. In other words, output must be prodigious and yet sustainable.

Yet agriculture does not merely provide food (and fibres and drugs and so on). Farming reflects the entire *Zeitgeist*, and it affects every aspect of human existence. It does this partly because it provides food, and food touches all aspects of human life; partly because it occupies so much space and is the principal shaper of landscape (and a major determinant of climate); and partly because it employs people, and the number it employs, and in what kind of set-ups – Stalinesque collectives, subsistence farms, modern intensive factory farms – largely determines the shape and the mood of the overall society. It seems to make sense in designing agriculture (or rather, since 'design' has an uncomfortably authoritarian feel to it, in working towards the systems we feel are desirable) to keep in mind all of the various end-points we want to achieve: what kinds of landscapes, what kinds of relationship between town and country, and so on. Agriculture affects all of these things, and if we simply allow it to be shaped by one single driver – and at the moment, the crudest kind of commerce is by far the most significant driver – then it is hardly surprising that the outcome has features that are unforeseen and undesired.

Thus the second great set of requirements for truly enlightened agriculture are social. Under this heading comes the general need for justice: for producers within each society (meaning farmers should be properly paid for what they do) and in the relations between societies. A world in which one set of nations operates as the dependants and effectively as the servants of other nations has long been considered unacceptable (this again was one of the more desirable changes of mind in the twentieth century), but the world's new commercial structures, including the notion of globalized agriculture, seem in many ways to be reinvigorating the old imperial relationships in a new guise. Agriculture that is essentially imperialist in structure and intent cannot be called enlightened. (More of this later.) In addition, as discussed in Chapter 9, enlightened agriculture should provide good jobs – and, of course, to get the best out of a landscape, adapting farming to its vagaries and exploiting its microclimates, requires detailed knowledge. In short, the social need to employ people, and the need to combine productivity with sustainability, go hand in hand. Here is another great serendipity. The third social and moral requirement is kindness to livestock, which is discussed later in this chapter.

Finally, for reasons both of biological security and of morality and aesthetics, we (humanity) must ask if we really want to be the only species left on this planet – apart, that is, from the domestic livestock, crops, ornamental flowers and trees that we grow for our own convenience, and the cockroaches, flies and mice that come along for the ride. I don't expect ever to see snow leopards on their home ground, or even wild tigers, but I would feel a great sense of loss if they no longer existed, as if the world as a whole had been sullied. I know of no formal surveys to show that others feel the same but my impression is that people in general do feel this. A sense of identity with other species is surely part of human nature. Agriculture is the most invasive of all human activities, and directly and indirectly does most to influence the fate of our fellow creatures. Attitude matters enormously. It seems altogether too perfunctory simply to categorize all our fellow creatures as 'biodiversity', and I hate the mentality that sees 'biodiversity' simply as a resource. As discussed in the next chapter, the materialism thus implied is probably necessary to some extent, but it is certainly not sufficient. The basic point is, though, that if we want other creatures to share this earth with us, then farming must be wildlife-friendly.

These, then, are the physical and logistical requirements: good, plentiful food for everybody for ever; a fair deal for producers; labour intensiveness – a maximal number of good jobs, giving rise to working rural communities; benign husbandry; and wildlife friendliness. These desired end-points will not arise by default. They must be expressly written into the strategy; and they must be underpinned by moral values that are broadly accepted and kept well to the fore. So what are the values on which enlightened agriculture (or indeed 'enlightenment' of any kind) is founded? I discuss the morality that underpins human values in more detail in the next (and final) chapter. But here is a shopping list of practicalities that emerge from it.

Values

The fundamental problem for all organisms is to stay alive; and the fundamental problem for all social organisms (and all organisms are social up to a point) is to rub along with other organisms. The set of rules and ideas that we call morality and the behaviour (ethics) that emerges from morality have to do with both of these sets of problems: the things that we owe to ourselves, as individuals; and the things that we owe to other individuals – of our own society, our own species, and other species.

The things that we owe to ourselves are not simply selfish. At least as a metaphor, I like the idea that we do not belong to ourselves. We did not create ourselves; we are, as it were, stuck with ourselves, and have to make the best of what we are stuck with. It is for this kind of reason that Roman Catholics regard suicide as a sin: we simply do not have the right (they say) to rob God of one of his own creations. Various notions are in line with the idea that in the end, whatever else we do, we must also look after ourselves. These ideas include happiness, of course. More broadly, they include ideas of personal fulfilment, autonomy, dignity and personal security – safety from random attack. These ideas can be extended to the society with which we identify; and are manifest in the concepts of autonomy, self-determination, national security and general peace.

The thing that we owe to other individuals (or other societies) is to acknowledge that they too seek fulfilment, autonomy, dignity, security; and – the key concept – that they have a right to such fundamentals. The idea of human rights of course is now a powerful component of modern politics (which can also be seen as one of the truly significant political advances of the twentieth century). Some people, including me, would extend the concept of rights to other species too, of which more later. With the concept of autonomy (of individuals and societies) goes the idea of cultural diversity. People can hardly claim to be autonomous if their society is simply a facsimile of somebody else's (unless, of course, the people as a whole expressly desire to become like somebody else, as opposed to having somebody else's customs thrust upon them. But historical examples of a spontaneous desire to imitate are surely rare).

When we put the two kinds of requirements together – the physical and logistical requirements, and the underlying values – a number of themes emerge that between them define the shape of enlightened agriculture. These include: the broad structure of enlightened farming in any one country – including the idea of the 'mosaic', animal welfare and a short supply chain – and at the global level the importance of national self-reliance (incorporating the ideas of autonomy and cultural diversity.

The Broad Logistics: Cows in the Meadows and Sheep on the Hills

As emphasized throughout this book, if farming is to be as productive as possible and yet be sustainable, it must first and foremost march to the drum of biology; and this means that it must acknowledge and play to the strengths of the landscape, the climate, the crops and livestock, while of course it is geared primarily to the human requirement for good food. As outlined in earlier chapters, human beings need a diet that supplies enough but not too much energy – and we fare best when this energy comes mainly in the form of unrefined carbohydrate; but we also need some, though not a lot of, high-quality protein; a mixed array of fats; an even more miscellaneous list of minerals and materials classified as vitamins; and (it seems) a possibly vast catalogue of arcane materials that have loosely been classified as nutraceuticals or functional foods which are as yet largely unidentified but seem to include, for example, plant sterols. All this is supplied in passing if the diet is high in cereals and pulses (and also tubers), includes modest amounts of meat and other animal products, provides as much fresh fruit and vegetables as a person can reasonably eat, and is as mixed as possible.

So how can farming best be arranged to meet such needs – and to go on doing so, century after century, without degrading the ground – indeed improving it as the seasons pass? The general answer (by and large) is to give the best, most suitable land to pulses, cereals and tubers (that is, to arable farming); to fit in horticulture in every spare pocket – and be prepared to spend a lot of time and effort on it, and to invest capital for example in greenhouses; and to allow the livestock to slot in as best it can. The most fundamental principles of biology show that animals and plants work well together (they have after all been evolving side by side for the past few billion years). In short, farms in general should be mixed: even the most committedly arable areas would in general benefit from at least some livestock, as all traditional farmers knew. With industrial chemistry it is possible to override the need for rotations of crops (including periods of grass and livestock) but still, in principle, rotations remain desirable. The areas that are truly marginal – too high, too steep, too rocky, too dry, too wet – can be ideal for ruminants, notably sheep and cattle but also goats (or the quasi-ruminant camels, llamas, guanacos, alpacas and vicunas). Some cereal and pulse can be grown expressly for livestock –

but in general, only enough to keep them going through the winter, so they can make better use of the grazing in the summer.

Of course, there will be some areas of the world – like the vast grasslands of the prairie, the pampas, the savannah and the steppe – which, if managed for food at all, are best used more or less exclusively for sheep and cattle. On some soils, any ploughing is likely to do more harm than good. Perhaps these areas can best be regarded as a bonus for livestock production. All farming should have an eye to wildlife, and the grasslands should surely be managed not exclusively to maximize beef production (the world doesn't need that much), but also to encourage native grassland fauna, from prairie dogs and black-faced ferrets to bison, gazelles, rheas, strand wolves, Przewalski's horses, and strange, lonely birds like the great bustard.

If the world as a whole were managed according to such principles then it should be just as sustainable as is wild nature, since it would in effect reflect wild nature; and the ratios of food that it provides would perfectly match human nutritional requirements (which are geared to nature) and gastronomy (which has evolved in response to what traditional farming finds it easiest to produce). In short, when agriculture is expressly designed to feed people, all the associated problems seem to solve themselves. In essence, feeding people is easy. Nature determines the overall pattern. Human ingenuity in the form of traditional husbandry over 10,000 years has long since shown how to manage those aspects of nature that we need to deal with; and science, properly applied, can supply useful insights into the underlying mechanisms, and a few useful tweaks.

That's it in a nutshell. The problems and the hand-wringing begin when people in power decide that the principal role of farming is not to feed people but to supply wealth, and try to treat farming simply as a business like any other. Treat agriculture simply as another means for creating wealth, and we immediately move away from the idea of the mixed farm. We move towards monoculture and seek perforce to maximize livestock. In addition, science ceases to be the accomplice of enlightenment, and becomes the handmaiden of commerce and power politics. Thus humanity's most important enterprise is expropriated; and science, which is humanity's most powerful provider of material understanding, is subverted. The powers that be have changed the nature of the problem. No longer is farming designed to feed people. It is designed primarily to keep the powers that be in power. Somehow the world's most powerful governments and the powerful industries that work hand-in-glove with them have managed to convince the rest of us that unless they remain in power, doing the things they do, then the world as a whole will be in deep trouble. In truth, the complete opposite is the case.

Critics will probably accuse me of Marxist subversion. It's happened often enough in the past and this, or some variation of it, has become the standard knee-jerk response to all awkward ideas. In truth I'm a good capitalist: or at least, I can see the advantages of capitalism, and the snags in centralized economies. But capitalism takes many forms. Some forms are benign and truly democratic, and are underpinned by moral values independently arrived at. Others are brutal, authoritarian, hierarchical, and content to allow such morality as there is to emerge from the market itself: it is taken to be self-evident that whatever can be sold, must be right. Those who espouse the brutal forms tend rapidly to acquire wealth – since wealth is what they are expressly interested in; and in a world that is run by money, wealth means power. This is the evil in the world, not capitalism itself: the fact that people with a lust for wealth and power, in whatever context, tend *ipso facto* to become rich and powerful; that, plus the belief that wealth and power are innately good. I would be very happy with the capitalism of Thomas Jefferson, who saw an America founded on small farms, personal dignity and justice. But in the modern USA Jeffersonian ideals are perceived to be dangerously to the left. I am sure that enlightened agriculture can be created within a capitalist-democratic context, and that the prize of achieving this is so enormous (the continuing well-being of humankind and of our fellow creatures) that it has to be worth a shot. It's a sad world indeed if such a view of life is construed to be subversive.

The Mosaic

Agriculture has to work – it has to produce good food sustainably from the land allotted to it – but it also has to fit in with all the other requirements of the world. Notably, on the one hand we need cities, and roads, and airports; and on the other, the world would be a sad place without other species, and other species above all need wilderness. The overall requirement, then, is to design the world as a whole, so that all these different needs are taken care of as best as possible, and all the different elements work with each other as harmoniously as possible. Overall, the world would be conceived as mosaic.

Of course, yet again, the word 'design' is not appropriate, because no one (or at least, not I) would advocate the kind of world government and authoritarianism that would enable humanity truly to design the whole world to one single plan. But at least we need agreements and understandings that would make it possible for independent states and societies to produce something resembling a design, even though each should continue to treasure its own autonomy. There are many precedents for such collective endeavour. Most nations have signed up, for example, to the Ramsar agreement, which seeks both to protect the world's wetlands and to coordinate them – for wetland birds in particular commonly commute between habitats, feeding in one place and breeding in another, and they need an integrated world network of necessary places. Hence we have a world plan, but it is operated by individual nations operating individually. The Kyoto Treaty that should have committed all countries to reduce the emissions that lead to global warming also provides a model in principle, even if it has been undermined in practice. Such agreements in general are mediated through various agencies of the United Nations, which most individual countries accept should at least have influence. The World Trade Organization has also shown that it is possible to exert power globally, even in the absence of an authoritarian world government which is unrealistic and in many ways seems so undesirable. Thus the world already has the kinds of agencies that could bring enlightened agriculture into being, if enough people saw the need for it, even though the precise organization that would be needed is clearly lacking. Sufficient coordination is possible even in the absence of grand, overweening design.

To be sure, many people have thought about the relationships between town and country, cities and farms, farms and cities and wildlife, but (I suggest), they have not thought about them enough. The essential principles, to my knowledge, have not been properly spelled out. I suggest that a sensibly designed world would embrace the following elements and principles:

First, there is a case for very intensive farms dedicated solely to crops (although they should incorporate livestock as much as possible, since mixed farms are in general preferable for various reasons). In Britain we may love to hate the spreading prairies of eastern England: cereals to the horizon. Often, indeed (more often than not perhaps), these vast fields are created insensitively, with nothing in mind but profit, and they sit very ill, replacing landscapes and communities that were in many ways superior. Yet beauty remains in the eye of the beholder, and I am at least intrigued by this quote from the American commentator A. H. Reginald Buller, from his *Essays on Wheat* of 1919:

There is no more exhilarating sight in the West than the prospect of the binders at work on the sea-wide, sky-skirted prairie, with the golden grain gleaming under the August sun and above and about all the cloudless blue dome of heaven. And when the last sheaf has been cut and the binders are silent, how splendid is the view across the gently rolling stubble fields: stook beyond stook . . . for a quarter mile, for half a mile . . . stooks cresting the distant horizon, ten thousand stooks all waiting to be threshed and each with its promise of bread, the gift of the New World to the Old. The unbroken expanses of the prairie create within one a sense of freedom which is best known only to those who dwell far from crowded cities, who plow and sow and reap, and whose daily toil causes them to commune unconsciously with Nature and thus to absorb something of her simplicity and her charm.

This is not quite modern (who, nowadays, knows what a stook is? there is no need for them when the crops are brought in by combine harvesters), but it is certainly in the spirit of modernity; and Buller's vision is broader than that of mere cash. He is an aesthete, and anyone who has seen arable fields in their glory is aware of their grandeur. These wondrous seas of corn later took a more sinister turn, as the scale and the mechanization increased, as John Steinbeck so poignantly described in *The Grapes of Wrath*. Still, though, intensive food production can be aesthetically fine, and indeed exciting. Logistically, too, the greater the output from a given area, the greater the area that can be used for other purposes – including wilderness. Horticulture can reasonably be even more intensive – in capital (as in greenhouses), in energy, in technology, and in labour. Intensive horticulture can be beautiful too, as I have seen in particular in Israel, in greenhouses stuffed with melons, the air heavy with their scent; and in Malta where there was hardly space to squeeze between the tomatoes; and outdoors in rural China in hydroponic columns dripping with lettuces and capsicums. The overall feel is Arcadian; you can hear the pipes of the satyrs. In short, hyper-intensive food production is not all bad. Not at all.

Still though, there are conditional clauses. Arable fields can still be enormously productive – at least as productive as is necessary – and yet be wildlife-friendly. In Britain, traditional arable fields accommodated dormice and corncrakes, cornflowers and shepherd's needle, and they could in theory continue to do so. Ecologically speaking, the arable fields stand in for the grasslands that they have replaced. It's just a question of adjusting times of cultivation, sowing and harvesting, and the type and frequency of spraying (if any). Concessions need to be made, but not to a degree that is crippling.

Then again, hyper-intensive cultivation of plants is acceptable because plants are not sentient, as far as can be discerned. Melons are happy, insofar as melons are capable of happiness, growing with scarcely an inch between them. Animals, in absolute contrast, are extremely sensitive, and if in doubt we should certainly err on the side of kindness. There seems no case whatever for hyper-intensive livestock production of the present-day factory kind. The factory farms fail by the criteria of nutrition, gastronomy and common humanity. The battery hen and the hyper-intensive piggery owe their existence purely to the dominance of accountancy, and although the criteria of accountancy are necessary (sloppy accountancy leads to waste), they are not sufficient. Surely, no sane person could write as lyrically of floor-to-ceiling poultry as Buller wrote of horizon-to-horizon corn.

This leads to the second great desideratum: the need for farms that are as wildlife-friendly as possible to sit alongside those that are maximally productive (and so save space). There are many movements worldwide which, through advice and sometimes cash, are helping farmers to make their farms more wildlife-friendly. There are many modern approaches, technical and logistical: to plant hedges and copses, for example – not just anywhere, but in places where wild creatures truly can make use of them, and use them as stepping stones from one habitat to another. For instance, many British bats will not fly over open fields, but they will happily follow a hedge; and the hedge is most valuable when it leads them from wood to wood. The hedges, ideally, need to be about a metre and a half thick (although they rarely are). Of course, there are many ways of faking the improvements, and a great deal of nonsense is being perpetrated: hedges that lead nowhere (but attract grants anyway); grants to restore stone walls, with more grants to obliterate the native pasture between the walls in favour of high-protein, brightly coloured, totally homogenous ryegrass. Thus the form is retained, but not the substance. This is window-dressing; spin. But good things are happening too. Many farmers need no encouragement to welcome wildlife on to their land. They need only the opportunity; some relief from the pressure to maximize production at the lowest possible cost. Sometimes their essays into wildlife are spectacular. Many farmers in southern and eastern Africa have encouraged wild game on to their farms, from gazelles to rhinos; and collectively, they may well be making a significant contribution to those species' conservation. In my adopted Oxfordshire, some farmers are prepared to risk bankruptcy (literally) in their quest to bring the skylarks back. Skylarks are less spectacular than rhinos, but are important too. In general, the world movement to encourage wildlife on farms has been one of the more encouraging features of twentieth-century policy. But it runs against the mainstream, just a few eddies in the swelling tide.

Third, we of course need areas that are dedicated solely to human activities and comfort. For instance, I can think of no sensible role for wildlife in a concert hall (the flowers around the orchestra don't count). Yet (I suggest) there are very few buildings and open spaces, no matter how anthropocentrically conceived, that could not be friendlier to other species than they are – or that could not be used, where required, to grow food. Above all, we need more wildlife-friendly cities. Again, this is happening to some extent: the spread of the urban peregrine is another encouraging trend of the late twentieth century, beginning in Berlin and Prague and then Salt Lake City, New York, London, and so on and so on. London now has far more foxes to the square mile than the countryside does, and its outskirts have badgers and muntjac deer. Cathedrals traditionally drip with bats and kestrels. Then again, of course, there are many anthropocentric areas from which human beings at large are largely excluded: railway embankments, nuclear power stations, rocket launch-pads. Each and all of these can be havens for wildlife.

The city is the ideal venue for horticulture, too. Then, the produce could truly be fresh. In the days when petrol routinely contained lead the wisdom of urban vegetables could be doubted, fresh though they were, but now there seems no excuse for not growing them in towns – except, of course, the dominance of cash, and the price of real estate. But there is no natural law which says that cities have to be as densely packed with people and offices as is physically possible. The greatest cities have plenty of open space, as in London and Paris and indeed New York. There was plenty of gimcrack and cynical building in the nineteenth century, but at least in the grander places the planners never lost sight of the fact, as those of the twentieth century often seemed to do, that people actually have to live in cities. That is what they are for. The great sprawling horrors of the present developing world, cities of 25 million plus, will surely grow even bigger (especially as farms shed labour), but they could be made far more tolerable, and diets and gastronomy could be greatly enhanced, if serious areas within them were dedicated to horticulture. The world could be so much better, for people and other species, if architects and town planners would only think a little more broadly – or perhaps, to be fairer, if the governments and councils that employ them gave them a more thoughtful brief.

Finally, the world needs wilderness. Many other species have adapted to a greater or lesser degree to human habitats – farms and airports and cities – but most thrive best in truly open spaces. Pristine wilderness is hard to find these days; and whatever is left needs to be managed, precisely because the wild areas are so much smaller than once they were, and smallness imposes pressures of its own. Ecologically, a national park is not a microcosm of the continent of which it is a part, but is an island within it, surrounded by farm and sometimes by city; and island ecologies have many special features, quite different from those of continents. To some extent the management of wilderness seems an oxymoron, for management seems the antithesis of wildness. Yet this circle must be squared. Modern conservation is an exercise both in science and in aesthetics, and it can succeed on both fronts when given a chance. Wildlife-friendly farms and cities can extend the area available to some species, and indeed in the modern world may fill a vital conservation role. But wilderness is the core, which no human artifice can replace.

Most importantly, each of these elements should be good of its kind. A farm that is not particularly wildlife-friendly and yet is fabulously productive can be conservationally useful since (in principle), hyper-productive farms can provide the food that people need but in a smaller space. But a more extensive farm is excellent too, if indeed it is wildlife-friendly. Sadly, many farms worldwide are neither particularly friendly nor especially productive. In Britain, I seriously hate those landowners and their even more obnoxious managers who keep people off their land (by destroying the footpaths) but do not farm particularly productively and do not provide havens for wildlife either; and then have the gall to pretend self-righteously that they are 'guardians' of the countryside. With guardians like that, bring on the vandals. Many modern cities are bad for all living things, including human beings. Too often one feels that the creators of these all-pervasive entities don't really know what they are trying to achieve, or indeed have not given thought to the matter.

Animal Welfare

It's the role of science and perhaps of all philosophy to improve on common sense. Yet we abandon common sense at our peril; and in difficult issues, with many different threads, common sense remains our main guide, albeit as informed as possible. The welfare of animals is one such issue. The more information we can bring to bear, the better; but in the end, in deciding what we really *ought* to do, instinct and common sense are indispensable. Some philosophers have made excellent contributions to this discussion but many others have been less than helpful precisely because, as professional pedants, they put too much store by their own pedantry.

Human beings keep animals for food, and along the way we control their reproduction, and then kill them: all of which amounts to a huge and indeed an absolute invasion of their lives. Some (the vegans) say we simply have no right to do this, and condemn all animal husbandry. For my part, I am inclined again to appeal to the idea that we, human beings, are not of our own making. We did not design ourselves, and our opinion of how we would like to be was not asked. We are, in short, stuck with ourselves; and the world we find ourselves in is not of our making either. Every aspect of our evolved and inherited biology proclaims that we are omnivores: not out-and-out carnivores, like cats, but at least requiring some meat (or else we need very clever ways of getting around that need, or very good luck). We can acquire the meat we need either by hunting or by husbandry. For many reasons, and especially since there are so many of us, husbandry is preferable. Besides, if well done, husbandry can be much less destructive than hunting: the Pleistocene overkill, the loss and serious compromise of many a wild beast in modern times (from the great auk and the pink-headed duck to the wild Arabian oryx), and the plight of present-day fisheries and of the great whales, show how difficult it can be to hunt sustainably. Of course, we could in theory farm without keeping animals, as the vegans have demonstrated; but as discussed in Chapter 10, farming with livestock is more efficient in the end than farming without, and so in principle requires less room, and so in theory leaves more room for wilderness and wildlife, so that even vegans have blood on their hands. So indeed does every creature that must compete to stay alive. The only logical way to avoid incommoding other creatures is to commit suicide. In short, we are stuck with the need to raise animals, and having raised them, to slaughter them. Lacto-ovo vegetarianism does not obviate the need for slaughter.

Because we are stuck with the need to raise and slaughter animals, some industrial agriculturalists have effectively concluded that anything goes. If we are going to keep them at all (the logic seems to have it), we might as well raise them as rapidly as possible, by whatever means are most efficient; and if the animals don't like it we can simply say 'So what?' or perhaps alleviate their misery by turning down the lights, plying them with sedatives or (the latest recourse) engineering them genetically so that they lack sensitivity. This is one approach – and indeed is the prevailing one. But I (and many others) prefer what might be seen as common-sense morality. Yes, as human beings we are more or less obliged to raise animals for food, and then to slaughter them. That's the way life is. But this is a huge imposition, and we should see it as an expediency and as a privilege rather than as a right. We should not grant ourselves carte blanche. In short, having decided that we must keep animals, we should contrive to treat them (and in the end to slaughter them) as kindly as possible.

This is not straightforward for two sets of reasons. The first set are logistical and economic, and can be summarized with reference to animals that are on the edge of husbandry, neither quite in nor quite out: the red deer, and its variants such as the North American wapiti and the Chinese Père David's; and the eland of Africa, the world's largest antelope. Forward-looking agriculturalists have dreamed of raising all of these animals for food (and farmed red deer are already big business, especially in New Zealand). These big deer and antelope produce large amounts of fine, lean meat, and they thrive in difficult conditions in which more conventional livestock – cattle and sheep – are wont to languish. Thus eland can survive in the virtual absence of water, while red deer can get by on the wildest of Scottish (or New Zealand) mountainsides, partly because they thrive on low-grade fodder, and partly because they can run from bad weather. Père David's deer, traditional denizens of Chinese marshes, are, as Manchester University's professor of zoology Andrew Loudon is wont to comment, 'tough as old boots'.

The trouble is that such beasts cannot demonstrate their ecological advantages unless they are allowed to run virtually wild; but then it is extremely difficult to catch them when they are needed, and the catching needs to be carried out by experts (including expert marksmen) or else it becomes more traumatic than an abattoir. If we add in the perceived need for hygiene, then safe slaughter in the (semi-) wild becomes extremely expensive. In the case both of eland and of deer, the logistical problems have been overcome by keeping the animals in corrals. But eland flourish in the wild by grazing far and wide on native vegetation at night, out of the midday sun. If they are confined then they obviously cannot forage far and wide, and if the food is brought to them they are no more economical than cattle. Red deer, too, when incarcerated, are not significantly more efficient than sheep or cattle (but they are a lot friskier, and more difficult to handle). In practice, eland and red deer enterprises have succeeded – but not because the animals themselves are efficient exploiters of landscape since, in reality, they don't get the chance to demonstrate their own survival skills. Both flourish primarily as niche markets. In particular, since the 1970s, there has been a vogue for lean meat which venison admirably supplies.

The principles that are so clear in these marginal species apply to all livestock. All husbandry is a compromise. The freer the animals, the more difficult it is to make use of them at all; but curtailment of freedom seems innately harsh. So welfarists ask (although the more industrial farmers feel the question is nonsensical) how can we effect the best possible compromise?

There have been three main approaches, all with their strengths and weaknesses; and on the general principle that in difficult issues we need all the help we can get, it is surely wise to learn from all three.

The first general approach is that of common sense, and it takes us a very long way. Thus traditional texts of husbandry, from beef cattle to farmyard hens, and indeed domestic pets, are essentially commonsensical. They tell us to keep animals warm and dry and out of draughts, to give them the kinds of things they eat in nature and constant access to fresh water. Such directives are refined to provide the kinds of principles that inform most laws and codes of practice of animal welfare, and for example lie behind the seminal report of 1965 chaired by the British veterinarian F. W. R. Brambell. Most basically, such codes generally demand that animals should be spared from obvious pain and stress. Slightly more subtly, they recommend that animals should be allowed to fulfil basic functions: stand up and turn around; stretch their wings; and – preferably – walk, run or fly. The codes may further suggest that animals should at least have access to the outdoors, and should be able to do at least some of the things they would naturally do in the wild: scratch, peck or root in the ground for food; graze, as opposed to eating constantly from a stall; fly to a perch to roost; lay their eggs in protected places (such as nesting boxes).

All this seems to fall within the purlieus of common sense and common decency, yet it is absolutely at odds with the kinds of intensive systems that have grown up over the past fifty years. Intensively kept broiler chickens, ducks and turkeys, for example, are commonly kept by the tens of thousands in sheds, again with no room for the stretching of wings, in permanent twilight to keep them calm (or calmer than they otherwise would be). Pigs, whose intelligence when given a chance proves comparable with dogs', are similarly raised in dim light, packed in pens so tightly it's as if they were already canned. Sows in recent years conventionally spent the whole of their pregnancy in stalls, without being able to turn around; and then in farrowing stalls in which they lay on their sides while their piglets sucked with their noses under a steel bar that prevented their mother from lying on them. Britain has banned the farrowing stall, but its farmers complain that they are now undersold by imports of pork and bacon from countries all around the world with no such scruples. The implied cruelty of the practice is if anything compounded by the injustice, and indeed the self-immolation of British farming. Thus we find half-hearted rules unevenly applied. Again one asks, is this really the best we can do?

The common-sense approach does have its drawbacks, however; or at least it is open to criticism, some of which is spurious (more bad philosophy), but some of which must be taken seriously.

The spurious criticism springs from a misunderstanding of the concept of anthropomorphism: what it is, and its limitations. The trouble began, or at least came formally on to the world's agenda, with René Descartes in the seventeenth century. Descartes argued in his Gallic and logical fashion that thought depends on speech, and since animals do not have speech, therefore they do not think. Since they do not think they are not self-aware, and so are incapable of any finer feeling that involves any degree of instrospection. Of course, animals appear to respond adversely to pain – kick them, do they not whimper? – but they cannot be said to experience pain in the way that humans do since they are not self-aware, and so they do not know that they are feeling pain. In fact, said Descartes, animals are simply like the automata, the clockwork toys that were so popular in his day. They can behave for all the world like humans, but obviously they are not. We know that a clockwork toy does not care what happens to it (or at least, all our experience suggests that this must be the case), so why infer that an animal does? Later philosophers suggested that to draw direct parallels between the behaviour of animals (or clockwork toys) and ourselves is simply 'anthropomorphic'; and took it to be self-evident that anthropomorphism is muddle-headed and therefore bad.

In the twentieth century the science of animal (and to some extent human) psychology was largely taken over by the behaviourists. They began soundly enough after the First World War, but then made a philosophical mistake. Their initial intent was simply to turn psychology as far as possible into an exact science. In the first instance, this requires measurement. But the traditional subjects of psychology – thought and feeling – are not directly measurable. Therefore, at least in the first instance, thought and feeling could not be taken into account. The only aspect of a creature's (or a person's) psychology that was directly measurable was behaviour. So these new, no-nonsense, new-broom enthusiasts, beginning with J. B. (John Broadus) Watson and culminating in B. F. Skinner set out to devise a system of psychology as complete as possible based simply on measured behaviour. Hence 'behaviourists'. The basic ingredient of their system was the learned reflexive response.

It was and is remarkable how much of animal (and human) behaviour can apparently be explained in behaviourist terms and the approach was (and remains) both powerful and salutary. It seems to espouse the fundamental principle of 'Occam's razor': don't make explanations more complicated than they need to be (or – which is what Occam really said – invoke hypothetical forces that don't really need to be invoked). Yet behaviourism has faded somewhat this past two decades, for three kinds of reason. First, various biologists have pointed out that in reality, much of what animals (and people) do cannot be satisfactorily explained in behaviourist terms. Animals in laboratories with specific tasks to perform might indeed behave in very predictable ways. But whenever wild animals are studied closely, as more and more species have been, it becomes clear that their behaviour is flexible, extremely dependent on social and physical contingencies of many kinds, and that they do many things (including the transmission of information) that cannot be explained away as learned reflexes. Jane Goodall's studies of wild chimpanzees were extremely influential. So, as Herb Terrace of Columbus University put the matter when I talked to him at the Royal Society in the 1980s, 'The point is not that animals don't think. The task is to explain how they think without human language.' In truth, the idea that animals do think and reason was not entirely discarded in the centuries following Descartes. David Hume wrote in *A Treatise of Human Nature* in 1739–40 that 'no truth appears to me more evident, than that beasts are endow'd with thought and reason as well as men'. Voltaire made a similar point. Now this idea may be considered formally restored.

A second blow was struck by philosophers of language, and in particular by Noam Chomsky of Harvard. For behaviourists sought to argue that human language, too, was acquired essentially by accumulating reflexes: for example, by learning to associate the word 'cat' with the physical presence of the beast. Such learning does seem to account for our vocabulary but not, said Chomsky, for syntax: for our far more interesting ability to manipulate words to frame a virtual infinity of new sentences and new thoughts. In short, human language, one of the most striking of all behaviours, must clearly be acquired by mechanisms that the behaviourists had no conception of.

Finally, the fundamental philosophical flaw in the extreme behaviourist position became obvious. To be sure, it made sense for the purposes of experiment to leave out the concepts of thought and feeling. As Sir Peter Medawar said, 'Science is the art of the soluble'; and thought and feeling could not be 'solved' or even sensibly addressed by the methods and in the state of knowledge of early- and mid-twentieth-century science. But just because thought and feeling were left out of the experimental design, for reasons of convenience, this does not mean that they do not exist. The mistake is obvious. But it was made nonetheless, and some scientists with a particular mindset, and some of their followers (including the more naive politicians) make this kind of mistake all the time.

In truth, we do have to beware of naive anthropomorphism. Of course it is foolish to assume as folk philosophers and many a storyteller has assumed, that an eagle is 'proud' just because it has (by definition) an aquiline profile and swoops down upon its prey *de haut en bas*, or that a fox is 'wily' just because it has a, well, 'foxy' expression. Indeed, careless anthropomorphism can even be cruel. Medawar himself tells how, as a small boy, he tucked up a frog in a warm dry bed only to find it desiccated; for frogs, of course, like their beds cool and wet, the very opposite of what we (and most other land mammals) find pleasant.

On the other hand, the only way we can come to grips with anything that we do not understand is by comparing it, in the first instance, with what we do understand. The behaviourists, following Descartes, compared animals to automata: the clockwork toy in effect became their model of animal psychology. Yet this model is clearly inadequate, so where do we go from there? We need something altogether more complex, that we do nonetheless have some insight into. As Pat Bateson of Cambridge has pointed out, our own selves are suitably complex, and although we are less than brilliant at introspection, we do have some insight into ourselves. In the first instance, then, says Professor Bateson, it makes perfect sense to use ourselves as models by which to gain insight into other animals, including their putative thoughts and feelings. In short, he says, 'anthropomorphism, used sensibly, can be heuristic'. Obviously we should not suggest that monkeys or pigs or frogs or seals *are* human beings. But, *faute de mieux*, it can be helpful to assume as a starting point that they are similar, until otherwise demonstrated. Certainly, our fellow mammals, and birds, are more like us than they are like clockwork toys.

So it may be 'anthropomorphic' up to a point to assume that a chicken would like to be warm and dry, free from attack by neighbours it can't get away from, and free to stretch its limbs and find at least some of its own food, if only to alleviate boredom. But it is sensible anthropomorphism. Certainly, the argument which says a priori that chickens are *not* like human beings at all, and that they bear comparison only with mindless machines, has long been shown to be false. Common sense says that chickens and pigs and cattle should be afforded basic comforts, freedom from stress, and freedom for self-expression; and in this instance there is nothing worthwhile either in philosophy or in science to gainsay that commonsensical view.

Yet one criticism remains. Pigs and chickens are not human beings. It is reasonable to assume in a generally anthropomorphic way that they do think, and do have feelings as we do (if not quite so thoughtful or sensible), but it is clearly false to suppose that they are made happy or miserable by the same kinds of things that affect us. If we really want to be kind to other species, then we should find out exactly what they do like and dislike, or we could well make mistakes comparable with the young Medawar's mistreatment of the frog. But it is hard for human beings even to discover what other human beings think. How can we hope to probe the minds of other species?

In practice there have been two main approaches. The first is experimental, as demonstrated for instance by Marian Stamp Dawkins at Oxford and by British government scientists at Roslin Institute near

Edinburgh (which was involved in poultry before it got into the cloning of sheep, and still is). Among other approaches, Dr Dawkins has sought to measure the importance that chickens attach to particular components of their environment by seeing what stress they are prepared to endure in order to achieve desired ends. For instance, it is known in a general way that hens prefer to find a quiet place to lay their eggs, and will go to a nesting box if one is provided. It's also known that they dislike strange objects. In one set of experiments Dr Dawkins placed a slowly inflating balloon by the side of the nesting box: something that was not enough seriously to frighten the bird (these experiments were not cruel), but enough to make it wary. The question was, to what extent would it run the gauntlet of the balloon to get to the box? The answer (it transpired) was quite a lot. To a hen, a nesting box is important. What matters, though, is the general logic of the experiments: effectively inviting the animals to reveal, by their behaviour in standardized situations, what they care about and what they do not.

The Roslin scientists adopted a physiological route. For example, when animals are stressed, they produce particular hormones in greater amounts (such as adrenalin) and reduce their output of others. These responses are universal, and of course are shared by us; and we know, from introspection, that a rise, say, in adrenalin, is accompanied by subjective feelings of 'flight or fight', which can be pleasant or unpleasant. Basically, the Roslin approach was to measure endocrine changes in different conditions. Again, what matters here is the generalization: that it is possible to gain insight into another animal's responses to life, provided we begin with the common-sense assumptions that animals do feel, and do not assume a priori that we cannot extrapolate at all from us to them.

The other main approach is as demonstrated in the 1980s by the late Michael Woodgush at Edinburgh University. Professor Woodgush simply kept animals in conditions that offered them the kinds of things they are known to like but which also gave them a wide range of options. Then he simply watched what they did – what, indeed, they *chose* to do. In particular he offered pigs a choice of woods and open ground to spend their time in, and in general they preferred woods. Among other things, they like shade and cool; and perhaps they respond to other signals too, including the physical presence of trees. He enclosed them only informally, with walls of straw bales, and gave them plenty of straw to play with. Then he let them get on with their lives. He found that when the sows were due to give birth, they made nests of straw, just as they would in the wild. Conventional farmers fear that when sows raise their own piglets in free conditions they will roll on them and crush them. The farrowing crate was devised to prevent this. But Woodgush's sows never did this. In conventional systems, too, the boar is generally perceived to be a murderous creature (adult male mammals often are) and is removed as soon as he has impregnated the sow. But in Woodgush's set-up the boar could safely be left with his family. In the wild, pigs do live in family groups.

One important generalization to emerge from all these formal studies and many other observations in the laboratory and in zoos is that animals above all like to feel that they are in control of their own destinies. They don't mind a bit of stress if the stress is of their own making, or if they know they can get away from it. What makes an animal truly neurotic is knowing that there is nothing it can do to change its condition. Intensive livestock husbandry is sometimes justified on the grounds that a pig, say, that's warm and dry in a shed is better off than one scratching round in some drizzly wood. But the latter feels in charge of its own destiny. Philosophers who still feel that anthropomorphism is the cardinal sin will say that this is nonsense, but anyone who knows anything about animals, including scientists like Marian Dawkins and Michael Woodgush, curators of good zoos, traditional livestock farmers, and anyone who has ever kept a dog or a cat, know that it matters a great deal.

Clearly, then, the concept of welfare does mean something: animals are aware of their own circumstances, and they do care. Common sense tells us that hens do not enjoy cages, and cattle and pigs do not like to be crammed with hundreds or thousands of others in vast, anonymous herds; and common sense and a body of good science show us how we could, if we chose, keep them in more kindly fashion. So why don't we?

Many people, of course, simply don't give a damn. Some feel that the arguments presented here are simply spurious while others feel that if animals suffer, so what? Their lives are short anyway. Others do care, or say they do, but many modern farmers go on to argue that they have to keep animals harshly – often much more harshly than they would like – for economic reasons. They point out that the profit margins on their animals is already close to zero, and if they allowed them any more space, or freedoms of any kind, then their farms would immediately be running at a loss. Free-range animals in general take longer to mature and fatten than the intensive kind, and time is money, and turnover is all. Other things being equal, it costs nearly twice as much (not quite, but nearly) to raise a steer for beef over two years, as it does to raise it for one; and if the rate of growth is doubled then, in a given time, the farmer has twice as many animals to sell. With broiler chickens, every day counts. Furthermore, livestock farming that is kind requires more of the most expensive input of all: labour. For many commercial farmers in the present economic climate, kind husbandry is simply not an option, however much they might personally favour it. Many industrial farmers and politicians simply become sanctimonious. It is desirable (they say) to produce as much meat as possible for nutritional reasons; and vital in general to keep the cost of food as low as possible. Cruelty, in short, is justified on grounds of cost.

But to defend cruelty on grounds of cost is specious in the extreme; either that, or hypocritical. To begin with, as argued in Chapter 3, it's clear that people do not need vast amounts of meat. The urge to maximize production has nothing to do with human need, and everything to do with money. A little meat is desirable, for all kinds of reasons; but what matters in gastronomy is flavour and texture rather than amount, and what matters nutritionally is biochemical complexity, for the true benefits lie in the details. Flavour, texture and general biochemical complexity are increased when animals are raised more slowly, on varied diets, and belong to the kinds of breeds that best respond to such treatment (Angus beef, Berkshire pigs, and so on). Contrariwise, flavour and texture are demonstrably compromised when animals are raised in fear (the adrenalin released by fearful pigs for example leads to pale, watery pork). The breeds that grow fastest, too, are not likely to have the finest flavour or texture. In short, it makes excellent gastronomic and nutritional sense to produce relatively small amounts of meat to a very high standard, and it makes no sense at all to produce huge amounts of meat of indifferent texture and flavour. But that is what conventional farming is designed to do. Here is another serendipity: kind husbandry, which is desirable on moral and aesthetic grounds, also provides the kinds of food that people actually need and (given a choice) prefer.

Of course, kilo for kilo, kindly produced meat is more expensive. It takes longer to produce. It requires more labour. Yet the arithmetic is not quite so stark as I presented it a few paragraphs ago. Turnover is reduced and labour is increased in kind systems, but there are other gains. The farmer who raises pigs in woods and encloses them with bales of straw does not spend many thousands on concrete buildings. In short, capital costs are low. Disposal of excrement becomes easy; not the huge environmental burden of the typical intensive unit. Most simply of all, the livestock farmer simply ups sticks at the end of the season and begins again in some fresh field – which, of course, is what is meant by 'rotation'. Disease control is simpler too. The animals are healthier to begin with (unstressed, well fed, exercised) and are not left to wallow in their own parasites. There is no need for heroic douches of disinfectant, or (worse) for a steady trickle of antibiotics, creating fearsome strains of resistant microbes. Most of all, perhaps, there is hands-on, attentive husbandry. The poultry farmer who practises kind husbandry does not, as the keeper of broiler chickens does, go round each morning with a bucket picking up the dead birds. Any birds that look off-colour are dealt with before they get to that stage. Furthermore, today's markets are designed to deal with the mass products of conventional (generally intensive) livestock farms. Beasts that are raised kindly these days are typically slotted into 'niche' markets. If the whole system of retail was geared to smaller, kinder production units, then the cost would come down.

Keep It Simple, Keep It Fresh

All nutritionists and gourmets, anyone interested in human health, and all ecologists with an eye to pollution and the need for sustainability, emphasize how desirable it is to shorten the food supply chain: to reduce the time and the distance between production and consumption, field and fork. Then again, although I do not believe that crime and corruption are the main sources of the world's food ills – the legitimate structures and strategies are mainly to blame – we clearly have to be alert to them. Villainy exists, and will be manifest wherever it is given a chance. As outlined in Chapter 5, villainy flourishes at present (to an extent that can only be guessed) largely because the present ways of producing and distributing food are so enormously complex that effective policing is impossible. Even small cracks in the edifice can be disastrous (Britain's foot-and-mouth epidemic may well have begun with a single import, though of course nobody knows). Laws and policies are innately bad if they cannot be enforced. For reasons of safety alone, then, and in the general cause of honesty, we need to keep the food supply chains as simple as possible. In enlightened agriculture, this is a high priority. In the present system, despite a plethora of ad hoc regulations (including the largely fatuous sell-by dates in supermarkets) the chains are encouraged to grow in distance (this is what globalization means), while storage time and general complexity of distribution are primarily decided by commercial convenience.

Autonomy, Diversity, Self-reliance

Looking at what's required for agriculture to become enlightened worldwide, we come to a whole raft of interrelated concepts that are of supreme importance to all individuals and societies and moral codes: self-determination; autonomy, both of individuals and of societies; dignity; individuality; and cultural diversity. These ideas have personal connotations – we all of us want to feel that we are ourselves, and that what we want to do is important and should not be overridden by the whim of others – that we and the society to which we belong are in control of our own lives.

These ideas also have the most profound political implications. Many countries were still trying to build empires in the twentieth century, and some succeeded in doing so. But the twentieth century is mainly memorable as the time when the concept of empire was finally discredited. Empires were always built for the benefit of the conquerors, but in past centuries the conquest was typically justified by layer upon layer of self-righteousness: the need to convert the pagans and infidels, to obliterate their ancient cultures and sometimes indeed to wipe them out for their own benefit, if not to save their corporeal lives then at least to rescue their souls. At least by the late twentieth century these ideas no longer washed. Some at least of the people who in the past had been immolated for their own good, including the native North Americans, the Maoris, Australia's Aborigines and many peoples of Africa fought back, at least to an extent (given that they were already sadly depleted). The United Nations, theoretically the pooling of the world's political will, acknowledged that all such people have a right to be themselves and that this right should be asserted.

In practice, of course, these rights are routinely overridden, both at the personal and the societal level. Some people or countries still seek to dominate others for their own reasons, and if they are obliged to find an excuse at all they can generally find one. Yet the modern world is run above all not by crude military powers but by commerce, which in turn is underpinned by and depends upon a particular swathe of technologies. The modern world is run, in short (as described in Chapter 9), by the feedback loop of capital–science–high tech–capital. The logic of this loop leads us to the global market; and the global market in turn leads us to global-scale production. Whether the global market and global-scale production are good or bad in other contexts – steel, computers, cars, mass-produced furniture – I do not care to argue. In some contexts – such as scholarship of all kinds – globalization is certainly very much to the good. In agriculture, however, as I have argued throughout this book, globalization is various kinds of disaster.

Strategically, it is surely perilous in the extreme for most countries to buy in as they are doing to agricultural globalization. A country that grows food primarily for the global market compromises its own ability to feed itself. If it does that, then it has to buy in its basic requirements from elsewhere. Some countries choose to do this from a position of commercial and technical strength, or because they are effectively obliged to. Israel, notably, has more sunshine than it knows what to do with, but very little water. It is also rich, very well connected commercially, and has a high proportion of world-class scientists and technologists, thanks to the ancient Jewish culture of learning. It makes little sense for Israel to grow wheat, since wheat has relatively low cash value tonne for tonne and requires too much water. But it makes perfect sense to grow strawberries and avocados, measuring out the water drop by drop (drip irrigation controlled by computer) and selling produce specified to the last molecule and timed to the hour to rich European markets that are only a couple of thousand miles to the north and west. Commercially and historically, after all, Israel is virtually part of Europe and is well represented in all European capitals. By not growing all of their own staples, the Israelis make themselves hostages to fortune. But the cash rewards more than compensate. As one Israeli agriculturalist put the matter to me: 'By buying wheat, we are effectively buying water. By selling strawberries, we are selling sunshine. For us it makes sense.'

Contrast Israel's position with that of say Angola, where 70 per cent of the people live by agriculture, or Rwanda, where the proportion is 90 per cent. Neither country has any commercial clout in the world markets. Yet, as things stand, under the logic of globalization, Angola and Rwanda are in theory being invited to sell strawberries or French beans on the high streets of London or indeed Paris (where surely they would get short shrift) at the expense of the maize, sorghum, cassava and several hundred variations on a theme of pulse that in fact keep them alive. Most countries are far more like Angola and Rwanda than they are like Israel. Those who really care about the people who live there should surely be urging that they focus their farming primarily on feeding themselves. That way they will at least be well fed, whatever else happens in the world. They will also keep their own people in work; and they will not be dependent on hand-outs from the US or the UN. That is a good basis for going forward, or indeed for maintaining the status quo which, when left alone, is in many ways enviable. To begin the new century as they are being invited to do with a denuded countryside, a debased and pointless agriculture geared to cash crops that are in world surplus, vast and swelling cities with no legitimate livelihood on offer – that is the real horror. But this is what globalized agriculture implies, and indeed is urging. We have seen many offences against humanity this past century or so but none is greater than this.

In short, it makes sense on all levels – ecological, nutritional, gastronomic, financial, social and strategic – for almost all countries in the world to become self-reliant in food. Most are perfectly well able to do so. 'Self-reliance' means simply that each country should strive to produce all the basic foods that it needs, so that it could feed its own people in a crisis, notably in times of political or economic blockade. It stops short of total self-sufficiency, which implies that a country produces absolutely all its own food, including the kinds that it cannot easily grow at home in open fields. Some countries could easily be self-sufficient. China and Australia, for instance, contain all the climates needed to produce virtually anything anyone might want. Britain (or anywhere else) could be self-sufficient if we spent enough. We could easily grow all our own bananas, for instance, if we built big enough greenhouses. But many other countries grow bananas of many kinds much more easily than the British could do, and so it makes sense to buy them in. The point is, though, that we do not depend on them. If the Atlantic is again blockaded (it's hard to imagine the circumstances, but it's the principle that matters) we would not starve for lack of West Indian bananas, although the thought of them might give us another reason for wishing the blockade would end (as it did in Britain, in 1945).

The principle of national self-reliance is simple, commonsensical and historically justified on strategic grounds in all countries that are amenable to analysis. It also fits perfectly with the biological need to make the best use of each country's landscape; with nutritional and gastronomic needs; with the social and economic need to maintain high employment; with the political and ethical desire for autonomy and self-determination; and the general desire for cultural diversity. It is also, of course, totally at odds with present world policy, which is for globalization, industrialization with minimal employment, standardization and homogenization. I bet that the only people who really prefer the world to which this is leading are those who benefit directly from it: the politicians who consolidate their power; the big industrial companies who supply the required technologies; the scientists who work for them; the big farmers who remain; all the so-called 'experts' in fact.

One complication. For convenience, I have been talking in terms of *national* self-reliance in food. But of course, many now doubt whether nation-states as now conceived are always the ideal units. The straight lines on the map of Africa (drawn by the former colonial powers) cut through ancient tribes and alliances, and bring traditional enemies together. A country like modern Turkey seems to make no sense as a political unit. At least three separate countries are ruled after a fashion at various distances from Ankara. The island of Britain is at least three countries, but although the Welsh and the Scots demand greater autonomy neither seems inclined truly to go it alone. Some modern commentators have argued for the reintroduction of the city-state, like ancient Athens and Sparta and Renaissance Florence and Venice. Others would like to see the whole of Europe bound in one federation, like the United States.

All these issues arise for political and historical reasons, but agriculture raises another complication since, as argued in Chapter 2, some regions really are better suited to some crops or livestock than to others. Florida, for example, has a natural bias towards fruit, Texas towards cattle, the Loire Valley to grapes, and so on. Then again, it can make very good sense, for the sake of avoiding pest or disease epidemics, to introduce discontinuities deliberately. When all such considerations are taken into account, it may well be that current national boundaries do not necessarily define the ideal units of agricultural organization. Sometimes the ideal farming regions will be parts of nations, sometimes they would embrace several nations, and sometimes simply straddle national borders. So be it. Whatever the regions, they would still be designed on lines determined by considerations of biology and human values (culture, aesthetics), and individual societies (whether or not they are nations) could still pursue the general principles of autonomy and self-reliance. In conception and in practice, such a structure would be worlds away from the present reality and the future vision of a globalized market and a globalized farm, driven by nothing but the ruthless principles of commerce.

Enlightened agriculture, in short, is common-sense agriculture: rooted in good husbandry; traditional in structure, yet making all the use it chooses to of the very best science and (where appropriate) the highest technology; guided by biological reality (ecology, physiology) and by the human values of kindness, autonomy and justice. It uses the mechanisms of markets and money and accountancy because these are well established and convenient. But these financial devices, like science and technology, should be seen only as tools: players in the greater need to serve humanity and our fellow creatures well, within the bounds of physical possibility. Although commonsensical, such a vision is totally at odds with modern policies, in structure, in intent, and on virtually all points of detail. To be sure, it is the job of science and perhaps of all scholarship to improve on common sense. Yet we abandon common sense at our peril; and in this instance, which surely is the most important that can be imagined, the abandonment has been altogether too precipitate.

12

From Where We Are to Where We Want to Be

We (humanity) are in a strange position. We are perfectly capable of feeding ourselves well – to the highest standards of gastronomy and nutrition – and of going on doing so for ever. We have the resources to do this – although the contrast between the plight of people in some of the world's poorest countries, and the fertility and extent of their land, not to mention their oil, is in many cases staggering. We have the know-how – rooted in craft, but abetted by science. Yet in order to unleash and realize our own potential we will have to think radically in all areas: politics, economics, moral philosophy, aesthetics, religion, science. It is the case, too, that the political and commercial forces that now dominate the world, and indeed are increasing their hold, are on virtually all fronts at odds with the philosophy and strategy that are now needed. Biology and human values must have priority, and economics must be adjusted to them – if, that is, we want a world that's sane, tolerable and safe. At present, farming itself and all that goes with it (the biology, ways of life and human values) are thrust into an economic mould – the MICG model of capitalism – that might or might not be suitable in other contexts but certainly is not appropriate to agriculture. Those who defend the economic status quo are wont to insist that no other way is 'realistic'. In truth, it is the defenders who are living in a fantasy world, and on borrowed time. The world as a whole needs to call their bluff.

The following is a shopping list of the things we need to address and (with luck) to get right.

The Big Ideas

The most fundamental idea is as outlined in Chapter 1: that in addition to all our aspirations, pretensions, and manifestations as philosophers, artists, scientists, warriors, what you will, we are also animals: evolved creatures; children of biology. By the same token, this world is not just an address, but a habitat: the collective nest. Biology should be a leitmotif that runs through all political thinking. To be sure, the ideas of biology are gaining a political foothold through the world's Green parties, and strength to their arms. Yet this is not enough. The Green parties fight primarily on a specifically environmentalist ticket. Green issues in practice are perceived only to be matters of pollution and of wildlife conservation. If we really take our lives seriously, and those of our descendants, then the habit of thinking biologically must pervade all political decisions. In particular we need to think in terms of biological time: not just of the next decade or even the next half-century, but of the next few thousand years. A thousand years must be seen as a proper unit of political time. Agriculture in particular must be conceived as a biological pursuit, because that, emphatically, is what it is. It is extremely dangerous to perceive the world's agriculture simply as a tool of political and economic whims, and the world's farms as a theatre for acting them out. The fabric of the world and the creatures that live in it are not that resilient or forgiving. The political vision that now prevails – MICG capitalism – fits just as awkwardly upon the natural world as the politics of Stalin did upon the people and landscape of Russia, and in the long run is likely to cause at least as much misery and physical destruction. Neither Stalin's collectivism nor the industrialized and corporatized global market are geared to biological reality, or to what is socially desirable.

Then there is the grand, prevailing fact of capitalism. It's a mistake to suppose that capitalism is intrinsically bad, but the form that now prevails, an essentially monetarist version that might be called hyper-capitalism, is extremely damaging. Quite simply, it has had the morality leached out of it. The perversion of capitalism is reflected in the rise of the corporations. The US was founded in capitalism and freedom of trade but its late-eighteenth-century founders already recognized the dangers of corporate freedom. The early American constitution contained laws specifically to restrain the power of big companies. They were not supposed to exceed the boundaries of their own state. Their charters – their right to exist – were granted only for a limited time, and only within their own particular state. Each corporation had a mission statement (as it would now be called) that specifically stated, in paragraph 1, that the company existed to serve the needs of the state. The company directors were not allowed to hold shares in their own company. The companies were absolutely forbidden to put funds into political parties. All these restrictions lasted until the end of the nineteenth century when, at a stroke, they were rescinded – which was done in very dubious circumstances, as Thom Hartmann records in his excellent *Unequal Protection: The rise of corporate dominance and theft of human rights* (Rodale Press, 2002). The sophistication of those eighteenth- and nineteenth-century capitalist politicians was remarkable, and from a modern perspective is incredible. Today's American political parties, Democrat as well as Republican, cannot function without corporate support. As Thom Hartmann indicates, it is game, set and match to the corporations.

Of course, the advocates of modern capitalism acknowledge that outright corruption is a serious flaw. Yet, as with the modern food supply chain in particular, the chief criticism of today's capitalism in general is not corruption per se. It is that the corruption is institutionalized. The market is too complex and too arcane to be adequately policed; and increasingly, as with Enron, the accountants who are meant to do the policing may in effect be employees of the company itself. More generally, modern traders typically feel that the only practical way to operate is in the manner of professional footballers: play to the referee's whistle. Whatever the referee doesn't spot is fair game. Indeed when the occasional tycoon is caught with his hands in the till people at large are wont to comment 'Good luck to them!' Of course, too, when the trader owns the referee, the whistle can be mighty slow a-blowing.

In much of Western agriculture the Enron principle prevails on the grand scale, and with legitimate status. Thus US farmers are among the world's most vigorous defenders of free enterprise, yet they happily accept annual subsidies that far exceed the income of many small countries. They seem to see no contradiction in this. Why? Because they regard the Government itself as a part of their own enterprise. The Government is effectively on the board. At the same time, of course, the US is urging poor countries to compete on the world market, even though those countries are poor to begin with and typically have no subsidies to help them compete overseas. Again, apparently, the US sees no contradiction.

The market for agricultural produce is supposed to become global, and to supply this market each region, and each country, will specialize in the commodities it grows best. Each region will be increasingly monocultural and so the global market will be served by what, in effect, will be a global farm; and the whole will be controlled by a few corporations, to whom we will all be beholden willy-nilly. This is the current vision, or at least the logical end-point of the current vision. Accountants are on hand to demonstrate that the whole process is maximally efficient. Few seem to realize that the accountancy is tautological: a circularity. The accountancy demonstrates the cash efficiency of the globalized market and farm because that is what it has been set up to do. If some hypothetical, benign, dispassionate Martian did the accountancy, and added in the cost of human misery and ecological degradation caused by modern farming strategy, and compared the global farm with what could be achieved through a strategy of enlightenment, the reckoning would be very different.

Notably, we are asked to believe that poor countries can lift themselves out of poverty by selling food to the rich. Many countries are already well on the way. To take just one among scores of possible examples, Senegal now devotes half its agricultural land to peanuts for Western margarine, at the expense of its home production (*The Ecologist*, November 2002, p. 41). Yet the rich countries which are supposed to provide the market for all this produce, and more to come, in truth have very little spare buying capacity. Britain, for example, could already meet its food requirements half a dozen times over. Established agricultural countries the world over are queuing up to supply us and some (like New Zealand) are already very annoyed that we no longer buy as much from them as we once did. How much more does Britain really intend to spend on food? Can it really buy in more than it already obtains from New Zealand, Australia, Canada, South Africa, the Caribbean, California, Florida, France, Spain, Italy, Holland, Greece, Hungary, Denmark, China, not to mention its own beleaguered farmers? Can we really spend enough on the produce of Africa to make a significant difference to that continent's economy? Where are the figures that demonstrate this? I have looked for them, and cannot find them. I don't believe they exist. I have commented in various contexts in this book that farming is currently run on a wing and a prayer. This seems to apply to the whole vast drive towards globalization. Even the most basic arithmetic that should be in place to show how globalized agriculture might work does not seem to have been done. A few fat cats cashing in on the cheap labour of Angola and

Rwanda, a handful of happy labourers forcing a smile from the glossy brochures, are not demonstration enough.

All in all, the net effect of the corporatization of agriculture is to make some people very rich indeed, while most of the people who previously were making an acceptable living from it (including the majority of the farmers) are thrown out of work (at best to be re-employed as labourers). It's a zero-sum game: wealth transferred from the many to the few. The proper word for such a process, as John Ruskin the 'high Tory' (as he described himself) and Karl Marx both pointed out, is theft. I would not want to go as far as their contemporary Pierre-Joseph Proudhon who declared that all property is theft ('*La propriété c'est le vol*') because the world seems to run more smoothly if people do own things and respect what others own. But when a few people expropriate the property of many then that is theft by any judgement. Yet the present mass transfer of wealth and livelihood is not only legal but is sanctified. Indeed it is called progress.

The general point, most simply put, is that in a finite world we can envisage a future in which a few people are rich and a great many are very poor; and a future in which everyone is at least well fed, which is the basis of health and general well-being. But we cannot envisage a future in which everybody is rich; and the more we encourage the pursuit of unlimited personal wealth, the more we condemn the majority to be poor. The present 'war on poverty' is a huge paradox. In effect, it amounts primarily to hand-outs from rich to poor; yet the way of life and the economy that created the rich societies that can afford such apparent beneficence does much to create the poverty, or at least to exacerbate its ill effects. This is simply because the world is finite. The point is not even one of biology, but merely one of physics. Yet the driving premiss of the prevailing world economy is that wealth is good and extreme wealth is better and everyone should strive for it and those who achieve it should be admired and that 'charity', the virtue that St Paul singled out as the greatest, means hand-outs *de haut en bas*. Here, in practical terms, is the most fundamental nonsense of all.

In truth, no society can run on an economic algorithm, whether Marx's formulaic version of socialism or modern monetarism, which aspires to run the whole world on market lines. All societies need economies (of course), but behind societies that are capable even of stability, let alone of justice, economics must be secondary to underlying morality, and to the human values that underpin that morality. It's morality that holds societies together.

Matters of Morality

So what is this 'morality', and what are the worthwhile human values? Or as the Cambridge literary critic F. R. Leavis put the matter: 'What for – what ultimately for? What do men live by – the questions work and tell of what I can only call a religious depth of thought and feeling.' (He wrote at a time, and in a university, where 'men' meant humankind.) It's not my place to sermonize but it's clear that this is precisely the question that the modern world has written out of the act, and needs above all to address. In truth there are two issues: first, what indeed do we 'live by'; and second, how can we ensure that the moral principles and values that we decide are worth holding dear, in fact come to the fore? After all, in the economic system that now dominates, that of the unfettered market, the notions that once were classed as 'morality' have been purged as a matter of policy. The prevailing value is the freedom of the market itself.

As a matter of history, all moral standpoints have been roughly divisible into two. The first, which can be called 'absolutist', maintains that there are absolute standards of rightness and wrongness to which we should all subscribe. The second, known as 'consequentialist', insists that the goodness or badness of any act (or thought) should be measured only by its outcome. The best-known example of consequentialism is 'utilitarianism', attributed mainly to the late-eighteenth-century English philosopher Jeremy Bentham, which is summarized as 'The greatest happiness for the greatest number'.

Absolutist standards are most easily maintained in societies that believe in an all-powerful God, or gods. Then, what is 'right' is, by definition, what God decrees. Holy texts, interpreted by dedicated priests, are taken as God's word. This straightforward approach runs into trouble, however, if people stop believing in God. If God is indeed the sole maker of rules, then with God out of the way, it seems to follow that anything goes. Different thinkers have responded in different ways to this unfortunate twist. Dostoyevsky, in mid-nineteenth-century Russia, never satisfactorily resolved the conflict: with God dead, people were free, for example, to commit murder. However, those of Dostoyevsky's characters who did commit murder (or seemed to condone it) were riddled with guilt and generally finished up seeking punishment and/or reinventing God. In eighteenth-century Germany, Immanuel Kant sought to identify fundamental principles of ethics, 'categorical imperatives', which, he suggested, even God ought to obey. That is: if God disobeyed the categorical moral rules, then He was wrong (which a religious fundamentalist would find a very strange idea indeed). Most moral philosophers agree that Kant ultimately failed in his quest to define absolute ethical principles, although he did provide some very useful insights, one of which was that no one should adopt any rule for themselves unless they thought that that rule should be applicable to all humankind. (We should not, for example, suppose that it is OK for us to seek riches but that other people should be content to be poor.) George Eliot was among many English nineteenth-century intellectuals who lost their faith in a literal God (and the widespread scepticism of Victorian times had very little to do with Darwin, as is commonly supposed), and she wrestled until the end of her life with the notion that there are absolute standards of right and wrong, even though God does not exist. The idea that there are such absolute standards, even in the absence of God, pervades English literature, as reflected in the literary criticism of F. R. Leavis and, for example, the novels of Iris Murdoch.

There can clearly be serious clashes between the absolutist and the utilitarian views of ethics. For example, a gang of hypothetical Nazis might derive enormous pleasure from beating up a gypsy. A dozen Nazis are thereby made happy, while one gypsy is miserable. The happy clearly outnumber the unhappy, but even the most dyed-in-the-wool utilitarian would surely concede that this does not justify the beating-up. Other factors are clearly important in deciding right and wrong, including the means by which happiness is achieved. There are acceptable ways of being happy, and unacceptable. But once we admit this, then we see how utilitarian positions, taken at their face value, fall short. Nazis, we feel, don't deserve to be happy, because they are innately evil. But by what standards do we judge their evilness? We might try to rescue simplistic utilitarianism by suggesting that more people are made unhappy by Nazis than are made happy. But in principle Nazis might outnumber their victims, so even this duck-out clause might not apply. In any case, it's clear that we cannot really distinguish right from wrong merely by counting heads.

In short, anyone who claims any morality at all must at some point acknowledge the existence, or at least the need to presume the existence, of absolute standards – even if they do not believe in a literal God who has laid down the rules. So where do we go from here?

For my part (and morality in the end is a very personal thing, so I feel free to state my personal position), I admire the moral philosophy of David Hume. He declared, in his *A Treatise of Human Nature* (1739–40) that 'Morals excite passions, and produce or prevent actions. Reason of itself is utterly impotent in this particular. The rules of morality, therefore, are not conclusions of our reason.' Then again: 'We speak not strictly and philosophically when we talk of the combat of passion and of reason. Reason is, and ought to be only the slave of the passions, and can never pretend to any other office than to serve and obey them.' Paraphrasing, I take this to mean that in the end, morals are a matter of feeling: of emotional response, and of attitude. We have to *feel* that a particular thing is right or wrong. If we don't feel it, then the concept of rightness or wrongness has no meaning. Contrariwise, all the argument in the world is pointless without underlying feeling. It is salutary too that moral philosophers may be right-wing or left-wing like the rest of us, and although either may argue their case at tremendous length, in the end they each finish up justifying their initial predisposition. The arguments themselves are not a complete waste of time – they do clarify positions, and can help people to change their minds – but the initial prejudice ultimately shines through.

I conclude from Hume (as I think Hume himself concluded) that moral philosophy is a very limited pursuit, not least because it trades in intellectual argument and does not directly engage with the underlying feelings. At least, if it does so engage, then it ceases to be 'philosophy'. If we really want to get to grips with morality, we need to go for the feelings themselves: to identify, cultivate and refine those emotional responses and the attitudes to life that are appropriate. If academic moral philosophers are not suited to this task, then who is? And what kinds of emotions and attitudes emerge?

In truth, whether or not we accept the existence of a literal God, the fact is that the people who have focused most clearly on the essence of morality – on the underlying feelings and attitudes – are the prophets. Typically, whatever God they believe in, or the nature of their spirituality, prophets seek to discover truth by revelation, and they strive to achieve this by measures that include isolation, prayer and meditation. Although sceptics are wont to emphasize the differences between religions (with arguments like, 'They all say different things, so they can't all be true') the most striking, not to say stunning, fact is that the morality of all the great prophets, of all the great religions, is so similar: Moses, Christ, Muhammad, the Buddha. In truth, the prophets of religion don't have a monopoly on this approach to morality. Some more secular, or at least only quasi-religious thinkers also emphasize that ethics must in the end be rooted in emotional response and attitude, and seek primarily to define what those attitudes should be. In this category we might mention Lao-tzu and indeed Aristotle. Both of these, however, are arguing in the manner of prophets even if they are not conventionally recognized as such; and both arrive at the same kinds of conclusions as the great religious prophets. The essence of what all of them said was summed up beautifully by the Hindu mystic Ramakrishna, which of course is appropriate because Hindus tend to be the most eclectic of all religious thinkers.

The essence of all morality, said Ramakrishna, can be summarized under three headings: personal humility; respect towards fellow, sentient creatures; and a sense of reverence for the universe as a whole. That's it. I am modifying what he said somewhat, since to my knowledge he spoke of respect for other people, which I would prefer to extend to animals (and as a Hindu, Ramakrishna surely would not object to this); and he spoke of reverence towards God, which I prefer in this non-theistic age to equate with the universe (or indeed with 'nature'). The point, either way, is that we are not in charge. The universe is not of our making. It deserves awe. We may prod at it here and there, but awe is the inspiration for all the probing, as all true scientists emphasize.

Such a simple morality, simply spelled out (Ramakrishna was a genius, for it requires genius to be that simple) seems to me to tell us all we need to know about effectively everything that has to do with human relations and politics as a whole and with the particular issues of this book, including agriculture and science. Just to take one small example from science: if we were personally humble, and had respect for fellow creatures, and any sense of awe in the face of nature, we would not for one second contemplate the kind of liberties that are now being taken with animals in the name of biotechnology, and we certainly would not be keeping them as cruelly as has now become conventional. Such excesses would simply be unthinkable. By the same token, it would be unthinkable to arrange an economic dogfight between the farmers of Africa and those of Europe and America and everywhere else and see who wins, and call that progress. Such a strategy would strike us as intrinsically obscene. If the economic system seemed to demand that we should do such things then we would, without hesitation, demand that the economic system should be rethought. That is the stage we are now at.

If there is no God, though, where does the revelation of the prophets come from? Who is doing the revealing? Of course I do not presume to provide the definitive answer. I would certainly not be so arrogant as to declare – as dogmatic, fundamentalist atheists do – that there really is no God to make the rules. I do have a penchant for biology, however, and for evolution in particular, without which, as Dobzhansky observed, biology does not make sense. I like the idea that our deep feelings, our attitudes, are at bottom evolved. Scientists and non-scientists alike have interpreted Darwin's idea of evolution far too crudely (although Darwin himself was not crude) and concluded that natural selection is bound to produce aggressive, competitive creatures, 'red in tooth and claw' (an expression that in truth derives from Alfred Tennyson's *In Memoriam*, dating from the 1830s, more than two decades before Darwin wrote *Origin of Species*).

But modern evolutionary theory, thanks largely to the late William Hamilton of the London School of Economics and Oxford, shows that human beings, like all animals, are capable of altruism and self-sacrifice. Observably we are social creatures; and we have, built into us, all the attributes of sociality, including innate kindness, generosity, and indeed respect. We need to get on with our fellow beings, and these attitudes are the prerequisites. As creatures that need to survive, too, we know intuitively not to take too many liberties with our environment, and know full well that the universe is intrinsically, in the end, beyond our ken. This is not, as some scientific zealots are wont to insist, 'superstition'; it is good plain sense that natural selection has provided us with. In short, humility, respect and a sense of reverence are part of human nature, at least as much as any of the qualities that are less obviously attractive. Unfortunately, the evolutionary thinking that leads to this optimistic view of ourselves has not become fashionable, largely because the notion that we do have in-built predilections (however favourable those predilections may be) is held (by some) to imply 'biological determinism'. In truth, the modern evolutionists are merely saying that there is, indeed, such a thing as 'human nature', which Hume clearly endorsed (in the title of his first major work), and which George Eliot (and many other writers) took to be self-evident. The fact that we all partake of human nature absolutely does not negate the idea that individually we also have free will.

In short, I am perfectly happy to take Ramakrishna's irreducibly simple distillation of morality as it stands. It needs no further support. It expresses what we all, deep down, know to be the case. The idea that this irreducibly simple but nonetheless powerful morality is actually built into us, through our own evolutionary past, I find very encouraging indeed. It is pleasing (it doesn't prove anything, but it is nonetheless pleasing) that evolutionary theory and the meditations of prophets should lead in the same moral direction.

In practice, prophets operate in the purlieus of religion (though a few who seem more secular deserve prophet status), so we might hope that the world's religious leaders, the priests and mullahs and rabbis and gurus, would provide the moral leadership which the world so obviously needs, and which the modern monetarist market so obviously lacks (and indeed has officiously abandoned). In practice, some representatives and leaders of the great religions are outstanding moralists. From fairly recent times, the names of Mahatma Gandhi, Archbishop Tutu and the Dalai Lama suggest themselves. In a more secular vein, the world has cause to be grateful for example to Nelson Mandela and Vaclav Havel; and we could of course extend the list. What all have in common is that they present morality, as Hume suggested, primarily as a matter of feeling: of emotional response and attitude. The sad fact is, though, that religious leaders as a whole do not come across in the modern world as the natural leaders of morality. Fundamentalists of all religious persuasions all too obviously are bent on conflict. Clerics summoned to take part in TV debates too often regale us with details of their own theology. Religion as a whole often presents itself badly, frequently appearing simply to be cruel. Morally speaking, then, despite the prominence of religion worldwide, the world as a whole remains effectively rudderless. The morality of the modern market has stepped into a void. For the world as a whole, this is a precarious state. The one great hope in the end is humanity itself. We are innately moral creatures. Common observation suggests that this is the case, for most people behave generously when not harassed or threatened, and would prefer to behave generously; and biological theory, which surely is not trivial, seems to justify such optimism.

If human nature really is as benign, deep down, as I am suggesting is really the case then the grand task facing humanity becomes conceptually straightforward – albeit immensely difficult in practice. First, we need to create conditions in which people feel they can afford to be nice. We prefer to be nice (I suggest) but often are anything but because we fear that other people may take advantage, and we don't want to be taken for a sucker (which can after all be lethal). We are also knocked off course by the sheer pressure of life – not the least of which is the constant exhortation, in the modern economy, to compete, which implies being active for more hours in a day than other people, and being prepared to do other people down. Society as now constituted is not designed to bring out our niceness, but to reward aggression. But life doesn't have to be like that.

Democracy, Common Sense and Science

The essence of democracy in the end (is it not?) is that societies should be governed according to what the people as a whole want. If it really is the case that human beings, taken individually, are fundamentally benign, then the creation of true democracy should, logically, create a benign society. Given that human beings are basically nice, that notion ought to be encouraging. All we really need to do to create a fine world, in which human beings can live well effectively for ever, is to make democracy work. The trouble is, that this is immensely difficult. The formal notion of 'democracy' is commonly ascribed to the ancient Greeks of the sixth and fifth centuries BC, but although it has taken many forms since then, few have been convincing. The versions that the United States, Australia and Britain now pride themselves on are in many ways crude, and historically are steeped in blood. Democracy (like socialism, or Christianity, or Islam, or indeed capitalism) is a grand idea but it is very difficult indeed to translate into acceptable and stable practice. But (like socialism and the rest) it is also too important to give up on.

Modern evolutionary theory again suggests fundamental reasons why democracy is so hard to get right. Bill Hamilton can properly be seen as one of the doyens of modern evolutionary theory, and the other is John Maynard Smith. In particular, Maynard Smith brought game theory to bear. Game theory was devised from the 1920s onwards primarily by the Hungarian-American mathematician John von Neumann, and has been much used by military and other strategists. The point is to regard every situation in which different groups interact – competing or cooperating – as a 'game', and then seek to analyse, mathematically, what the optimum strategy of the 'players' should be if their intention is to win or at least to arrive at some stable compromise.

One of Maynard Smith's most basic ideas is that we can think of any one society as composed of hawks and doves. Hawks are aggressive, and so will dominate. Doves are passive and never retaliate. An all-dove society is not stable, because it is always liable to be invaded by a hawk, and once the hawk invades, it will take over. Having taken over it will reproduce and hence multiply. However, hawks flourish only insofar as their aggressiveness remains unchallenged. It is fine and dandy for a hawk to tell a dove to get off its favourite bar-stool, because doves immediately give way. But if the incumbent of the bar-stool of whom this request is made is also a hawk, then a bloody battle ensues. As hawks multiply within any one society, so each of them is more and more likely to encounter other hawks. It follows, then, that the number of hawks is self-limiting. If there are too many, they spend their days punching the lights out of each other. The doves always get pushed around, but numerically they are bound to outnumber the hawks. Hawkish behaviour simply does not pay if the hawks are in the majority.

So we finish up with an interesting paradox. In one sense the doves are more successful, biologically, since they are numerically superior. On the other hand, they are the ones that are shoved around, and seem constantly to be drawing the short straw.

Although this is a very simple example of game theory in action, it seems to me to describe the state of human societies with uncanny accuracy. Most of the people in any one society are nice, partly because most people are nice by nature but partly because game theory ensures that niceness will lead to numerical superiority. But the nice majority will inevitably be ruled by a tough, hawkish minority. In other words, paradoxically, we almost invariably find that nice people are ruled by nasty people. That, in a nutshell, is what's wrong with the world's politics.

We need to find a way of ensuring that the niceness of the majority prevails. Since they are the majority, this should not on the face of things be difficult. In practice, however, we come up against another catch-22. Nice people do not seek office. Part of their niceness is to be self-effacing. Occasionally nice people do get thrust into positions of power ('have greatness thrust upon them' as Malvolio put it), but only under rare and special circumstances. Occasionally we find societies with people at their head who are truly admirable, like Nelson Mandela. Much more usually the people who get to be in charge are hawks. It is also the case, of course, that human beings are not robots. Most of us are neither out-and-out incontrovertible hawks, nor dyed-in-the-wool irredeemable doves. Most people are able to behave either hawkishly or dovishly, depending on circumstance.

If doves are promoted to leadership, they commonly metamorphose into hawks. As the late-nineteenth-century historian and politician Lord Acton commented, 'Power tends to corrupt and absolute power corrupts absolutely . . . Great men are almost always bad men'. Thus it is that extremely nice people are so often ruled by conspicuously nasty ones. Milosevic in Yugoslavia and Saddam Hussein in Iraq are two obvious examples from recent times. It can be very disappointing. The most vicious leaders sometimes begin well.

There is an intriguing twist in this line of thought. The ultimate exercise in hawkishness is indeed to be the leader: president, dictator, tycoon. But this requires a certain measure of organizational skill. The easier way to be a hawk is simply to let people do their own thing – run their farms and grocer's and carpenter's shops – but then offer to bash the lights out of anyone who does not pay them a kickback, or a tithe. In other words, hawks can thrive most straightforwardly as bandits.

Here, however, hawks come up against other hawks. There's nothing worse for bandits than to descend upon some village and find that other bandits have already stripped it bare. So now, an extra layer of complexity is introduced. No longer do the bandits simply demand payment, on pain of death or serious damage. Instead, they offer a deal: the farmers and the shopkeepers pay the bandits to perform a specific task, which is to keep other bandits at bay. Hence of course the old-style Mafia which (before they focused on drugs) grew rich by offering 'protection'. The protection they offered was against people like themselves.

In this way, banditry becomes respectable. After all, it's worth paying somebody half your income if, in return, they prevent somebody else, even nastier than themselves, from taking all of it. When bandits achieve respectability, they no longer call themselves bandits. They become professional warriors, and are looked up to because they are necessary and are good at what they do, as well as being extremely threatening. Hence, for example, bandits become Samurai. Or indeed, in Europe, they became barons, who protected their own bands of peasants against other barons who, if provoked, would prove even more vicious (or so at least the peasants were given to believe).

Life being as it is (hawks do exist), the old-style Mafia offered value for money. Societies where the old-style Mafia rules still tend to abound with small shops and farms, and this may in large part be because the corporations have been kept at bay. The corporations and the governments who support corporations may suggest that those countries are 'backward', but the people who are able to stay in business presumably would not agree. Those who pay the Mafia are indeed protected. It's in respectable societies, devoid of visible godfathers, where the small farmers and traders have been swept aside. It isn't obvious to me, though, that the corporations that do the sweeping are different, morally or economically, from the Mafia; except that the Mafia does, or did, offer a palpable service to the people at large while it isn't obvious (at least in the context of food production) that the modern corporations justify their presence at all. Yet people who work for the modern corporations (having lost their jobs as independent traders) feel beholden to them. This is the ultimate con-trick, as Bertolt Brecht commented: the poor don't realize that the rich need them far more than they need the rich.

Of course societies need organizers, and organizers do tend to become leaders, if only because the rest of us, doves that we are, find it convenient up to a point to be told what to do. Still, it is not so easy as might be assumed to distinguish between the leader-qua-organizer, who is paid and necessary, and the predator. This was acknowledged by the laws of nineteenth-century America that limited the power of corporations and ensured that those who organized those corporations remained merely their employees. Now the distinction between organizer and predator has in effect been lost, not simply by neglect but by statute. We could, in short, do much to restore democracy simply by reinstating laws that operated until just a few decades ago – which for the most part are still on the statute books; laws that distinguish expressly between the leader-qua-organizer and the leader-qua-predator.

True democracy would have practical advantages, apart from the abstract advantage of justice. It would reintroduce common sense: folk wisdom; the kinds of things and ways of thinking that we all take self-evidently to be correct. Common sense, contrary to much modern lore, has enormous advantages, and much of what is now wrong with agriculture, and hence with the world as a whole, derives from our peremptory abandonment of it.

To be sure, scientists and specialists of all kinds claim that their role in life is to improve on common sense. Indeed, science does do this. This is precisely its role. Common sense says the world is flat (it looks flat, and it feels flat) but science has shown beyond reasonable doubt that it is not, with pictures from space delivering the *coup de grâce* to the few remaining flat-earthers.

But life is not lived on the basis of isolated facts. If we want to make decisions that have to do with survival (and justice) then we have, at any one time, to keep many different balls in the air. Thus, as discussed in Chapter 8, in assessing even simple matters of risk, we all of us consider many different angles. The mathematical chance of being killed in any one venture is only part of the issue. It matters just as much what we think we will gain from the exercise (pleasure, thrills), and who is in charge. We risk hang-gliding because it is fun, and because it's our choice. But we don't want to eat additives or GMOs (or at least some people don't) because we don't want other people to take risks on our behalf unless we give them a specific mandate to do so (as we do, for example, to airline pilots).

Experts, though, have been defined as people who know more and more about less and less, and they have the unfortunate habit of demonstrating that this is indeed the case. Thus, on the specific matter of risk in the context of food, scientists these past three decades in particular have been assuring us ('the public') that everything they do is perfectly safe, and for our good. Very often this has turned out to be wrong. The recent creation of BSE through sheer bad husbandry was grotesquely wrong. The broad point is that in this, as in so many contexts, narrow expertise was allowed or indeed encouraged to override common sense. Traditional farmers, relying on common sense, simply would not have fed cow flesh to cows. Traditional farmers, relying on common sense, do not plant novel crops for which there is no obvious agronomic need. It's only experts, driven by their own narrow perceptions, who insist that common sense must routinely be overridden. Common sense can indeed be improved upon, and must be improved upon. But we abandon it at our peril. A world run on truly democratic lines would surely be more commonsensical; and a commonsensical world would be far safer. The corollary is that the current mistrust of experts, particularly in the whole field of food production, is absolutely justified. Common sense has been abandoned to the point of nonsense.

The fate of science in the modern world is tragic, both for the world as a whole and for science itself. Its role is indeed to improve on common sense; but it is deployed, instead, as the antithesis of sense. Increasingly, research is promoted only if it seems likely to generate immediate wealth. Even the most obvious and pressing social problems are left untackled if there is no pot of gold at the end of the rainbow (and usually, these days, the rainbow has to be very short). Science could indeed help to create truly enlightened agriculture, designed expressly to provide good food without wrecking everything else. Indeed, it is a vital player. But at the moment it promotes those very practices that push enlightened agriculture further and further into the margins: literally, as the most fertile lands are given over to the industrialized, corporatized, global farm. In this guise, tragically, science has become the enemy of humankind and of human values. That really is a betrayal: of science itself, and of humanity.

Again, the problems now are worse than in the past. Again, fundamental principles that were obvious to earlier generations have been abandoned in the cause of hyper-capitalism. Thus, until well into the 1970s, it was widely accepted that science was indeed for the common good of humankind, and not for the narrow profits of a few. Governments paid for most of it, and were proud and pleased to do so. One of the most valuable biotechnological creations of the 1970s was that of monoclonal antibodies, announced in 1975 by the Argentinian-British molecular biologist Cesar Milstein and the German immunologist George Kohler (for which they received the Nobel Prize in 1984). Although monoclonal antibodies clearly had value in all branches of biotechnology, from pure research to medical diagnosis, Milstein refused to take out a patent precisely because he felt an invention of such universal value should be available to all humanity. Even in our own time, Britain's Sir John Sulston has insisted that his laboratory's description of the human genome should be freely published. Sir Kenneth Blaxter, as director of the prestigious Rowett Research Institute, Aberdeen, wrote in 1977: 'It seems wrong that . . . the science related to producing food has to be used in a competitive fashion: the essence of science is its universality, and freedom from hunger should be the birthright of all mankind.'* But in the present climate, nothing is published without patent, and intellectual property rights have become one of agriculture's hottest topics, a cornucopia for lawyers. Instances abound of modern biotech companies seeking not only to protect their own creations, but also to claim those of traditional farmers who (innocent souls that they were) saw no need for any patent. Most grotesquely, an American company in recent years has sought to patent the name 'basmati', India's (rightly) most

* 'The Options for British Farming' in *Agricultural Efficiency*, The Royal Society, London, 1977.

famous strain of rice. Patents have their place but the modern freneticism, like the hyper-capitalism of which it is a part, is grotesque. The root point, again, is that morality has been leached out. Those who apparently seek to demand payment from traditional farmers for materials that those farmers developed themselves over many generations seem simply to have lost sight of what morality means. Banditry and theft do not become less reprehensible simply because they have become legal.

More broadly still, we surely need to reinstate, or properly to appreciate, the philosophy of Ivan Illich. His phrase, 'tools for conviviality', hit the nail on the head. Some forms of technology and science, adroitly deployed, are indeed liberating, increasing the autonomy both of individuals and of entire societies. Some, in the end, restrict our options, and give power to minorities. We need to distinguish more keenly between the two, and to emphasize the former.

This, then, is the grand task. Agriculture at its best is already brilliant, a wonderful blend of traditional practice, evolved and refined over 40,000 years, and of science, imposed in modern form only in the past two centuries, and truly significant only in the past seventy years. But its brilliance cannot be realized until we rest it on sound foundations; and to do that, we need to rescue and resuscitate the notions of morality and democracy, and to look again at the particularities of capitalism, religion and science. All this, of course, is a very tall order. Yet there are movements in the right direction; and if the different currents could be brought together, and given some coherence, then perhaps, just perhaps, humanity may yet be in with a fighting chance.

Straws in the Wind

'Straws in the wind' perhaps is not the right expression, for some of those straws (to extend the metaphor) are more like forest giants, crashing along mighty rivers; and some of them are more like the wind or the river itself – entire philosophies, changes of mood or of *Zeitgeist*. But 'straws' will do. For the various philosophies and practical manoeuvrings that can be seen to be moving in the right direction are failing to make the impact they should, and to reverse the tide of industrial, corporate globalization, because they do not cohere. Between them, the various trends and movements are powerful: virtually all of what needs to be done is already being done, here and there, in one form or another.

What's lacking is coherence. The different movements remain as straws because they lack an underlying, coherent philosophy. They need to be marshalled.

One powerful straw is or are the many signs that people at large, in all walks of life, in all countries, at all levels of income, are disaffected with modern hyper-capitalism and all that goes with it. The doubters are not necessarily revolutionaries, although they are often treated as such. Those who actively protest are too easily classed as hooligans, and hooligans are too easily lumped with terrorists, in the suppression of whom modern governments have given themselves carte blanche. More general, though, and perhaps in the end more powerful, is the idea promoted by Derrick Bell in *Ethical Ambition* (Bloomsbury, London, 2002), that big business really ought to be ethical, and (as Machiavelli noted in a slightly different context 400 years ago) for the most part would work better if it was. In the same vein, only 30 per cent turned out to vote in the 2002 US midterm elections; and it's been suggested that the 70 per cent who did not include a great many who, by staying at home, are registering positive disgust. Their attitude to both the warring parties is that of Mercutio towards the Montagues and Capulets: 'A plague on both your houses!' Both American parties, after all, have come to depend entirely on the support of corporations who (many perceive) are much of the problem. But those who do take the Mercutio view of modern US politics have the Constitution on their side, at least as spelled out originally by Thomas Jefferson and others: the notion that capitalism is grand, but must be contained within an even grander moral framework. It must not be allowed to break out of that framework and redefine the rules for itself. Of course, those who give voice to the ideals of Thomas Jefferson in the modern US are liable to be called 'communists'. This is a nice irony, or perhaps is a nasty one; but the irony nonetheless is on the side of the protesters. All America really needs to do (one feels) is return to the principles that it has so often so vigorously defended.

On the practical front, too, there are signs that democracy might at last be made to work. Many take comfort in particular from the Internet, which enables individuals to communicate with individuals, and groups with groups, immediately, in depth, and across the world. I commented in Chapter 1 that human beings have dominated the earth not simply because, as individuals, we are brighter than individual animals of other species, but mainly because we have learned to pool our thoughts, through the medium of verbal language. Each of us, however inadequate we may be individually (and we are all inadequate in some respects, even those we think of as geniuses), can partake of the intelligence and thoughts of the whole world. Democracy can be seen as the kind of politics that comes about when humanity pools its thoughts. But the pooling of thoughts is in practice difficult. In early medieval Europe even the greatest and best-connected of scholars were more or less reduced to wandering about in the hope that in some hermitage or monastery, somewhere, they might meet a kindred spirit. Letters are slow, go astray, and are intercepted. But the Internet joins us all up, like a kind of meta-nervous system, providing the connections between all our individual nervous systems. We could say, albeit in high-fallutin' vein, that the net is enabling humanity at last to realize our true evolutionary potential: our ability not simply to think, but to think collectively. More to the point, it could at last enable humanity to make democracy, in its broadest sense, a practical proposition. Since we have cause to believe in the fundamental goodness and good sense of human beings (or so I maintain), this must be seen as a huge step in the right direction.

There are many practical movements, too. As discussed in Chapter 10, the organic and vegetarian movements contain a huge amount of value, even though they do not encompass all the required principles of enlightened agriculture. Many individuals and flourishing societies, too, are specifically concerned with the welfare of animals, including (in Britain) the Universities Federation for Animal Welfare (which supports research into animal welfare, including basic studies of physiology and psychology) and Compassion in World Farming. Even the law may be catching up. Europe as a whole continues to tighten its restrictions on factory farming. Some American states too, including Florida, are tightening up. There are some fine movements even within governments, and in international agencies too – such as the Sustainable Agriculture and Rural Development Initiative (SARD) within the UN Food and Agriculture Organization, based in Rome.

Paradoxically, or perhaps not so paradoxically, the greatest political hope in the immediate term may lie in consumerism. The corporations claim that they have achieved their power only with the consent of consumers, and in the end this is true. It is not true that consumers have always consciously 'demanded' what the companies provide, as the companies like to claim. To a huge extent (as discussed briefly in earlier chapters) companies contrive to manipulate both the producers and the consumers, squeezing the former and leaving the latter with little or no choice, and generally pushing the whole food supply chain in the directions that can generate most profit. It is true though, at least in principle, that if consumers ignore their blandishments, if they simply don't buy what the corporations offer, then even the biggest must wither on the vine. Less dramatically, if we elect to buy some things, and not others, then in principle we can mould the corporations' efforts; and if the corporations find that their re-moulded efforts are no longer profitable, that's too bad. Indeed, the whole, modern, neo-monetarist, hyper-capitalist economy rests, in the end, entirely on the complicity of consumers, even though it does not reflect their deep desires as accurately as its advocates insist.

As citizens we may be largely disenfranchised, as voters and potential taxpayers we have the option only of voting for one party from a very short shortlist every few years and *faute de mieux*, but as consumers in a capitalist world we still have power. We can use that power to call the bluff of the big companies, and the governments that so complaisantly support them with our taxes. If only we, as consumers, identify food that really is good ('the future belongs to the gourmet') and enterprises that really are worthwhile, and are prepared to pay for them, and leave the rest to wither away, then we could indeed change the world. In passing we would expose one of the central lies (or at least misconceptions) of the modern food industry: the idea that the world needs farming to be industrialized, and corporatized, and globalized, and needs the high tech (including most of modern genetic engineering) that makes all this consolidation possible. In truth the world needs farming with traditional structure, abetted but not transformed by science, and consumers who know what's really worth paying for probably offer the greatest hope in bringing such a strategy into being.

Of course, we cannot support small and benign producers and traders unless they exist; and the good news here is that even in the rich world, despite the best efforts of governments and corporations to see them off, they still do. I know several companies that distribute 'real' meat: not necessarily organic, but raised nonetheless to the highest standards of husbandry. The breeds for the most part are traditional, bred to do well on traditional pastures, and to taste good: Gloucester Old Spot, Berkshire, and Tamworth pigs; Longhorn, Aberdeen Angus, and Dexter cattle; Norfolk Bronze turkeys; and so on. Small farmers in general are fighting back, too. In mainland Europe they have always been a powerful force and they have formed formal societies both in Britain and the

US, the strongholds of corporatized, industrialized agriculture. Small shopkeepers are also becoming a political force in Britain. Everyone needs to gang together – co-operatives, but not corporations.

Traditional markets still exist, too, like the open and covered markets in hundreds of towns and villages throughout Europe, including Britain, which of course are standard in the rest of the world. Both in Britain and the US these traditional markets are now supplemented more and more by farmers' markets, in which local producers contrive to sell directly to their customers. In truth, we should not be deceived by this. Sometimes small farmers are able to take time off to become retailers largely because their own farms are languishing, and they have time on their hands. In addition, a prime task for the future – perhaps *the* prime task – is to ensure that farmers once more are able to make a living simply by producing good food. They should be able to leave the processing and the sale to the butchers, bakers and grocers. As Mark Overton noted in *Agricultural Revolution in England* (Cambridge University Press, Cambridge, 1996), the trades that abetted farmers in seventeenth-century East Anglia included 'brewer, butcher, carrier, chandler, grocer, innholder, mercer, miller, oatmeal maker, wool chapman'. By 'diversifying', as Britain's farmers are now encouraged to do, farmers are simply nudging themselves along the supply chain, and taking other people's jobs. That is certainly not the ideal. But it is still a good thing to restore the contact between consumer and producer, as the farmers' markets do.

These, then, are some of the forces that are opposing the overall, destructive trend towards the industrialization, corporatization and globalization of farming: the antithesis of the MICG model. Most of these forces are traditional: notions, practices and movements that have been around for a very long time, and have survived the modern political and commercial assault. Some are new, including the concerted effort of some scientists to improve animal welfare. Because these movements often run against what has become the commercial and political mainstream, they are often opposed from on high. Small American farmers, for example, may find it hard to get their produce into the stores. Those who argue against the excesses of modern science and on behalf of traditional craft are often derided, and derision can be hard to fight off. Small businesses, including many of Britain's abattoirs, find themselves closed down for reasons that seem to have nothing very much to do with anything, except to protect the interests of the big ones. But those who are disaffected with the status quo and can offer alternative ways forward just have to keep plugging away. It's up to individuals, of course, to decide how far they are prepared to go in defence of their ideals and ways of life: bloody-mindedness, civil disobedience, active protest. Mahatma Gandhi recommended a judicious mixture of passive resistance and reconstruction, proceeding side by side – conceived locally but repeated a million times. Certainly we should not be giving our elected leaders and their attendant experts an easy time. The notion that those in charge know what's best for the rest of us is well and truly discredited. We may draw encouragement from the success of the French sheep farmer José Bové, who helped to pull down a McDonald's in his own small home town, has published a bestseller (*The World Is Not for Sale – And Nor Am I!*), and has become a national hero.

The French, after all, have a culture of food in a way that the English don't. The British have plenty of excellent food, but for some reason we don't take proper pride in it, or try hard enough to keep home-grown recipes alive. All French people are traditionally gourmets. Many French politicians see the paradox and absurdity of their present position: that as modern statesmen and women, locked into the European Union and the grand cause of global capitalism, they are obliged to pursue measures that they know strike at the heart of their own (primarily rural) way of life, and threaten to destroy the gastronomy that has been one of their great gifts to civilization. If only those politicians had the courage of their convictions, and could bring themselves finally to acknowledge that human values are more important than corporate profits, movements such as José Bové's could be of global significance. Certainly we should be grateful for the existence of such charismatic figures, who actually care.

Of course, if consumerism is truly to be a force for good, then consumers must know what is worth paying for. At present (to judge from what's on sale in the supermarkets) people are prepared to pay highly for out-of-season fruit and vegetables even though they have been whisked from the other side of the world and hence (often) are several weeks old by the time they hit the stores. If we truly favoured freshness we would perforce be supporting local production, which in turn militates against globalization and monoculture. If we were prepared to pay for kindly raised livestock, factory farms would die even if they remained legal.

Consumers, though, also have to resist derision from on high. Many, for example, resist GM crops and are duly told off by politicians and scientists alike, who appear on television, wagging their fingers. There is nothing wrong with GM produce, we are told. Some object to GM rapeseed oil and yet, we are assured, oil from GM rape is chemically identical to oil from rapeseed bred conventionally. There is no reason to doubt that this is true. It is possible to think of reasons why genes that are added to a rape plant to protect against herbicides might affect the quality of the oil. But there is no reason to suppose that such added genes are bound to do so. Since GM rapeseed oil is identical to conventional rapeseed oil then, its advocates insist, anyone who is prepared to pay more for the conventional kind is obviously foolish. People who insist on conventional crops are not buying better quality. They are merely paying a premium for the way the crop is produced. In truth, they are buying history. That, to the modern advocate, is the height of absurdity. 'The public,' the cry goes up, 'needs educating!'

Yet in all contexts except that of food, we constantly buy history. We make a virtue of it. Furniture made in the workshop of, say, Thomas Sheraton would typically cost thousands of pounds. Reproductions that are good enough to fool even the experts do not approach the original in value. We value the original precisely because its history is important to us. We commonly apply the same principles to machinery (as in vintage cars), houses (the bijou cottage) clothes (the designer label) – indeed, everything. We are not guided simply by crude chemistry or even by functionality. Yet for food, the most important material commodity of all, the powers-that-be apparently believe that chemistry is all that matters, and are aghast if the rest of us take a different view.

Yet in practice the notion that it's worth paying for a food's history has already achieved widespread and formal status. People have always had a special affection for home-made cakes, and farmers at their celebratory dinners favour beef from the Scottish hills. The scent of the heather is there in spirit even if chemists could detect no actual molecules. 'Ethical trading' is already a powerful movement: many people are prepared to pay more for coffee or tea, say, when they know that the farmer who produced it has been properly rewarded. Besides, food produced by traditional means is often superior to the mass-produced kind, at least when we have learned to live without the tang of the tin can. Only sometimes (as in refined rapeseed oil) is the modern food chemically indistinguishable. But even then, when there is no detectable chemical difference at all, history still matters.

On the broader front, at least whimsically, there is a parallel to be drawn between the idea of enlightened agriculture and William Morris's Arts and Crafts Movement of the late nineteenth century (itself inspired by John Ruskin). Morris pointed out the intrinsic virtues of craft: that the thing created is an expression of human ingenuity, skill, and indeed passion, and as such has an intrinsic value that goods that are merely manufactured cannot have, however cleverly they are made. Morris is easily derided (not quite the original beard-and-sandals, but certainly one of the pioneers), but he draws on a deep well of humanity nonetheless. All art, including performance art, is valued for just this quality: that it reflects, and in the end is dependent upon, the individual who made it. Art and craft demonstrate what human beings are capable of. We take farmers and cooks for granted but that's because (I suggest) we have developed a very poor sense of values. These are the crafts that should be valued above all, not only because they are so self-evidently important, but because, when done well, they are brilliant: as brilliant in their way as the designs of Morris, or the carvings of Grinling Gibbons, or the masons who built Europe's great Gothic cathedrals, or the calligraphers who adorned the world's great mosques. Ordinary people can be good farmers and cooks, and we just don't value ordinary people enough. In truth (as the pioneer of artificial intelligence Marvin Minsky was wont to point out) the ordinary talents of ordinary human beings are far more extraordinary than the apparently outstanding talents of the few.

By invoking the memory of William Morris, or John Ruskin, or even Thomas Jefferson, I am bound to be subject to the criticism levelled at all who suggest that the strategy of the modern world is in any way unacceptable: that my objections and solutions are effete and nostalgic, and that I am an enemy of progress. Paradoxically, however, it's the out-and-out, global-industrial zealots who are out of date. It is commonly acknowledged that we now live in a 'post-industrial' society. Since we still use the things that industry provides (oil, steel, ships, cars), this means in the main that we have moved beyond the nineteenth-century factories with their tangible smoke and their visibly corrosive fogs, their noise and their immediate physical danger, and (to a large extent) into the clean, low-energy, apparent austerities of biotech and electronics. It means, too, that a higher proportion of people now derive their incomes from 'services', from legal advice to driving taxis.

Agriculture, though, seems to have fallen behind the broad trend. While the world as a whole is moving beyond nineteenth-century manufacture, agriculture is still trying frantically to move into it; still trying to re-enact the processes that took Britain, and then the rest of the world, into the Industrial Age at the end of the eighteenth century. Enlightened agriculture can be seen as post-industrial agriculture: not gratuitously nostalgic, as defenders of the status quo tend to insist, but ahead of the game; leap-frogging the 200 years of heavy machinery and pollution that the Industrial Revolution brought to manufacturing. Some wise person (I wish I could remember who) observed that we can envisage a post-industrial society, but we cannot envisage a post-agricultural society. Enlightened agriculture can properly be seen to be modern, for it belongs to the age of biology; while the corporate, highly mechanized, industrial-chemicalized, homogenized, monocultural kind of farming that we are still so frantically being urged to develop can properly be seen as yesterday's news, as crude in its way as the sulphurous hell-holes of the early nineteenth century that excited the wrath of social reformers like Marx's friend and collaborator Friedrich Engels. In short, enlightened agriculture emphatically is not retrospective. It should appeal above all to those with a penchant for progress, when progress is sensitively defined.

Enlightened agriculture has another, perhaps more rigorous, claim to modernity. For millennia, philosophers have drawn parallels between the organization of societies and that of living organisms. This is far from foolish, for an organism is a miracle of organization, with many billions of components combining to form the whole. Most philosophers have concluded that living bodies are hierarchical in structure: that there is some organizing spirit within each of us that runs the rest. Many have suggested that somewhere in our brain there is a 'homunculus', like Pinocchio's Jiminy Cricket, in ultimate control. Modern biology has reflected this general notion too. The nucleus, containing the DNA, is conventionally held to run the cell and hence to control the whole organism: the boss in his office commanding the minions.

Increasingly, however, biologists perceive that organisms are not run in this hierarchical fashion. The DNA is a big player, but it is in constant dialogue, from day one and throughout life, with the rest of the cell, which in turn interacts with all the other body cells, and hence with the environment at large. It is impossible to say, in this dialogue, who or what is actually 'in charge'. Clearly, too, the brain contains no 'homunculus'. In the nervous system as a whole we see no hierarchy. Instead we see the 'neural net'. Each bit of the nervous system communicates primarily with the bit next to it, and somehow the whole coheres. The neural net, in general form, is remarkably like the markets that Adam Smith envisaged. Organisms also contain long-range, overall systems of coordination – like the nervous system itself, and the endocrine (hormonal) system. But in the main they are held together, with wondrous efficiency, by a Smith-like invisible hand.

The natural economic structure of enlightened agriculture is that of the neural net. It is old-fashioned in the sense that it reflects the vision of Adam Smith. But it is also ultra-modern. By contrast, the hierarchical structure of the MICG model – everything run by a few corporations, as Egypt was run by the pharaohs – is grotesquely out of date. Corporations may or may not be useful but society as a whole certainly does not *need* them, any more than bodies need homunculi.

Where Do We Go From Here?

I suggest (presumptuously, but why not?) that the concept of enlightened agriculture really does mean something. It's what humanity has to build and develop as quickly as possible if we are to create a tolerable world for ourselves and our fellow creatures, and one that can go on growing and developing not simply for the next few decades but for many thousands of years to come. In the end, the fate of everything depends on how we farm. The idea of enlightened agriculture is rooted as deeply as it seems possible to root any human endeavour: in the physical facts of the universe; in biology; in the most refined form of moral philosophy (for such, I suggest, are the insights of Ramakrishna). Enlightened agriculture should, indeed, become the major player in all human affairs, the bedrock from which the rest of politics takes its lead. It is in agriculture, after all, that *Homo sapiens*, the inescapably flesh-and-blood creature, and Aristotle's conceptual 'political animal', need to become one.

There is a paradox here. Truly to achieve its ends, and to be true to its own spirit, enlightened agriculture should be a people's movement. It is intended above all to be democratic, and to foster a world whose guiding principle is '*vive la différence!*' – for diversity, of wildlife and of cultures, can be seen as the world's greatest asset. But if the movement is to be more than a vague idea it has to be able to survive and indeed compete in a world that is currently dominated by corporations and governments, and so long as those institutions continue to dominate it needs to be as well organized and as politically adroit as they are. Thus enlightened agriculture needs to become a discrete and identifiable entity – to be a proper noun with capital initials: 'Enlightened Agriculture'. To this end it needs in the first instance to establish a formal society, a clearing-house of ideas; and in the fullness of time this society should acquire the gravitas and clout of an agency – comparable with that of the United Nations Food and Agriculture Organization or the World Conservation Union (IUCN). Yet Enlightened Agriculture in practice should have the structure of the neural net; and any attempt to create a central body, however necessary this may seem for the purposes of survival, again raises the spectre of the hierarchy. But such is the dilemma of all democracy. If we take democracy seriously, we just have to be aware of the traps.

A prime practical question, then, at least for starters, is how to create the society, and then the agency, for Enlightened Agriculture. Clearly, to bring it about, radical change is needed. The present trend – monetarization, industrialization, corporatization and globalization (MICG), channelled through the World Trade Organization – is running in the diametrically opposite direction. The minds of people at the top, who have the power to make irreversible changes to other people's landscapes and ways of life before those people have a chance to react, need to be changed. Indeed we need nothing less than revolution, albeit of the bloodless kind which, invariably, proves the more enduring.

Mahatma Gandhi was the greatest of the twentieth-century revolutionaries, and if we are talking of hearts and minds and not simply of political boundaries, he was also the most enduring. To achieve change he recommended action both of a negative and positive kind. On the negative front he advocated, not violence, but non-cooperation. In the specific context of British India, when the British said 'Do this!', the followers of Gandhi did not fight back, but neither did they do as they were bidden. On the positive front, Gandhi urged people to do their own thing, whatever the authorities might be saying.

Gandhi's approach readily translates into the present context. Above all, we have power as consumers: not so much power as the processors and retailers pretend, as they muscle their way into our town centres, but a significant amount. Simply by refusing to buy what's on offer we could have an enormous impact.

At least as important, though, is to build the foundations of something different and better – and as is roughly outlined above, and everybody knows, there are a lot of good things going on. Small farmers and shopkeepers are fighting back, and so are processors of the traditional kind, truly wedded to good food. There are scores of movements in animal welfare, nutrition, gastronomy, conservation, fair trading and ethical investment, and although they are not all pulling exactly in the same direction, they make common cause. There are plenty of fine financial models, too: notably the world's many trusts, such as Britain's National Trust and Woodland Trust, or the great legacies of American philanthropists such as Rockefeller and Kellogg (which already contribute a great deal to world agriculture). In the first instance, clearly, there has to be more cohesion: farmers communicating between countries – Denmark to Kenya, Myanmar to Wisconsin; nutritionists talking to farmers and cooks, and so on and so on – and everyone talking to the world at large.

In all this, the Internet surely has a huge role to play. It is not the exclusive domain of the Western middle class: far from it. It is emerging truly as an appropriate and 'convivial' technology even in the remotest villages. A simple website entitled 'Enlightened Agriculture' could be enough to set the ball rolling. When it is rolling, then surely it will gather momentum. As the society grows into an agency it can establish centres of learning (or at least take root in those that exist already) and commission research in science, economics and sociology, even if it did not carry out such research itself – liberating science from the fealty it now owes to corporate commerce. It could spread the crafts of traditional farmers, which so often are wondrously subtle yet at present are so insouciantly swept aside, all around the world, to whoever can make use of them. It should also help to create the kind of alternative commercial framework that would make it possible for the multifarious enlightened players (farmers, processors, retailers, researchers, communicators) to make a living, until enlightenment becomes the norm, and such support is no longer needed.

This putative movement would not need the blessing of governments. It would merely need governments to stay out of the way (though this is something they seem to find hardest of all – in 'democracies' as much as in totalitarian states). Neither does the movement of Enlightened Agriculture need the blessing of corporations – indeed it offers a direct challenge to the corporations within the specific, but vast, area of food production. On the other hand, the growing movement should not be a priori exclusive. Politicians and industrialists who want to get on board and have something serious to offer should certainly be welcomed. Some people in high places will surely see that this is where the future lies, and will want to be part of it. I do not intend to start such a movement. It would need full-time energetic organizers almost from the outset, which doesn't sound like my job description at all. I am merely suggesting that the movement of Enlightened Agriculture, with a brief to put the world back on course, could and should come about; and there are plenty out there with the talent and the will to help make it happen. It's just a question of what people think is really worthwhile, and what they really want to do.

The stakes could hardly be higher. We are talking about the difference between a world that could endure effectively for ever, in peace and conviviality, and one that could be in dire straits within a few decades.

Afterword

I sent the hardback text of *So Shall We Reap* to the publishers in the Spring of 2003. Then, like everybody else, I watched with stupefaction as the war in Iraq unfolded and is still unfolding – a lesson in the extraordinariness and primitiveness of mainstream politics. I also got involved with more and more meetings and symposia on farming; and spent a couple of months in Brazil, Costa Rica and Panama, and then a month in various parts of India, looking at politics, agriculture and forestry. Some of the people I have talked to both at home and abroad have tried to persuade me that the central thesis of *Reap* is wrong. Many, including some whose hearts are in the right place, who really do want to do good, believe that despite appearances the world is now on the right lines. These enthusiasts truly believe that farming must be integrated with the world economy as a whole – as the economy now is; and that the central task for the world's governments and industrial companies is to increase disposable wealth – cash – and that this indeed is the *sine qua non*. They believe furthermore that the agrarian way of life is anachronistic, destructive and cruel and should be swept aside with all possible speed.

Some of those who feel that the world is on the right lines have their own vision of the future. They see it in terms of middle-class western ways: city, suburb, white-collar job, pension scheme, car. Food production, like everything else, should be in the hands of corporations, regulated only gently if at all by governments and the law, and run by *experts*, in science, medicine and management, who 'know what they are doing'. Others have no specific vision for the future, but feel in a general way that what is happening now is 'progress' – more and more technology, to replace hard labour, and more circulating cash – and that everything will work out for the best if only we give the free market its head. We cannot say *where* the world is heading: one corporate expert told me with some irritation that to have any 'vision' at all of how the future might or should turn out was simply 'childish'. But this particular brand of zealot believes that the free market can do only what people at large want it to do, so that whatever happens must reflect public desire, and the outcome whatever it may be must be democratic and by definition 'good' (since 'good', in this philosophy, is equated with the satisfaction of short-term desire, at least of those who can afford to pay).

Both kinds of enthusiast (and others who are mixtures of the two) argue vehemently that the visions presented in *Reap* are ridiculous – at best pie in the sky, at worst vicious and imperialistic. A spokesman from Britain's National Farmers' Union felt that the points raised in *Reap* were simply beneath his lofty attention, although others in comparably powerful positions have at least given it the time of day. 'Unrealistic' is a favourite pejorative. To advocate any deviation from the present course of corporatization, urbanization and globalization, is to be self-indulgent, other-worldly, gratuitously nostalgic, 'romantic', elitist. Deviation is also perceived to be defeatist, since it denies the possibility that significant numbers of people who are now subsistence farmers could in reality, in the foreseeable future, be transformed into salaried suburban commuters.

But I have also met a great many who agree with the thesis of *Reap*. Some agree with it absolutely. Others accept its sentiments but then shake their heads and suggest, as the out-and-out critics do, that present ways of doing things and the official policies that underlie those ways will be too difficult to change. Besides, these reluctant critics assert, the people who now are peasant farmers would really rather be something else. At the first opportunity the farmers and especially the farmers' children take off for the cities. The city may not in reality have much to offer, but it does offer a great deal more than the countryside as it has now become. In short, even if the powers-that-be did agree that we should take agrarian economies seriously, the people most directly involved in those economies would rather get out. To impose agrarian economy, therefore, would be high-handed, imperialistic, inhumane, and non-democratic.

Despite the critics, though, both the out-and-out opponents and the regretful head-shakers, I have become more and more convinced by world events, the various meetings I have attended and by my American, Asian, (and continuing British) travels, that if anything I didn't argue the point quite strongly enough in *Reap*. If I were starting the book again I would say the same things. In particular, and above all, I would stress the need to gear the world's food production to biological realities, working to the strengths of landscape, crops, and livestock, and gearing output to nutritional need and local cuisine.

But I would reinforce the political and economic arguments. The more I have travelled and talked (and watched world events unfold) the more it becomes obvious that present-day economic and hence political ambitions are ludicrous. At this point in history humanity as a whole could be flourishing, while in reality we are constantly on the point of war (indeed it suits the present US regime to pretend that the world is already in state of war – against an all-purpose enemy called 'terrorism'). Specifically, we could all be well fed both by the standards of modern nutritional theory and by those of the finest cuisines – yet starvation continues and the 'diseases of affluence' grow daily more grotesque. I suppose I wrote *Reap* to some extent in the spirit of, 'Wouldn't it be nice if . . .' Now I feel that the Enlightened Agriculture that *Reap* advocates is a burning and urgent necessity.

I can summarize my newly emerging or newly consolidating thoughts in a few brief paragraphs:

1 The World Economy as Soap Opera

The people who run the world (governments, corporations, experts of all kinds) act as if they believe that the economy of any one country is an integrated whole; and that soon, if we put our best feet forward, the economy of all the world will be a single whole and interdependent entity.

Yet it is obvious, wherever you live, that this just isn't so. Even in the best organized countries, as Britain and the US would claim to be, several more or less distinct economies run side by side. There's the legitimate economy, in which visible income is taxed. There's the black market, which includes much of the catering industry, since waiters live by largely untaxed tips. There's crime, which is huge in all countries: drugs underpin one of the world's biggest industries and in some are the biggest employers. Prostitution often has ambiguous status, with prostitutes enjoying little protection from the law and often falling foul of it, but expected nonetheless to pay taxes. Similar to crime in outcome but generally different in its *modus operandi*, is corruption, which takes many different forms, is significant in all societies, and often makes a nonsense of all other endeavours. Within the legitimate economy there are obvious, vast anomalies. Typically, the very rich are those in the business of manipulating money, even if they are almost unqualified to do so and have only a narrow education, while people in recognized trades and professions – doctors, teachers, academics, nurses, musicians, plumbers – typically rub along in a fairly mundane fashion unless they find some way of tapping in to the wealth of the money manipulators.

Within Third World countries, as India still is, the different strands often seem hardly to interact at all. India has a billion people and the middle class is vast. Although some of the middle class defend Indian traditions and spiritual values, most look towards the west (and make a virtue of this). The IT professionals in Bangalore typically earn only a tenth of the western wage for an equivalent job – $6000 per annum against $60,000 in California – but still they are well off in a country where 60 per cent of the people – 600 million – live on the land, mostly as subsistence farmers. As a westerner it's hard to adjust to the Indian economy, as a hotel in Delhi may easily cost $120 a night, while a taxi driver may gratefully work for a whole day for $10 (which includes the cost of the petrol). In the villages food is good (excellent masala dosai in the south – stuffed pancakes made with rice and bean flour) and costs only pennies (proving once more that feeding people really is easy if you grow what is easy to grow and know how to cook). But the farmers who grow the rice and beans might work for a week for $10. Yet the same countryside may contain high-class western-style chalets that again charge $100 or more per night.

My point is that the different strands of the economy in all countries, and especially in Third World countries and in the world as a whole, do not form one coherent entity. Different people pursue their own economic destinies as separate as the different storylines in a television soap opera. The present manipulations of western money people have absolutely nothing to do with the day-to-day lives of poor people in western cities, and still less to do with the struggle for survival on the subsistence farms which are still the world's greatest employers. At least, the manipulations of western wealth have nothing to do with poor people until the poor people find themselves dispossessed, their land 'claimed' by some company that probably doesn't recognize that the farmers even exist, and yet has plans of its own for 'development' – and possibly is in possession of some grant, ultimately financed by western taxpayers, to help it on its way.

To be sure, some people in high places do seem vaguely aware that the world's economies have this soap-opera quality, with many strands and little direct interaction. Some feel that this is simply the way of the world, and *vive la difference*. Thus, people in many countries feel that traditional societies of all kinds, cut off from cash economies of any kind, should simply be helped and encouraged to go on living in their own way. At best this is benign – genuine handover of large tracts of forest to indigenous peoples, for instance. At worst it means that 'native' or 'aboriginal' peoples are 'placed' on reservations, with very few rights if they leave them. Others apparently feel that the only economy worth taking seriously is that of the rich west, and that the rest will just have to sort themselves as best they can. Enthusiasts for the global market are often benign in intent (or so it seems to me) and feel that if only everyone in the world was equally free to trade in a free global market, then they would be integrated into one economy, and the world could move ahead in a freshly integrated mode, with the hope (at last) of justice for all.

This last vision can be made to sound good (while other models can be made to sound imperialistic and old-fashioned) but anyone who has been to rural India or rural Brazil (or anywhere in the Third World) can see how fanciful it is. Often there is roughly a hundred-fold difference between the earnings of poor farmers and those of the ordinary westerners: after all, the average working Brit earns £18,000 per year which is about $30,000, while the UN estimated in 1999 that a billion people worldwide get by on not much more than $300 per year. There is no built-in mechanism in the free market that can begin to bridge that gap. The much-bruited and vaunted idea that poor farmers can acquire western-level incomes by selling their crops into western markets is ludicrous. Western standards are largely arbitrary but are nonetheless stringent, and very few Third World farmers could afford to meet them. Entrepreneurs might meet western market standards – but only by turning the indigenous small farms into monocultural high-tech estates, and throwing the traditional farmers off the land. Indeed, those entrepreneurs are often westerners themselves. Besides, all the markets for food in the western world are already saturated several times over (not least by the western farmers themselves – who also have a right to make a living).

In short, the idea that the world's economy *can* become one simple integrated whole, a perfectly worked piece of literature as opposed to a rough-and-ready soap opera, is a nonsense, at least in the foreseeable future. Models based on the idea that the world's economy is already integrated are rooted in a huge lie; and models that envisage integration in the immediate term, or even within the next century or so, seem bound to create a great deal more misery than is evident already. In particular, they will take from the poor even that which they have.

2 The World Economy must at bottom be Agrarian

Most people – four billion out of the world's total of six billion – live in the 'Third World'. Sixty per cent of that majority work on the land, mostly in subsistence farming. Modern western scientists and politicians, and the corporations they support, speak as if subsistence farming was some kind of dying trade, already an obvious anachronism; as if the world as a whole self-evidently depends upon the western, high-tech farming that is so rapidly sweeping the old ways aside; and as if this sweeping aside was good – the stuff of 'progress', which is 'inevitable', and 'development', which is self-evidently good. In truth, the way of life that the powers-that-be so disdain, and are so anxious to do away with, is the *norm* for most of humanity. Ours is a world of small farmers. The people of influence who take it to be self-evident that they are at the centre of the world's affairs, are the minority. The idea that they (self-evidently) represent the future is very much their own conceit – and when you look at it closely is again ludicrous.

Of course, subsistence farming on small farms cannot represent the human ideal. Even when things are going well, such a life is extremely hard. Neither are crops or income guaranteed; so the traditional farmer's life also tends to be precarious. In addition, the countryside in most places in the world is in general less well-off than the cities. It is far harder and therefore more expensive to supply rural communities with the infrastructure that urban people take for granted (or at least used to take for granted), such as electricity, plumbing, transport, schools and a doctor. Thus we have a double whammy: that country people tend to be poorer than city people (and generate less cash); yet it is far more expensive to supply them with the basic necessities. Rural life has often been boring, too, to the point of distraction, especially for young people who are needed to carry things on. Perhaps above all, for all of these reasons, farming has often enjoyed very low kudos – as I am told, by people in a position to know, is the case in Brazil and most of Africa. A farmer (as opposed to a rich estate-owner) is not widely perceived as a good thing to be; and it is hard to overestimate the importance of social kudos to social creatures like us.

Simple arithmetic suggests, though, that whatever the problems and however hugely they loom the world as a whole *has* to establish agrarian economies. I hate to quote Margaret Thatcher, but her famous line 'There is no alternative' applies to the agrarian economy in spades. People need food and shelter, of course, but equally they need jobs. Forced unemployment erodes the soul. Those who believe that present-day economies are on the right lines are wont to suggest that brand-new industries of all kinds are sweeping in to take over from farming. This hardly seems to be happening in Africa, but India is taken as the great example to us all. According at least to a recent article in *The Hindu*, about 60,000 people now work in India's new IT industry, largely centred around Bangalore. India has three million BScs, after all. Soon, perhaps, IT could employ 100,000. But if that industry employed much more than that, there would be enough to serve the whole world. The whole global market would be saturated. In addition, IT is a one-off: apart from tourism, it is (or has been) one of few genuine growth industries in the past few decades. But 600 *million* people work on the land in India. For every one person involved in Indian IT, there are ten thousand farmers. It is astonishing, but it is the case. Alternative industries that could in theory employ dispossessed farmers in reality fall short of what would be needed by four orders of magnitude.

For the foreseeable future *nothing* except farming can even begin to employ the majority of the world's farmers – which means, the majority of the human species. Unless we are content to see many hundreds of millions of people permanently out of work, then we have to build up agrarian economies, rooted in labour-intensive agriculture. We already see the first signs of what could come about if we do not. According to the United Nations, *one billion* people are now living in slums – and you have to go to places like India to see what 'slum' really means: basically a rubbish heap, stretching for miles into the distance, with shelters of scrap metal or sticks and sacks or whatever comes to hand, polythene bags and dead pigs and open sewers and all the rest. Much is done to improve ailing cities, but it is a losing battle. The only feasible way to halt and reverse the tide is to ensure that the countryside is tolerable: that basic infrastructure is in place; that people can work there. For the foreseeable future (which in reality is surely the next few centuries) farming is vital as the great sump of spare human labour. I am reliably informed by people on the spot that many Asian communities already see agriculture in this light. This isn't just a Third World problem, however. Britain has remarkably few farmers left but is still losing more than a thousand farm workers a month. Their passing goes unnoticed – although if it happened in banking or manufacture, it would make the national news. But the run-down of farming is somehow seen to be inevitable – and in any case, Britain has been assiduously industrializing and urbanizing itself for more than two centuries and does not take the countryside seriously. The spasmodic outbursts of country people have largely to do with fox-hunting, and reinforce the urban belief that traditional country life has had its day.

But again, there are several outstanding serendipities. Agriculture is at bottom a craft industry, of the kind that really can benefit from hands-on labour. It is of course much kinder to raise livestock in small herds, in natural surroundings, with tender loving care; and animal welfare ought to be seen as desirable in itself. The meat, eggs and milk should taste far better – time was when the milk from every herd had its own characteristic flavour – and have a far finer texture. Smaller amounts of meat and dairy produce are more satisfying than large gobbets of inferior stuff (and here I speak from first-hand knowledge). Meat thus produced should be far safer, too. Possible improvements in safety these past few years – for example by superior hygiene and vaccines – are largely obviated by the official drive to produce as much as possible as cheaply as possible, which in the context of livestock in particular is intrinsically hazardous. All these considerations apply to crops, too (apart from those of welfare, since we generally assume that plants are unaware). More hands on deck can produce better flavour, texture, safety – all set in a far more agreeable society.

Kudos, vital as it is, is entirely a matter of perception. In many societies farmers have been despised precisely because they are poor and live such hard lives. This does not have to be so. The true value of modern technologies, including the highest of technologies, is to make life easier for people – and life on small farms could be very agreeable indeed. Ingenious technologies of many kinds can take much of the back-break out of it (although some hard labour is eminently satisfying) and IT, which essentially is cheap, can put anyone in the world in touch with everyone else. Rural communities need not be 'cut off', and therefore 'backward'. With the internet, it is easy to live in the sticks and be an intellectual. I have come to know plenty of people who manage it.

Indeed, I have now come to see that one of the most pressing tasks for all humanity is to work out what would be an ideal ratio of rural to urban people – basically the ideal proportion of farmers – in each community. Indians to whom I recently put the question thought that half-and-half was about right (reducing present farmers by about 10 per cent). In Britain, one might intuitively suggest, 20 per cent would be reasonable – increasing the present quota of farmers by about twenty-fold. Since World War II the NFU, which finds this book so risible, has presided over a roughly 20-fold decline in the proportion of farmers – perhaps the only union in the history of the movement actively to collude in the collapse of its own industry. I suggest that it has misread history, and misjudged as absolutely as can be conceived the real problems and needs of the world.

Then again, and finally, it is simply the case that the world is an uncertain place politically. All countries that buy in to the WTO philosophy must depend absolutely on trade. But a very great deal can go wrong with trade, and the people of virtually all nations (or regions within some nations) would surely sleep easier in their beds if they knew that if the worst came to the worst, they could at least feed themselves. Self-reliance can be no bad thing. Of course, if you say this in a country like Britain, there are guffaws from on high (not least from the NFU). Do I seriously suppose that the Atlantic will again be blockaded by German U-boats, as in World War II? No, I don't seriously suppose that. But can the scoffers of self-reliance also be absolutely certain that Britain will always find it as easy to pick and choose from the world's produce as it does now? What happens as China grows significantly richer, as is the case, and is able to become a major bidder as well as a major producer? And what would happen if the greenhouse effect bit even harder in China than it is doing so far, and reduced its agricultural output without (in the short term) reducing its buying-power? Would the US still respect its 'special relationship' with Britain if China (say) was offering to bid more for US surpluses? Wouldn't it be against the spirit of the WTO to favour Britain? Will the US escape the depredations of global warming enough to retain its status as the world's bread-basket? Is it safe in such uncertain times as this to run down the agrarian economy anywhere in the world – let alone in those countries where farming is still by far the greatest employer?

3 The Nonsense of GDP

Throughout this book I argue that farming has suffered from its position both as the poor relation in political thinking, and as the political shuttlecock. That is, agriculture has rarely been designed expressly to feed people, and has been expressly designed to employ people only *in extremis*. Always farming is shaped to fit in with the particular political conceits and ambitions of the day. At present, globally, the idea of free trade prevails (at least in theory), overseen by the WTO. Europe this past half-century or so has laboured under the yoke of the Common Agricultural Policy. The job and intent of CAP has not been to feed the people of Europe well, and to make the best use of Europe's landscape and agrarian skills. It has been to promote the central notion of the European Union, which is that a European common market would be a good thing; a notion that in turn derived in large part from the belief that if different nations depend on one another for trade, they are less liable to go to war. The CAP has become horribly complicated, because the French and Germans in particular (and also the Italians, but they have less clout) are still largely agrarian, and have accepted the run-down of traditional farming less complaisantly than the British. So the CAP remains a procrustean exercise – of ramming traditional *modus operandi* and ways of life into a set of rules drawn up for reasons that had nothing to do with good farming and everything to do with politics and bureaucratic convenience.

Personally, I like the idea of a more united Europe. Absolutely not am I a little Englander; and despite many lapses, Europe as a whole always seems to me to be eminently civilized. But behind both the WTO and the CAP – indeed behind *all* the prevailing economic models – lies the supremacy of GDP: Gross Domestic Product. GDP in effect represents the total amount of cash circulating in the economy at any one time. It is taken to be self-evident that the bigger the GDP, the better. 'Economic growth', means increase in GDP. Governments win elections by promising and delivering 'growth', which means they increase GDP. Economies that fail to 'grow' in this sense are said to be 'stagnant' and the nations that have such apparently unenviable economies are commonly said to be 'going nowhere'. In these competitive times (the world's economy is specifically designed to be competitive) nations with the greatest GDP, and the fastest 'growth', tend inevitably to prevail.

There is a huge snag, however, which many economists (including John Maynard Keynes) pointed out as early as the 1930s, which is that GDP has almost nothing to do with human wellbeing. Richard Nixon illustrated the point during his presidency. In effect, he defended industrial pollution. Industries that don't clean up after themselves cut their own costs, and so get richer quicker, and so contribute more to the GDP. The pollution itself can then be cleared up by specific anti-pollution industries – which also, by generating cash, contribute to GDP. Similarly, the American people (much more than any other bona fide First World country) suffer horribly from crime. Many are killed but many more curtail their own freedoms out of fear – a fear that seems perfectly justified. Yet crime supports several enormous industries: to provide both hardware (guns on both sides, and prisons, safes and reinforced houses), and the enormous industries of the police, the prison services and the legal profession. Crime stimulates the GDP enormously. Of course it isn't true that nobody benefits from it. Clearly, the industries of prevention and punishment are significant employers, and the law provides one of the prime routes to riches. Common sense says, however, that if the American or any other society had less crime most people would be far better off – even if the anti-crime industries suffered as a result, and the GDP was thereby reduced.

The modern food industry illustrates the same point beautifully. There are many examples (pollution and its cleaning up is one of them) but the point is made most strongly by meat. As discussed in Chapter 4, western agriculture in particular has sought to increase its total output – and hence its earning power – by raising meat production. This in turn has encouraged greater consumption of calories in general and of fat in particular (which, within the fast food industry, is accompanied by enormous quantities of sugar and salt). All this leads to 'diseases of affluence', including obesity, gallstones, diabetes, stroke, and coronary heart disease. But again, there is a phalanx of complementary industries anxious to put things right. Slimming is a major industry, and so too – matched only by the professions of law and accountancy – is medicine. GDP flourishes like the green bay tree. Intensive livestock generates ever-swelling rivers of cash, and so do the various 'health' industries that are designed to counter the all too obvious ill effects. But the intensive livestock industry employs as few people as possible (one per half-million chickens is fairly typical) and has thrown an enormous amount of traditional farmers out of work and a lot of people die before their time because of it.

As outlined in Chapters 8 and 9, science has become sucked up in the manic drive to maximise GDP. Many technically excellent scientists who aspire to do good have come truly to believe that it must be helpful to maximize productivity. But it just isn't so. The world as a whole is not short of food *per se* (or at least it wouldn't be, if what there is got to the places where it is needed). Increased productivity at present serves either to increase consumption (by those who can afford it, though they generally have enough already); or to create glut, which reduces world prices, which enables food traders to cash in in the short term (for the reductions are rarely passed on to the consumers). Food industry scientists are not fools, and know all this perfectly well. Still, though, the ones I have spoken to take it to be self-evident that they must be doing good because their efforts are helping to swell the economy – to raise the GDP. Once you see that GDP has so little to do with wellbeing, you realize that so much of modern science, and the often excellent people who practice it, have been seriously duped.

But there are ways of gauging human wellbeing, which at the most basic levels have to do with infant mortality, general length of life, literacy, and so on; and these measures can be made more and more subtle, even after such basics are taken care of. Many now argue that the world's economies as a whole must be based on these new measures, rather than on GDP. This matters to all of us, but particularly perhaps to the majority of humanity who live in the Third World. 'Development' is an important concept, and there is something to be said for the idea of 'progress'. But when both of these ideas are equated with 'economic growth' defined in terms of GDP, then the result can be very damaging in the short term, and disastrous for the longer future.

4 The 'Triple Bottom Line' Redefined

The Earth Summit in Rio in 1992 generated the idea of the 'triple bottom line'. This decrees that the performance of agriculture in particular must be 'sustainable': a difficult word to define, but it means in general that whatever we do today should in principle be do-able in ten years' time, or 100 years – or, I would say if we are thinking sensibly, in 10,000 or a million years. Sustainability in turn, said the Summit, should be considered under three headings: environment; society; and the economy.

This is fine as far as it goes – obviously far superior to a 'bottom line' that acknowledges only the increase in cash. But it does not go far enough. For 'environment', we should read 'biology'. The principle of Enlightened Agriculture – truly sustainable agriculture – demands that we should of course gear farming to the needs of the environment as a whole (including climate, landscape, and the needs of other species) but should also be gearing husbandry specifically to the physiology of livestock, crops and human beings; and to the psychology both of livestock (welfare) and people (gastronomy). For 'social' we should truly be looking at the whole spectrum of human existence, including the fact that most people in the world are farmers of one kind or another, and that there is no realistic alternative to this in the foreseeable future. 'Economy' is obvious too – but if we continue to equate economic success with GDP, then we (the world) will simply dig the hole deeper. True measures of wellbeing must be built in to the equation.

5 The College of Enlightened Agriculture

You may gather from the fact that I have added this Afterword that I do not regard this book as the finished, once-for-all article. But no book of this kind can ever be finished. I would like it to be seen as a discussion document. Agriculture is not the only item on the world's agenda, but it must be at the top of it: the *sine qua non*. We, (humanity) need a forum, to take the discussion forward: either actual (bricks and mortar) or virtual (out on the internet) or (preferably) a mixture of the two. We also need to improve education. People newly qualified with masters' degrees in agriculture tell me that they never really learnt the basics – for example the key differences between ruminants (like cattle and sheep) and non-ruminants (pigs and poultry). They learnt only the specifics: how to maximize growth in each 'class' of landscape and (most important these days) how to qualify for grants, subsidies, and tax breaks. In short they never learnt what agriculture is really *about*. People who are not specific students of agriculture, of course, generally learn nothing at all. They just have a vague idea (as so many ministers of agriculture obviously do) that farming is economically and politically complicated, and that farmers are stroppy and the sooner it can all be handed over to big business, the better. Such ignorance, such a lack both of knowledge and of 'feel', both inside and outside farming, is already a disaster, and is leading rapidly to the greatest conceivable tragedy. Yet if we recovered that 'feel' (and had proper respect for people who still retain it) life for all humanity could at least be secure: and security is not a bad starting point.

A place or a medium for discussion and education is by definition a college. I have no skill or taste for organization or administration but I would like very much to see the College for Enlightened Agriculture come about. The first task is to bring together all the many groups who are already thinking along appropriate (though often separate) lines and to hack out a truly coherent strategy. My own discussions so far suggest that this would not be too difficult. After all, the proper task of agriculture – to feed people – is a lot easier, conceptually, than that of trying to run it according to the criteria of the CAP or of GDP in general, while accommodating the particular demands of corporations. There is good evidence that hundreds of millions worldwide are clamouring for change. So the game isn't over yet. But we need to dig deep, and fast.

Colin Tudge, Hook Norton, February 2004

Further Reading

I did not write this book as a PhD thesis or as a textbook, and so have not provided references for every passing comment. Besides, many of my best insights have come from conversations, especially with scientists and farmers in far-flung places, and would be impossible to catalogue. The following is a highly selected list of books (with some reports and articles) to which I owe a special debt, and which I believe would bring particular benefit to others. Novels and scientific papers cited in the text are not listed here. Immodestly, I have included my own books where they bear on the present subject.

1 The Nature of the Problem

Peter Bunyard, *The Breakdown of Climate* (Floris Books, Edinburgh, 1999).

L. T. Evans, *Feeding the Ten Billion* (Cambridge University Press, Cambridge, 1998). *A very fine factual account of the status quo, and a history of agricultural science.*

Colin Tudge, *The Day Before Yesterday* (Jonathan Cape, London, 1995). Published in the US as *The Time Before History* (Scribner, New York, 1996). *Agriculture in the context of human evolution.*

UNEP/Earthscan, *Global Environment Outlook 3 (GEO. 3)* (Earthscan Publications Ltd, London and Sterling, VA, 2002). *Summarizes the basic facts and figures.*

2 Farming

Stephen Budiansky, *The Covenant of the Wild* (William Morrow, New York, 1992). *Domestic animals chose us as much as we chose them.*

Juliet Clutton-Brock, *A Natural History of Domesticated Animals* (Cambridge University Press for the Natural History Museum, London, 1987).

William Cobbett, *Cottage Economy* (Oxford University Press, Oxford, 1979). *First published in 1822.*

Simon J. M. Davis, *The Archaeology of Animals* (B. T. Batsford, London, 1987).

David R. Harris (ed.), *The Origins and Spread of Agriculture and Pastoralism in Eurasia* (UCL Press, London, 1996).

D. R. Harris and G. C. Hillman, *Foraging and Farming: The evolution of plant exploitation* (Unwin Hyman, London, 1989).

Mark Overton, *Agricultural Revolution in England* (Cambridge University Press, Cambridge, 1996).

Redcliffe Salaman, *The History and Social Influence of the Potato* (Cambridge University Press, Cambridge; first published 1949, reprinted 1970). *A classic.*

Colin Tudge, *Neanderthals, Bandits, and Farmers* (Weidenfeld & Nicolson, London, 1998; Yale University Press, New Haven, 1999). *How farming really began.*

3 Good Farming, Good Eating and Great Gastronomy

Felipe Fernandez-Armesto, *Food: A history* (Macmillan, London, 2001).

Dorothy Hartley, *Food in England* (MacDonald, London; first published 1954). *This book is truly a classic, to be enjoyed for its literary merit and humour as much as for its scholarship.*

G. B. Masefield *et al.*, *The Oxford Book of Food Plants* (Oxford University Press, Oxford, 1969).

Colin Tudge, *Future Cook* (Mitchell Beazley, London 1980). Published in the US as *Future Food* (Harmony Books, New York, 1980). *Explores the idea that good nutrition, good farming and great cooking are perfectly compatible – and indeed depend on each other.*

Colin Tudge, *The Food Connection* (BBC Books, London, 1985). *A rapid survey of nutritional theory as it then was – and, bar a few details, still is.*

4 Meat and Appetite

Michael and Sheilagh Crawford, *What We Eat Today* (Neville Spearman, London, 1972). *Another ageing title; but Professor Crawford was among the first to draw attention to the role of meat as a source of micronutrients, notably essential fats.*

Marvin Harris, *Good to Eat* (Allen & Unwin, London, 1986). *One of several books in which Professor Harris probes the anthropological origins of food taboos and preferences. Ingenious and often counter-intuitive.*

Kenneth Mellanby, *Can Britain Feed Itself?* (Merlin Press, London, 1975). *Another old title, but clearly spells out the relationship between food and farming, and between arable and livestock.*

Julian Wiseman, *A History of the British Pig* (Duckworth, London, 1986).

5 Food Unsafe

John Burnett, *Plenty and Want* (Penguin Books, Harmondsworth, 1966).

Eric Schlosser, *Fast Food Nation* (Penguin Books, London, 2002). *Bestseller on the influence and perils of modern American commercial food.*

6 Craft, Science, and the Growing of Crops

Rachel Carson, *Silent Spring* (Penguin Books, Harmondsworth, 1995). *Another classic, which should be read by everyone who seeks to get to grips with the modern environmentalist movement.*

J. R. Postgate, *The Fundamentals of Nitrogen Fixation* (Cambridge University Press, Cambridge, 1982).

David Walker, *Energy, Plants, and Man* (revised edition, Oxygraphics, Brighton, 1992). *A fine, clear account of photosynthesis, the basic process on which plants, and therefore animals, ultimately depend.*

7 Better Crops, Better Livestock

National Academy of Sciences, *Underexploited Tropical Plants with Promising Economic Value* (National Academy of Sciences, Washington DC, 1975).

Vitezslav Orel, *Gregor Mendel, the First Geneticist* (Oxford University Press, Oxford, 1996). *The best basic biography.*

Robin Pistorius, *Scientists, Plants and Politics* (IPGRI, Rome, 1997). IPGRI stands for International Plant Genetics Research Institute; and it is one of the worldwide network of institutions under the aegis of CGIAR, the Consultative Group on International Agricultural Research. The CGIAR institutes between them maintain the genetic base for all the world's major food crops, and breed new types. CGIAR's many publications (traceable through their website) are essential reading.

N. W. Simmonds (ed), *Evolution of Crop Plants* (Longman Group, London, 1976). Another old title; but another classic.

Colin Tudge, *Food Crops for the Future* (Basil Blackwell, Oxford, 1988). The basics of plant breeding, including the early promise of GMOs.

8 GMOs

Graham Harvey, *The Killing of the Countryside* (Vintage, London, 1998). A fine analysis of the commercial and political ills and confusions that lie behind the modern farming scene.

Colin Tudge, *The Famine Business* (Faber & Faber, London, 1977; St Martin's Press, New York, 1977). The precursor of this book.

Colin Tudge, *In Mendel's Footnotes: Genes and Genetics from the 19th century to the 22nd* (Jonathan Cape, London, 2000). Published in the US as *The Impact of the Gene* (Farrar, Straus & Giroux, New York, 2001). Includes general accounts of the science of genetic engineering.

Ian Wilmut, Keith Campbell and Colin Tudge, *The Second Creation: Dolly and the Age of Biological Control* (Headline, London, 2000; Farrar, Straus & Giroux, New York, 2000). Includes a general account of the science and technology of cloning.

9 Of Cash and Values

Michael Edwards, *Future Positive: International Co-operation in the 21st Century* (Earthscan Publications Ltd, London and Sterling, VA, 1999). Essential guide to modern global trends, and how to change direction.

Susan George, *How the Other Half Dies* (Penguin Books, Harmondsworth, 1976). Classic exploration of insouciant capitalism.

Oxfam, *Rigged Rules and Double Standards: Trade, globalisation, and the fight against poverty* (Oxfam, Oxford, 2002).

Paul Richards, *Indigenous Agricultural Revolution* (Hutchinson & Co., London, 1985). Explores the key idea that traditional farmers have an extremely acute appreciation of their own landscape and problems, and that the way forward is to build on their traditional know-how.

Amartya Sen, *Development as Freedom* (Oxford University Press, Oxford, 1999). *Professor Sen won the Nobel Prize in Economic Science in 1998 and is Honorary President of Oxfam. This book is a key treatise in modern development literature.*

10 Alternatives off the Shelf

Philip Conford, *The Origins of the Organic Movement* (Floris Books, Edinburgh, 2001). *A modern book that will surely acquire classic status.*

11 Biology, Morality, Aesthetics

Michael Allaby and Colin Tudge, *Home Farm* (Macmillan, London, 1977). *Self-sufficiency in the context of global agriculture.*

Jeremy Cherfas *et al.*, *The Seed Savers' Handbook* (Grover Books, Bristol, 1996). *Dr Cherfas and others have laboured nobly for many years to maintain seeds of traditional varieties in the teeth of various commercial and bureaucratic efforts to kill them off.*

Kenneth Mellanby, *Can Britain Feed Itself?* (The Merlin Press, London, 1975).

John Seymour, *The Fat of the Land* (Faber & Faber, London, 1961). *A classic in the modern 'back to the land' movement.*

Sir George Stapledon, *Human Ecology* (Faber & Faber, London, 1964). *Another classic, by a key player in modern enlightened thinking.*

Keith Thomas, *Man and the Natural World* (Penguin Books, Harmondsworth, 1983). *Not specifically about agriculture, but very much about the extraordinarily diverse attitudes to wildlife and landscape which influence agricultural policy.*

12 From Where We Are to Where We Want to Be

Derrick Bell, *Ethical Ambition* (Bloomsbury, London, 2002). *Bell argues not simply that big business ought to operate ethically, but would be more efficient if it did.*

Satish Kumar, *You Are Therefore I Am: A declaration of dependence* (Green Books Ltd, Dartington, 2002). *Satish is a Buddhist intellectual based at Schumacher College, Dartington, England. His latest (largely autobiographical) book spells out the kind of moral approach that is necessary if humanity is truly to live peaceably, which surely is necessary for long-term survival; and in particular discusses the philosophy of his fellow countryman, Mahatma Gandhi.*

Index

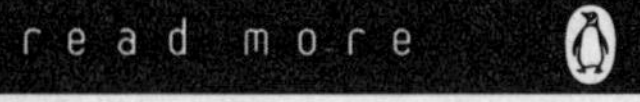

PENGUIN PRESS

HOW WE CAN SAVE THE PLANET

MAYER HILLMAN

'For thirty years Mayer Hillman has been busily turning conventional political thinking on its head . . . he has come up with solutions that are hard to dismiss'
Guardian

Climate change is the single biggest problem that humankind has ever had to face. Yet politicians cannot agree a framework for tackling it effectively and meanwhile we continue with lifestyles that are way beyond the planet's limits. Here Mayer Hillman explains the real issues we should focus on: what the government is doing, what role technology can play and, above all, why we must act *now* to protect our planet for later generations. Challenging, stimulating and practical, *How We Can Save the Planet* is the essential guide to help you understand how we can safeguard our future.

Shows how you and your community can make changes, and why government must take the lead.

Introduces a radical rationing scheme to reduce our individual carbon outputs to a fair and ecologically safe level.

Gives helpful short- and long-term guidelines for the home, travel and leisure to enable us to live within this ration.

Provides a wealth of information and contact details for relevant organizations, companies and websites.

Action must be taken! YOU can make a difference NOW.

read more

PENGUIN PRESS

GLOBALIZATION AND ITS DISCONTENTS

JOSEPH STIGLITZ

The International Bestseller

'A massively important political as well as economic document . . . we should listen to him urgently. Will Hutton, *Guardian*

Our world is changing. Globalization is not working. It is hurting those it was meant to help. And now, the tide is turning . . .

Explosive and shocking. *Gobalization and Its Discontents* is the bestselling exposé of the all-powerful organizations that control our lives – from the man who has seen them at work first hand.

As Chief Economist at the World Bank, Nobel Prize-winner Joseph Stiglitz had a unique insider's view into the management of globalization. Now he speaks out against it: how the IMF and WTO preach fair trade yet impose crippling economic policies on developing nations; how free market 'shock therapy' made millions in East Asia and Russia worse off than they were before; and how the West has driven the global agenda to further its own financial interests.

Globalization *can* still be a force for good, Stiglitz argues. But the balance of power has to change. Here he offers real, tough solutions for the future.

Compelling . . . This book is everyone's guide to the misgovernment of globalization' Jamie Galbraith

Stiglitz is a rare breed, an heretical economist who has ruffled the self-satisfied global establishment that once fed him. *Globalization and Its Discontents* declares war on the entire Washington financial and economic establishment' Ian Fraser, *Sunday Herald*

'Gripping . . . this landmark book . . . shows him to be a worthy successor to Keynes' Robin Blackburn, *Independent*

BY THE WINNER OF THE NOBEL PRIZE FOR ECONOMICS 2001

www.penguin.com

read more

PENGUIN PRESS

FAST FOOD NATION ERIC SCHLOSSER

Fast Food Nation has lifted the polystyrene lid on the global fast food industry . . and sparked a storm' *Observer*

Britain eats more fast food than any other country in Europe. It looks good, tastes good, and it's cheap. But the real cost never appears on the menu.

Eric Schlosser's explosive bestseller, by turns funny and terrifying, tells the story of our love affair with fast food. He visits the lab that re-creates the smell of strawberries; examines the safety records of abattoirs; reveals why the fries taste so good and what really lurks between the sesame buns – and shows how fast food is transforming not only our diets but our world.

'Has wiped that smirk off the Happy Meal . . . Thanks to this man, you'll never eat a burger again' *Evening Standard*

Startling . . . Junk food, we learn, is just that . . . left this reader vowing never to set foot in one of these outlets again' *Daily Mail*

'This book tells you more than you really want to know when you're chomping that hamburger . . . Have a nice day? Listen – you should live so long' *The Times*

'A shocking exposé . . . *Fast Food Nation* could make a difference to the way we eat. For ever' *Evening Standard*